APPLICATIONS OF COMPUTER TECHNOLOGY TO DYNAMICAL ASTRONOMY

APPLICATIONS OF COMPUTER TECHNOLOGY TO DYNAMICAL ASTRONOMY

Proceedings of the 109th Colloquium of the International Astronomical Union, held in Gaithersburg, Maryland, 27–29 July 1988

Edited by

P. KENNETH SEIDELMANN
U.S. Naval Observatory, Washington DC, U.S.A.

and

JEAN KOVALEVSKY
CERGA, Grasse, France

Reprinted from
Celestial Mechanics, Vol. 45, Nos. 1–3

KLUWER ACADEMIC PUBLISHERS
DORDRECHT / BOSTON / LONDON

Library of Congress Cataloging in Publication Data

International Astronomical Union. Colloquium (109th : 1988 : Gaithersburg, Md.)
Applications of computer technology to dynamical astronomy : proceedings of the 109th Colloquium of the International Astronomical Union, held in Gaithersburg, Maryland, 27-29 July 1988 / P. Kenneth Seidelmann & Jean Kovalevsky (eds.).
p. cm.
"Reprinted from Celestial mechanics, volume 45, 1-3."
Includes index.
ISBN-13:978-94-010-6932-8 e-ISBN-13:978-94-009-0985-4
DOI: 10.1007/978-94-009-0985-4

1. Astronomy--Data processing--Congresses. I. Seidelmann, P. Kenneth. II. Kovalevsky, Jean. III. Title.
QB51.3.E43I57 1988
523'.01'13--dc20 89-34269

ISBN-13:978-94-010-6932-8

Published by Kluwer Academic Publishers,
P.O. Box 17, 3300 AA Dordrecht, The Netherlands.

Kluwer Academic Publishers incorporates
the publishing programmes of
D. Reidel, Martinus Nijhoff, Dr W. Junk and MTP Press.

Sold and distributed in the U.S.A. and Canada
by Kluwer Academic Publishers,
101 Philip Drive, Norwell, MA 02061, U.S.A.

In all other countries, sold and distributed
by Kluwer Academic Publishers Group,
P.O. Box 322, 3300 AH Dordrecht, The Netherlands.

TABLE OF CONTENTS

Proceedings of the 109th Colloquium of the International Astronomical Union, held in Gaithersburg, Maryland, 27–29 July 1988

APPLICATIONS OF COMPUTER TECHNOLOGY TO DYNAMICAL ASTRONOMY

Part 2 / MATCHING PROBLEMS WITH TECHNOLOGY

Part 3 / NEW METHODS OF MATHEMATICAL PROGRAMMING

Part 4 / MATCHING COMPUTER LANGUAGE TO THE PROBLEM

Part 5 / USE OF NUMERICAL VERSUS ALGEBRAIC COMPUTATION

Part 6 / REPRESENTATION AND TRANSMISSION OF RESULTS

THE NUMERICAL ORRERY, A SPECIAL PURPOSE COMPUTER THAT CALCULATES THE MOTIONS OF SOLAR SYSTEM BODIES.

INTRODUCTION

P. K. Seidelmann
U. S. Naval Observatory
J. Kovalévsky
CERGA

After a half century of experience in the use of computer technology for the problems of dynamical astronomy, a wealth of experience and knowledge has been developed. Yet this wisdom has not been synthesized for future users. Presently, new technology is advancing with explosive speed. This includes personal computers, increased memory and speed, special languages, special purpose computers, array processors, graphic devices and algebraic manipulators to address the old problems in new and different ways. The objective of Colloquium 109 was to bring together the shared wisdom of the past and the new techniques of the present to provide the future solutions to old and new problems in dynamical astronomy.

International Astronomical Union Colloquium Number 109, "Applications of Computer Technology to Dynamical Astronomy" was held at the National Bureau of Standards in Gaithersburg, Maryland, on July 27 to 29, 1988. The meeting was attended by 120 scientists from the 22 countries; there were 40 oral presentations and 23 poster or computer papers. The meeting was sponsored by the U. S. Naval Observatory and the National Bureau of Standards.

The meeting was organized into six general topics: Matching Problems with Technology, Applications of Computers to Solving Problems, New Methods of Mathematical Programming, Matching Computer Languages With Problems, Use of Numerical Versus Algebraic Computations, and Representation and Transmission of Results. The Proceedings is organized into the same six general topics.

There were two invited keynote addresses. "The Automation of Dynamical Investigations" (a written version is not available) was given by Gerald J. Sussman (Massachusetts Institute of Technology) who developed the Numerical Orrery, a special purpose computer that calculates the motions of solar system bodies. This computer is pictured in the frontispiece of this proceedings. Sussman illustrated the need for computer programs to talk to other computer programs, not just to people. Not only can programs be taught to perform the

Celestial Mechanics **45**: ix–x, 1989.

decision and investigative process of the scientist, they can also write the paragraphs of scientific results. Once a computer research activity progresses from intuitive guesses to a systematic study, the computer program can free the scientist to pursue new original ideas. This concept is illustrated in the paper by Hut.

Sussman says "Knowledge" is knowing what not to compute. His main message is that we must change from trying to maximize the efficiency of the computer to maximizing the effectiveness of the scientist. The computer is relatively cheap, it is the trained scientist who is most valuable.

In the second keynote address, R. L. Duncombe (University of Texas) described the "Early Applications of Computer Technology to Dynamical Astronomy." His talk traced the worldwide introduction of punched card equipment and their applications through the use of the first electronic computers. Sussman and Duncombe, therefore, provided a forecast and review of the use of computers in dynamical astronomy.

This proceedings is intended to provide a consolidated general view of the past, present, and future of the Applications of Computer Technology to Dynamical Astronomy. Of necessity this requires that the papers be relatively short and provide references to the more complete discussions. Thanks to the representation from so many different countries the proceedings give a picture of the activities worldwide.

The Scientific Organizing Committee included K. B. Bhatnagar, V. A. Brumberg, G. Contopoulos, A. Deprit, J. Henrard, J. Kovalevsky, Y. Kozai, P. J. Message, S. Ferraz Mello, F. Tong and P. K. Seidelmann (Chair).

We gratefully acknowledge the financial support from the International Astronomical Union, Office of Naval Research, the U.S. Naval Observatory, and the National Bureau of Standards. Also the hard work of Dr. Fred Johnson, the Chairman of the Local Organizing Committee, Kathy Kilmore and Marla Holloway was vital to the success of the meeting.

EARLY APPLICATIONS OF COMPUTER TECHNOLOGY TO DYNAMICAL ASTRONOMY

RAYNOR L. DUNCOMBE
Center for Space Research
University of Texas at Austin
Austin, TX 78712

ABSTRACT. The review traces the progress from the early application of the Hollerith Tabulating Machine to the construction of astronomical tables by interpolation and to the evaluation of Brown's Tables of the Moon by L. J. Comrie in the late 1920's and early 1930's to the introduction of large programmable electronic calculators in the post World War II era. The early application of electromechanical punched-card machines to astronomical problems stemmed from the needs of the national ephemeris offices to evaluate theories of the Sun, Moon and planets and to subtabulate these positions to form the tables of daily values given in the national almanacs. These techniques were quickly applied to other astronomical problems such as the construction of star catalogs, the reduction of astrometric observations and the numerical integration of the equations of motion of objects in the solar system.

A suitable subtitle for this short review might be "Blame It On the Moon" because of all of the prominent celestial objects clearly visible in the sky the Moon has proven to be the most troublesome for the theorist and the ephemeris maker. In 1767 when the British Nautical Almanac was started by the Astronomer Royal Nevil Maskelyne, the position of the Moon rested on Mayer's Tables which were accurate enough to permit determination of longitude at sea within a degree by the method of lunar distances. Within a few years, however these tables departed from the true position of the Moon and in spite of repeated attempts to correct them, the predicted positions became less and less reliable. While the British Nautical Almanac was much sought after by sea captains of countries which had no comparable navigational almanacs, the experienced British navigator was beginning to suspect the accuracy of his derived longitudes. D. H. Sadler, former Superintendent of H. M. Nautical Almanac Office told this story of an incident occurring in the late 1700's. A British merchant ship bound for America met in mid-Atlantic with a treasure laden galleon bound for Spain. In the exchange of amenities, the Spanish captain sent over a small chest of gold. The British captain, searching for some suitable gift, sent his only copy of the British Nautical Almanac to the Spanish ship. As the story goes, the English ship with its small chest of gold, made port safely and the Spanish ship was never heard from again.

Celestial Mechanics **45**: 1–9, 1989.

A succession of lunar tables followed (Lalande, Laplace, Burckhardt, Hansen) each more accurate than the last but also more complex and more difficult to evaluate. The tedium of extracting positions from such tables is reflected in the plea of the Astronomer Royal John Bond about 1834 in requesting an increase in his staff of assistants: " I want indefatigable, hard-working, and above all, obedient drudges (for so I must call them, although they are drudges of a superior order), men who will be contented to pass half their day in using their hands and eyes in the mechanical act of observing, and the remainder in the dull process of calculation."

The International Congress on Astronomical Ephemerides, held at Paris in 1911, designated the distribution of calculations among the various national ephemeris offices. It was specified that with the exception of the Connaissance des Temps, all ephemerides would utilize Newcomb's and Hill's tables for the sun and planets, and for the Moon, Hansen's tables with Newcomb's corrections until Brown's new tables became available.

That day came in 1923 and with the introduction of Brown's Theory and Tables the drudgery of calculation was greatly exacerbated. In these tables over 1400 periodic terms were combined into 180 separate tables covering 660 pages. While the tables were well arranged, it required the continuous work of two experienced skilled computers, extracting and checking half-daily positions,to barely keep up with the motion of the Moon. L. J. Comrie, Superintendent, H. M. Nautical Almanac Office, who had experimented earlier with the use of Hollerith Electric Punched Card Machines in the construction of tables by interpolation [3], saw their application to the evaluation of Brown's Tables [4]. In 1928 actual punching of tables was started (a daunting task of punching 20,000,000 holes in half a million cards) and in 1929, using a Hollerith Tabulator and Sorter installed at H. M. Nautical Almanac Office, the evaluation of the Tables was carried from 1930 through the end of the century. Comrie pointed out that use of the Hollerith equipment had reduced the drudgery and fatigue, increased the speed by a factor of ten and had reduced the cost by seventy-five percent.

During this same period of time, Wallace J. Eckert, who had joined the Department of Astronomy at Columbia University in 1926, began experimenting with IBM Electric Punched Card Machines at the Columbia University Statistical Bureau [6]. With the addition of special circuits and control devices, he quickly demonstrated the efficacy of the electric punched card method for the solution of problems in astronomy and in 1934 he helped organize the Thomas J. Watson Astronomical Computing Bureau under the auspices of the American Astronomical Society, IBM and the Department of Astronomy at Columbia. The electromechanical machines utilized in the Bureau were ideally suited for batch processing but much less so for serial computations involving more than a few steps. Operations such as interpolation, sub-tabulation, formation of differences and checking the sums of products, for example, were easily performed but a stepwise numerical integration of the equations of motion of a minor planet, for instance, was tedious and time-consuming. For such sequential operations Dr. Eckert devised a piano roll like device which was advanced one notch at each step to alter set-up switches and plug-board wiring for the next operation. Carrying forward a numerical integration by this method might give a modern scientist the piano roll blues, but at the time it was far more accurate and efficient than computations with a desk calculator. In 1939 Dr. Eckert estimated that nearly all astronomical calculations being done by the punched card method were being performed at the Bureau.

At the beginning of World War II, the U. S. Nautical Almanac Office was faced with the requirement to produce an almanac for air navigation, similar to the Nautical Almanac but with the positions of the Sun, Moon and planets tabulated for every 10 minutes of GMT. Again the Moon was the problem. The computational burden was beyond the capability of the Nautical Almanac Office staff at that time so the Navy turned to Dr. Eckert [Fig. 1]. He accepted the position of Director of the Nautical Almanac Office in early 1940 and proceeded to establish the first punched card scientific computing laboratory in the U. S. Government. The initial complement of machines comprised the tabulator with special fractional wheels for adding degrees, minutes and seconds, a reproducing punch, a multiplying punch, a sorter and a key punch. Several years later a collator and an interpreter were acquired. The Electric Punched Card Method proved adequate for the task and the 1941 Air Almanac appeared in September of 1940 [Fig. 2]. Several innovations were introduced in the production process. To avoid the delays and errors of type setting, the printers copy was prepared on the tabulator from a set of cards on which the final data were assembled. The copy was then photographed and reduced to book size and the printing done from deep-etched plates made from the negatives. The proofs were keypunched and compared line by line with the printers copy cards to automatically proofread the data. In spite of alignment problems with the tabulator typebars, the copy was readable, and with introduction of the card operated typewriter, which IBM produced to Dr. Eckerts design in 1945, the results were as perfect as typeset copy [Fig. 3]. Computations for the Nautical Almanac and for much of the American Ephemeris were quickly transferred to the punched card machines, the most time-saving being the subtabulation of the half daily positions of the Moon to an hourly interval.

Now research on the motions of the principal planets, largely neglected since Newcomb's time, seemed more feasible. In addition, much of the drudgery of routine computation was removed by the punched card method. In discussing the observations of Mercury, Dr. Clemence had the help of three Works Progress Administration workers and it required all of their time from January 1940 through February 1942 to form the 57000 coefficients for the equations of condition. He formed the normal equations on the punched card machines, and accomplished in a few days a task that a skilled computer could not have done in three months time. Reduction of transit circle observations was done by hand on large forms with all entries being finally totalled on a manual adding machine. Transfer to the punched card method was accelerated by an incident on a hot August day in the late-forties when a colleague, who had operated the adding machine all day and had amassed an imposing pile of paper tape on the desk, reached over and turned on a high-speed fan. The result; instant confetti. Following the example of the transit instruments, the reduction of other USNO astrometric observations, visual and photographic, as well as the formation of star catalogs were transferred to the punched card machines.

With the end of hostilities, Dr. Eckert returned to New York to become Director of Pure Science at IBM. Dr. Herget became Director of the Cincinnati Observatory and set up a punched card computing laboratory which ultimately became the IAU Minor Planet Center. In 1948 Dr. Brouwer established a similar computing laboratory in the Astronomy Department at Yale University. These organizations cooperated with the Nautical Almanac Office during the late forties and early fifties, in applying their computing facilities to numerous aspects of dynamical astronomy, as attested by titles in the Astronomical Papers of the American Ephemeris such as; "First Order Theory of Mars," "Rectangular Coordinates of Ceres, Pallas, Juno and Vesta, 1920-1960," "The Motion of Jupiter's Fifth Satellite, 1892-1949," "The Secular Variations of the Orbital Elements of the Principal Planets," and "Solar Coordinates 1800-2000," for example.

Although punched card machines continued to supply computing capability well into the fifties, a quiet revolution had been taking place in sequential computing technology, starting with the Harvard Automatic Sequence Controlled Calculator in 1944. The use of vacuum tubes and relay networking then led to the development of several large sequentially operated relay calculators at Harvard University, Bell Labs and IBM. The first electronic sequence controlled calculator (ENIAC) was built at the University of Pennsylvania in 1946. It contained 20,000 vacuum tubes and was designed to compute ballistic trajectories faster than the projectile moved, but the storage capacity and program capability were restricted. When Dr. Eckert returned to IBM in 1946 the company had under construction a new Selective Sequence Electronic Calculator (SSEC) [7]. Dr. Eckert described the speed and mathematical capability of the SSEC in these terms: "With a little practice the average person could multiply two 14-digit numbers together to give a 28-digit answer in about 20 minutes; the machine does 50 such multiplications in a second, or 60,000 in twenty minutes. If the machine should make an error in an 8-hour day this would be only one error in over a million multiplications - less than one per lifetime for a computer with pencil and paper." This high speed, coupled with adequate storage, made the SSEC ideal for solution of a massive problem in dynamical astronomy that had hitherto proven intractable. So, at its inauguration in 1948, the demonstration problem was the simultaneous integration of the equations of motion of the five outer planets from 1653 to 2060.

The next astronomical problem for the SSEC was occasioned by the proposal in 1948 for a new time scale (Ephemeris time), defined by Newcomb's Tables of the Sun. It involved the Moon again. Brown's Theory of the Moon, with appropriate corrections to the longitude to make the independent argument of the lunar ephemeris conform with that of the solar tables, was evaluated for a period of 20 years (1952 - 1972). Observations compared to this Improved Lunar Ephemeris could then be used to derive ET - UT. The harmonic synthesis of over 1400 terms in Brown's Theory proceeded quickly and a comparison with positions derived from the Tables showed small but significant errors both accidental and systematic in the Tables. Other machines quickly followed. Cambridge University in England produced the Edsac in 1949 and in the next few years there appeared the Burroughs E102, the Monroe Monrobot, the RCA Bizmac, the Remington Rand Univac, the IBM 650 and 700 series, and others. Dr. Eckert estimated that in this period there were over thirty groups world wide developing electronic calculators.

In the field of scientific computation, the Age of Camelot began in 1954 when IBM completed the Naval Ordnance Research Calculator (NORC) [8]. In standard computations involving addition, subtraction, multiplication and division with 13-digit numbers the NORC produced 15,000 results in a second. It was straightforward to program and check, and easy to use. Herget quickly transferred all of the numerical integration of minor planet orbits to the NORC and this could be performed in a few runs per year. Both the SSEC and the NORC were decimal machines using four flip-flop elements representing 1, 2, 4, 8 taken one, two or three at a time to give all digits from zero to nine. To the astronomers and other scientists the decimal system was the way God intended man to reckon. To Dr. Herget, who prided himself on writing programs in machine language, this decimal feature was especially important. Within those four flip-flop elements, however, there remained combinations to represent six more counts. The IBM engineers felt the decimal feature was an extravagant waste of hardware and proposed using binary arithmetic only henceforth. To Dr. Herget this suggestion was blasphemous and he muttered that the engineer's paychecks should be made out in binary for a few months until they changed their minds. But efficiency and economy prevailed and when Dr. Herget later worked with an IBM 704 on Project Vanguard, he contented himself with detailing the orbit computation methods to be used and left the coding to others [Fig.4].
When the NORC was introduced in 1954, Dr. Eckert estimated that there were several thousand electronic calculators in operation, with the preponderance being in the U.S.A. In the years preceding and immediately following World War II, however, punched card data processing and electronic calculating techniques were introduced in the national ephemeris offices of many countries.

At the Astronomisches Recheninstitut (ARI) in Berlin, J. Peters began a cooperative program with L. J. Comrie in 1930 to prepare tables of natural trigonometrical functions for every second of arc [9]. In 1935, Peters detailed a conversion from rectangular to spherical coordinates, useful in astrometric plate reductions, using equipment at the Hollerith factory only a few hundred meters from the ARI (Berlin) [11]. In 1955, the Astronomisches Recheninstitut (ARI) in Berlin, used a paper tape controlled drum calculator designated G2 which was developed by a technical group founded by Prof. L.

Bierman of Max-Planck-Institut, Gottingen, for the solution of a problem in dynamical astronomy [1,2]. Dr. Peter Stumpff, then a student of Bierman, was requested by an ARI astronomer to compute the perturbed heliocentric orbit of minor planet Amor on G2. To simplify the task, Stumpff used unperturbed Keplerian ellipses for the perturbing planets (Venus through Saturn). The ARI astronomer refused to accept the G2 results and proceeded to compute the perturbed orbit by hand, using data from the the volume "Planetary Coordinates" for the perturbing planets. When Amor was finally detected the right ascension O-Cs compared to the ARI astronomers orbit were about 2 minutes. Bierman asked Stumpff to compute geocentric positions based on the G2 heliocentric orbit and the O-Cs were only 10 seconds.
At the Astronomisches Recheninstitut in Heidelberg, standard punched card equipment was in use, augmented in 1958 by an IBM626 computer used in the preparation of the annual volume "Apparent Places of Fundamental Stars" from 1960 on. In 1959, ARI personnel used an IBM650 at Darmstadt to compute apparent places for the polar stars of the FK4.

At the Bureau des Longitudes in Paris the subject of computer technology was first raised in 1956 when IBM offered courses in the use of the IBM650, but it was not until the early sixties that application to problems in dynamical astronomy occurred.

In Japan, the Hydrographic Office began using a FACOM 128 Relay Calculator [12] in the compilation of ephemerides as early as 1955.

In the USSR, at the time of the 1958 IAU meeting, delegates were shown a large scale, magnetic tape controlled calculator, similar to an IBM 700 series machine as well as several smaller calculators, but no applications had been made at that time to problems in dynamical astronomy.

At H.M. Nautical Almanac Office, preparation of ephemerides and astronomical tables continued in the late forties and early fifties by punched card methods, augmented in the late fifties by use of an HEC 1201 Calculator.

The U.S. Nautical Almanac Office continued the use of electric punched card machines initiated by Dr. Eckert, slowly upgrading the equipment. By 1955, NAO personnel had access to an IBM650 at the Naval Ordnance Laboratory, and in July 1957 acquired their own.

The growth of computing power, reliability and speed from prior to the punched card era to the large scale calculators of the late fifties caused a revolution in the field of dynamical astronomy. Problems which could not have been attacked by manual computation became tractable. The systematic accuracy of the computations improved, although the perils of programming errors remained to haunt the careless. But best of all, this was the beginning of the age of opportunity for the dynamical astronomer to assault those problems previously considered beyond calculability with freedom from the mind-numbing, tedious drudgery of manual computation. For much of this progress, which saw the transition from the manual evaluation of Brown's Tables to numerical integration of the equations of motion, we can truthfully say "Blame it on the Moon."

REFERENCES.

[1] Annual Report for 1955, Astron. Recheninstitut (Berlin), Mitt. d. Astron. Gesellschaft 1956, p. 59.

[2] Annual Report for 1956, Astron. Recheninstitut (Berlin), Mitt. d. Astron. Gesellschaft 1957, p. 96.

[3] Comrie, L. J. (1928) "On the Construction of Tables by Interpolation," M.N.R.A.S. Vol. 88, pp. 506-523.

[4] Comrie, L. J. (1932) "The Application of the Hollerith Tabulating Machine to Brown's Tables of the Moon," M.N.R.A.S. Vol. 92, pp. 694-707.

[5] Duncombe, R. (1979) "Punched Card Data Processing" in J. Belzer, A. G. Holzman and A. Kent (eds.), Encyclopedia of Computer Science and Technology, Marcal Dekker Inc., New York, N.Y.

[6] Eckert, W. J. (1940), Punched Card Methods in Scientific Computation, Lancaster Press, Lancaster, Pa.

[7] Eckert, W. J. (1948), "Electrons and Computation," The Scientific Monthly, Vol. 67, No. 5.

[8] Eckert, W. J. (1955), Faster, Faster. McGraw-Hill Book Co., New York, N.Y.

[9] Greaves, W. M. H. (1953) "Obituary of Leslie John Comrie," M.N.R.A.S. Vol. 113, pp. 294-304.

[10] Miller, J. C. P. (1951) "Obituary of Dr. L. J. Comrie," Nature, Vol. 167, pp. 14-15.

[11] Peters, J. (1935) "Verwendung der Hollerith-Lochkartenmaschinen bei Umwandlung von Tangentialkoordinaten in spharische Koordinaten," Vierteljahrschrift der Astronomischen Gesellschaft, 70 Jahrgang, 4 Heft.

[12] Tsukamoto, Y. (1957) "Ephemerides by Relay Computer," The Astronomical Herald, Japan Astronomical Soc. Vol.6, pp.89-93.

(Fig. 1) Wallace J. Eckert at the time he was Director of the Nautical Almanac Office (1940-1945).

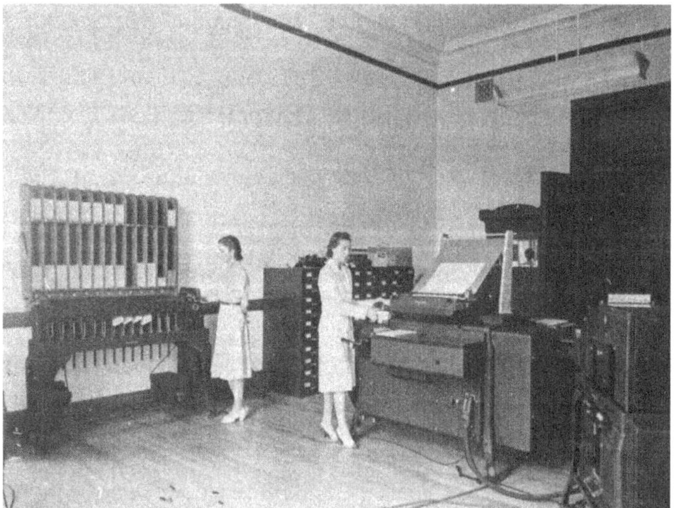

(Fig. 2) Punched Card Computing Laboratory, U. S. Naval Observatory about 1941, showing stand atop tabulator used to support large printers copy forms for the early Air Almanacs.

(Fig. 3) Card operated typewriter, designed by Eckert for Air Almanac printers copy production (1945).

(Fig. 4) Project Vanguard IBM 704. Standing, left to right: Joseph Siry, Paul Herget, Gerald Clemence and Hugh Chrisman.

APPLICATIONS OF COMPUTERS TO CELESTIAL MECHANICS

J. Kovalevsky
CERGA
Observatoire de la Côte d'Azur
Grasse, France

ABSTRACT. The development of digital computers induced major new developments in Celestial Mechanics. At present, one can hardly mention a project in Celestial Mechanics that does not use computers as the principal tool. One can distinguish many different manners of using computers in Celestial Mechanics. Among them, the following are presented and typical examples are given: algebraic manipulations for literal and semi-numerical theories, numerical integration of the equations of motion, determination of physical parameters, numerical checks of analytical results, studies of families of solutions, search for new conjectures, scanning the phase space. In all these approaches, and in others omitted here, major scientific achievements were obtained in the last 20 years and new problems can now be envisaged that were unaccessible even a few years ago.

1. INTRODUCTION

It is difficult to imagine what Celestial Mechanics would have been at the end of the 20-th century, if electronic digital computers had not been invented. A guess is that the trend of the years 1920-1950 would have been pursued, marked essentially by two schools.

(i) Theoretical developments initiated by Poincaré and Birkhoff directed towards regularization methods (Levi-Civittà, Lemaître), families of periodic orbits (Strömgren school), the consequences of the famous Arnold-Kholmogorov-Moser theorem, qualitative results on three or more body problems (Chazy) and relativistic Celestial Mechanics. This approach has indeed progressed in the last forty years and some important results have been obtained without the use of computers.

(ii) Painstaking analytical developments directed towards the improvement of the theory of the motion of some celestial bodies. After Leverrier, Newcomb and Brown, only partial improvements could be undertaken because of the practical impossibility of adding a significant number of terms without spending a whole life in computing them. So if some satellite theories could still be constructed or improved, no real significant progress would have been possible in practice.

Another guess is that Celestial Mechanics would have become an esoteric science

Celestial Mechanics **45**: 11–17, 1989.

involving only very few people.

In reality, Celestial Mechanics underwent considerable transformation using this remarkable new tool. Now, only very few papers published in the field do not refer, at least partially, to results obtained using a computer and most of them are devoted solely to results obtained by numerical methods. To describe applications of computers to Celestial Mechanics is practically equivalent to review the whole field of Celestial Mechanics - an impossible task in a few pages. So we shall only give some of the different types of applications and illustrate them by a few characteristic examples. By no means do they pretend to be the most important results obtained and the choice made must not be interpreted as an indication of its quality. To underline this, references were deliberately omitted, even if in some cases, it was impossible not to mention the authorship.

Many different ways of using computers or interpreting their results exist. We have identified six that are presented below.

2. NUMERICAL INTEGRATION

The most repetitive - and consequently the best fit for computers - algorithm is numerical integration of differential equations. So it is not surprising that the entry of Celestial Mechanics in the computer age was, in 1951, the renown achievement by W.J. Eckert, D. Brouwer and G.M. Clemence: the numerical integration of the motion of the four outer planets for a period of about 400 years. This work popularized Cowell's predictor-corrector method of integrating differential equations of Celestial Mechanics, and was later followed by a number of much more ambituous projects going along two main lines.

1. *Accurate representation of the motions in the solar system* - The main advantage of numerical integration is that one can complicate at will the equations of motion without affecting the algorithm. One may introduce a great number of interacting bodies, include developments of planetary gravity fields, introduce the effects due to general relativity, etc... Because one generally integrates the equations in rectangular coordinates, the conditions to apply in order to represent the observations are rather simple. And last but not least, the algorithm is well suited to prepare the determination of various physical parameters (planetary masses, initial conditions, etc...) governing the motion. For all these reasons, this technique is widely used to represent as well as observations permit, the motions in the solar system, in particular the Earth-Moon system and also of the artificial satellites. The best example is the series of J.P.L. Development Ephemeris (like the presently widely used DE200) that were and are used in US planetary missions as well as in astronomical ephemerides. This will continue in the future since new observations will produce new equations of condition to be used to update and improve older integrations.
2. *Investigation of the evolution of a gravitational system in the future or in the past* - This field was initiated in 1965 by Cohen and Oesterwinter who extended the numerical integration in the solar system to a million years. The most advanced results were obtained a few years ago by Milani, Nobili and some others after an integration over about 30 million years of the outer planets that identified periodic effects with periods of several million years and exchanges of kinetic energy and angular momentum between the planets with similar periods. Let us also mention a simulation by the same group of the motion of 410 minor planets over a period of 200000 years that identified several properties of the motion of asteroids.

These methods have a great interest in that they permit the discovery of some properties of motion. Of course, the knowledge of the actual initial conditions of motion is not sufficient to justify exactly the motion that is found. But generally - and this can be checked by changing by small quantities the parameters of the motion - the major trends that are found do not depend significantly on these parameters and it is therefore legitimate to consider them as being realistic. However, analytical confirma-tions are desirable for a complete proof.

Actually, numerical integration has many other applications in Celestial Mechanics. Some of them will be mentioned in the following sections. It has become the major tool of this science, but its effectiveness may be also somewhat overestimated. The results of numerical integrations are biased by various errors such as rounding or truncation errors. This has to be taken into account when interpreting the results and avoiding the word "proof". However, having in mind this restriction, it is a very powerful and efficient technique in Celestial Mechanics which will probably be even more so in the future, with the advent of faster computers which will yield both extended results in accuracy or in time.

3. ALGEBRAIC MANIPULATIONS

Classically, the theories of motion of the Moon, of the planets or of other satellites are expressed in terms of multiple trigonometric series of a some number of linear time arguments, the coefficients being dependent on some parameters of the motion related to the masses of the objects and the mean semi-major axes, eccentricities and inclinations of the orbits. The theory was obtained by applying certain algorithms on such series and repeating them a certain number of times. They involved various operations on such series, the most complex and time consuming being the multiplication. It was natural to implement these algorithms on digital computers and many prople have done it. One of the most general and easily transferable Poisson series manipulator was constructed in the early seventies by R. Broucke, but many others were and still are programmed by various teams working in Celestial Mechanics.

The main advantage of this analytical approach in comparison with numerical integration is that the expressions obtained may represent a manifold of solutions for a large domain of initial conditions and parameters rather than a single orbit or a finite set of orbits. The difficulty lies in the large number of terms that have to be kept in the final representation, a number that increases very rapidly with the required precision.

The most satisfactory solution is to obtain fully literal expressions. This was done in the early seventies by A. Deprit and his co-workers in the case of the main problem of the lunar theory and extended to some other perturbations by the Namur group. But the power of computers was insufficient to achieve accuracies comparable to the actual precision of observations. This situation may change in the near future.

In the meantime, for practical reasons, people are led to use semi-numerical analytical theories in which the coefficients of trigonometric terms are numbers instead of polynomials. Until now, and probably still for some time, this is the only practical approach to multi-body problems like the theory of motion of the planets in the solar system. The most prominent work along these lines is the planetary theories developed in Bureau des Longitudes (VSOP 82 and TOP 82) that take into account the eight major planets, some minor planets and relativistic effects. A similar type of solution was obtained in the Bureau des Longitudes for the motion of the Moon.

The problem of optimizing analytical expression manipulators is a vital one in Celestial Mechanics. While general literal algebra programs are very nice to use in moderately simple problems, they are awkward and inefficient when the number of terms and parameters increases. This is why, for each problem there is a corresponding specific computer algorithm; they have been systematically studied in particular in the Institute for Theoretical Astronomy in Leningrad by V.A. Brumberg and his team. In addition, the programs for large expression manipulations are almost necessarily machine dependent. This is why organizations engaged in the field have produced their own polynomial and trigonometric series manipulators for their problems. It is difficult to guess whether this trend will continue with future more powerful computers, and whether advances in analytical and semi-numerical theories in Celestial Mechanics will still imply progress in implementing optimized algebraic manipulation softwares.

4. DETERMINATION OF PHYSICAL PARAMETERS

The increased power of computers will permit improvements of the theories of the motion of celestial bodies and adjustments of a number of physical parameters so that the theories will represent as well as possible the observations, to their actual precision if possible. To achieve this goal, it is necessary to increase the number of parameters and include many more observations of various origins. The most striking example is the construction of Earth gravity field models from satellite observations (optical, laser, Doppler, altimetric) and ground based gravimetric data. The number of observations is well above 10^6 and more than 1000 parameters have to be determined. In the solar system, the masses of planets, a few more parameters pertaining to the Earth-Moon system and the initial conditions are systematically determined or improved for each version of the J.P.L. Development Ephemerides. Old observations to the most recent astrometric, range (radar and lunar-laser) and space probes data are used. Despite the efficiency of present computers, this global determination of the dynamical properties of the solar system is such a considerable computing task that it has not been duplicated. For instance, the ephemerides that are based upon the analytical theories of Bureau des Longitudes are obtained by comparing the expressions to the J.P.L. numerical integration as a representative interpolation of observations rather than to the observations themselves.

5. NUMERICAL STUDIES AND RIGOROUS PROOFS

As a branch of Mechanics and applied Mathematics, one of the objectives of theoretical Celestial Mechanics is to rigorously prove some results. It is clear that numerical methods cannot be used to make such proofs, but it appears also that computers may be very helpful.

If numerical integration is used to illustrate a result under the form of a solution of some differential equations, the result is a certain curve wandering in a family of solutions but is none of them. The residual round-off errors and their cumulative effects are such that at every integration step, the result of the computation deviates from the solution with the same initial conditions. The expected error ε of the result is a measure of the cross-section of this family of solutions and is therefore the approximation with which the expected result is proven. For instance, whatever is the quality of the integrator, the numerical integration of the two body problem never produces an exactly periodic orbit. Similarly truncated analytical representations of the solution of differential equations obtained by algebraic manipulations are affected by errors due to neglecting terms in the developments. In addition, if the property to be proved concerns a family of solutions depending on one or several parameters, it is impossible to be sure that all the cases have been treated and that there are no exceptions overlooked because the corresponding values of the parameters were not tested.

However, such numerical tests, if positive, may determine that the assumed result is verified to an accuracy of ε in so many different cases. Then one of the following situations exists.

(i) If the result was already proved in a rigourous manner, the numerical results confirm it and give to the author of the proof some confidence that he was not wrong in his deductions. In most cases, this check will not be mentioned in the published paper, but will have been very useful to the authors who use it. Conversely, if the numerical results deviate significantly from the expectations, the assumed result or conjecture is false. While numerical methods cannot prove the veracity of a result, they can disprove it in building counter-examples sufficiently far from the conjectured situation.

(ii) Similarly if the result is in the form of a truncated series, as in the case of analytical or semi-analytical theories of motion, numerical integration may be used to assess the accuracy of the expression. For instance, this recently permitted to E. Bois and myself to find that while a second order theory was needed to describe the rotation of HIPPARCOS during a few minutes while a first order solution was sufficient for many days for a fast rotating satellite.

(iii) Conversely to the first case, numerical studies of some families of solutions may indicate that the solutions seem to present some specific properties. Such conjectures are very useful, because they show theoreticians what to study and what results they should try to prove.

Let us give some examples. A first group relates to stability problems. The mathematical proof of the stability of a dynamical system of at least three bodies is very complicated. There are a great many definitions of stability, but none of them corresponds totally to the subjective idea one has for the notion of stability. When transformed into a mathematical algorithm, they appear very restrictive and assign instability to cases that one would not normally call instable (Hill, Liapounov, Poincaré, etc...). Within the capabilities of numerical methods (that is for an interval of time during which the integration errors are kept under control), a much better feeling for stability can be obtained, although not a full proof. For instance, several experiments were conducted by P.E. Nacozy on the Sun-Jupiter-Saturn system showing stability over 100000 years. If the masses are increased up to a factor of 29, the quasi-periodic structure of the system is conserved, though with larger amplitudes. By exagerating the disturbing effects of the masses and still finding stability, one gains confidence in the results. But for factors larger than 30 instability appears (ejection of Saturn from the solar system). In this example, unstability is proved, while stability is only suggested.

Another example that we wish to give refers to a conjecture, called Brown's conjecture, according to which the family of long periodic orbits around the triangular Lagrange point L_4 terminates in an asymptotic orbit passing through the colinear point L_3. After some unsuccessful attemps to prove or disprove it, accurate numerical integrations performed by J. Henrard showed that the family terminates on a double asymptotic orbit around L_3. This was later analytically confirmed by B. Garfinkel.

More recently, after the discovery of the interacting orbits of the so-called coorbiting satellites or of shepherd satellites, the properties of the orbits were first studied by numerical integration, and only later were they explained analytically. Most of the theories of the Kirkwood gaps in minor planet distribution are deduced from a systematic survey of numerically integrated orbits around the resonant regions. One could give many other examples: numerical integration is now a major tool for discovering properties of the motion before they are studied and hopefully proved more rigourously.

6. FAMILIES OF SOLUTIONS

Numerical methods are particularly efficient for exploring the solutions of a system of differential equations depending upon a parameter. They permit construction of a great number of solutions and determination at least qualitatively of how they behave as a function of initial conditions or various parameters. Among the manifold of solutions, periodic solutions play a specific role that is acknowledged since Poincaré has shown their importance. In the case of systems of interest to Celestial Mechanics, they are called periodic orbits and are particularly studied in the simplified, but still representative configuration, called the restricted three body problems.

The construction of such orbits by numerical methods started long before the computer era; between 1900 and 1914 with F.R. Moulton, G.H. Darwin and particularly E. Strömgren and his school of Copenhagen and continued between the two world wars. But the advent of electronic computers gave a considerable impetus to this field of research. Families of periodic orbits were studied as a function of the ratio μ of the masses of the main bodies and of the Jacobi constant, firstly for planar orbits, then in the three-dimensional space. A great number of results have been obtained on the structure and stability of various families (bifurcation points, generation and termination of families as a function of the parameters, asymptotic-periodic orbits, close approach, periodic orbits, etc...). Only a few of these properties have been analytically shown. Numerical results remain our main source of knowledge in this domain.

Similar studies are in progress for more complex equations, in particular those which model the galactic potential and a large spectrum of possible studies by this method is still open.

7. SCANNING THE PHASE SPACE

Studying the solutions of systems of differential equations of interest to Celestial Mechanics is not restricted to the description of families of periodic or non-periodic orbits in the configuration space. It is often advantageous to represent the solutions in the phase space and also their intersection with some surface of the phase space. In this latter case, an orbit is represented by a succession of points whose distribution on the surface of section characterizes the orbit. In this representation, many properties of families of orbits are clearly distinguished. Even if analogous properties can be obtained by area conserving mapping, this approach needs much computer time.

The numerous numerical studies performed have brought many very important results about the structure of the phase space such as islands, semi ergodic and chaotic motions, Hénon attractors, etc... that would have been impossible even to imagine before the advent of fast computers. The whole field of the structure of the phase space is the consequence of the availability of efficient computing techniques, even if some results like the famous Arnold - Kholmogorov - Moser theorem give a deep theoretical insight to some of the results. The recent extension to point dynamics and to Celestial Mechanics of the theory of chaotic motion, and its consequence that Celestial Mechanics has also some non-deterministic aspects, open many new applications to numerical investigations of the phase or configuration spaces.

8. CONCLUSION

Many other examples could be given of applications of computers to Celestial Mechanics, showing that practically in every domain of this science, computers now play an important and often a major role. This trend can only be amplified in the future. Even if some subjects may become less investigated or even dropped, many others, that are up to now unaccessible because they need more powerful computers than those presently available, will be candidates for extended studies in the future. This is the case of all analytical and general semi-numerical theories of the motions in the solar system that have not yet caught up with the accuracy of observations. This is also the case for the study of the evolution of the solar system and its subsystems The long term objective is the description of the dynamical state of the solar system near its origin. For this, it is desirable to be able to extend the studies up to few 10^9 years ago, with a sufficient accuracy despite error propagations. Such a limit is far from being accessible at present. The extension of the study of ergotic or chaotic motions among real systems is another field that will need more powerful computers, especially for cases in which double or multiple resonances will be encountered. Actually most of the still open problems described by P.K. Seidelman or V.A. Brumberg and J. Kovalevsky in the IAU meeting in New Delhi will have to be attacked using fast and powerful computers.

APPLICATIONS TO STELLAR AND GALACTIC DYNAMICS

R. H. MILLER
Department of Astronomy and Astrophysics
University of Chicago, Chicago 60637, USA

ABSTRACT. Computation and a wealth of new observational techniques have reinvigorated dynamical studies of galaxies and star clusters. These objects are examples of the gravitational n-body problem with n in the range from a few hundred to 10^{11}. Relaxation effects dominate at the low end and are completely negligible at the high end. The gravitational n-body problem is chaotic, and the principal challenge in doing physics where that problem is involved (whether computationally or with analytic theory) is to ensure that chaos has not vitiated the results. Enforcing a Liouville theorem accomplishes this with collision-free large-n problems, but equivalent recipes are not in common use for smaller n. We describe some important insights and discoveries that have come from computation in stellar dynamics, discuss chaos briefly, and indicate the way the physics that comes up in different astronomical contexts is addressed in numerical methods currently in use. Graphics is a vital part of any computational approach. The long range prospects are very promising for continued high scientific productivity in stellar dynamics.

1. Introduction

Numerical experiments provide a laboratory where the dynamics of stellar systems can be studied. They play the part that laboratory experiments play in other areas of physics. We would like to go around to the other side of a galaxy to see what it looks like. We would like to kick it to see if it bounces. We would like to do all the things a physicist would do with an unfamiliar object. That is a different motivation from that which leads to the construction of high-precision ephemerides.

Numerical experiments play a dual role in stellar dynamics. First, they provide the kind of check on analytic theory that laboratory experiments do in other areas of physics. But, second, they allow more realistic situations, with more of the physics included, than can be attacked by conventional analytic theory. Many properties of the structure and dynamics of star clusters and of galaxies have stubbornly defied understanding and interpretation by means of analytic theory. Computation promises to help.

Those of us who do numerical experiments in galaxy dynamics are constantly surprised by the variety and number of totally unexpected responses. Nature is much more imaginative than astronomers. Who would have thought of pulsars or quasars had they not first been observed? After the fact, those unexpected responses seem quite reasonable, but one has learned some dynamics in the process of trying to understand them. New discoveries sometimes result. The often unexpected results bring a feeling of excitement and discovery to the subject.

Holmberg (1941) made a charming attempt to study collisions of disk galaxies with

Celestial Mechanics **45**: 19–26, 1989.

the help of a specially built analog computer using light sources and photocells. The famous integration of the 5 outer planets by Eckert, Brouwer, and Clemence (1951) belongs to celestial mechanics. The gravitational n-body problem was prominent in Pasta and Ulam's (1953) list of nonlinear problems (another famous problem in this set is the "Fermi-Pasta-Ulam" problem that figured prominently in early work on chaotic systems), and von Hoerner's (1960) work on star clusters is well known. Lindblad's (1960) models of spiral patterns seem to be the earliest attempt to study a problem in galactic dynamics by digital computation. The floodgates opened shortly thereafter. Early work set the stage so we had a good set of tools ready and waiting to go when finally access to adequate computing power, on a fairly steady and reliable basis, became possible in the mid-1970's.

More recent numerical computation has led to remarkable insights and understanding. Its principal value has been in helping to understand ways in which galaxies and star clusters respond to various kinds of disturbances. It is a good example of Hamming's maxim that "The goal of computation is insight, not numbers."

2. Scientific Results

Some of the accomplishments large scale computation has already yielded in galaxy dynamics include:

1. The demonstration that the gravitational n-body problem is chaotic.
2. The demonstration that thin disk galaxies are not dynamically stable forms. The important instabilities in thin disks are not axisymmetric. This work led Ostriker and Peebles to their famous suggestion that galaxies are surrounded by a considerable amount of dark matter that is needed for dynamical stability. That matter is important in cosmological theories.
3. Confirmation of density wave picture for spiral galaxies.
4. Demonstration that barlike structures, rotating "end over end" (about a short axis) in space are the dynamically preferred forms for rapidly rotating stellar systems.
5. Discovery of a strong contraction within each of the participating galaxies during close approach in a galaxy collision, and the identification of the underlying physical process as excitation of the principal radial pulsational mode of a galaxy.
6. Demonstration of effects on the internal dynamics of galaxies as a consequence of being in a galaxy cluster and the first demonstration of the timescale on which a cluster affects the internal dynamics of a galaxy. Tidal braking of a galaxy rotating in the force field of a cluster of galaxies is one example of cluster effects. Puffing up envelopes, stripping material, and even disruption (in some extreme cases) are other examples.
7. Experimental verification of theoretically expected growth rates for unstable disturbances in both disk and three-dimensional galaxies. Stability thresholds have been studied and confirmed as well. Essentially all theoretical models with analytic stability limits have been tested experimentally.
8. Experimental verification of the theoretically expected growth rate for disturbances in the expanding universe. The growth rate was confirmed to 2–3%.
9. Protogalactic collapses that start from systems that join smoothly into their surroundings lead to "galaxies" with flat rotation curves (density going as r^{-2}). These agree with rotation curves for most spiral galaxies. Protogalactic collapses that start detached from their surroundings undergo more violent collapses and reach a density that falls off like

r^{-3}. These r^{-3} galaxies project onto the sky to produce the usual brightness profile of elliptical galaxies (de Vaucouleurs profile). The difference seems to be connected with the amount of sloshing around during the collapse. Galaxies that get shaken up by one means or another (collision with another galaxy; tidal stretching by force field of a cluster of galaxies) also change over to an r^{-3} density profile, strengthening the connection with sloshing. This finding may help to clarify the reasons that regular galaxies take two forms (spiral and elliptical).

10. Discovery of dynamical instability in spherical galaxies with strongly anisotropic velocity distributions (specifically, those with much larger velocity dispersion in the radial direction).
11. Discovery that a galaxy's nucleus orbits around its center of mass. Nuclear regions of galaxies are not static steady-state systems.

3. Chaos

The gravitational n-body problem is chaotic. Chaotic systems show exquisite complexity. Chaos is a property of the physical system that interacts in strange and essentially unpredictable ways with both numerical computation and analytic theory. It complicates both discussion and physical interpretation. The principal intellectual problem in using n-body calculations is to design programs to work around chaos, to allow valid physical inferences in spite of the fact that the detailed trajectory of the computed system through the phase space differs from that of any possible physical system. It may be plausible to assume that chaos does not vitiate conclusions drawn from analytic theory or from numerical experiments, but it remains an unproven assumption.

Attempts to draw valid physical inferences from computations that represent chaotic systems are virgin territory. Numerical experiments on the gravitational n-body problem were the first computational area to encounter this problem. There is no body of experience to draw on in finding ways to work with chaotic systems. Most current studies of chaotic systems are aimed at exhibiting the chaos itself, and at studying its properties as a problem that is interesting in its own right. They are not designed to develop methods that help draw valid physical inferences in spite of the chaos.

Computed results may look quite reasonable, but basing physical inferences on them is an act of faith. Some functionals can be computed reliably in spite of chaos: think of the first ten integrals (centroid position, centroid velocity, angular momentum, energy) and particle number. Others cannot. A detailed trajectory through the phase space is the best-known example of something that cannot be computed reliably. The Liouville theorem reflects properties that are sensitive to chaos. That suggests that an exact Liouville theorem is the best means to cope with chaos. That, too, is an unproven assumption. Happily, it is fairly straightforward to make the numerical calculation have an exact Liouville theorem. It can be made to conserve phase occupancy as accurately as it does the total number of particles.

Chaos is typically much more troublesome in galaxy or star cluster dynamics than in celestial mechanics. Rapid separation of initially adjacent phase trajectories, a signature of chaos, seems to be helped along significantly by close encounters. That presumably does not happen very often in the solar system. Close encounters, if they happen at all, only take place between low-mass objects like asteroids. Abstract three- and four-body problems can be quite another matter.

4. Various Methods

It is to be expected that any problem as charming, and with the scientific interest of, the dynamics of galaxies and of star clusters would inspire quite a variety of approaches. A few are listed here. More detail on some of these methods is given in Miller (1988), and several are described, without much commentary, by Sellwood (1987).

4.1 DIRECT METHODS

Historically, direct integrations of the n-body differential equations were the first method used. The physics is considered to be embodied in the differential equations, which are treated as carefully as possible. This approach has been brought to the highest level of development by Aarseth (1985). Aarseth's most important contributions were treatment of each particle on its own time step and regularization. The chaotic property of n-body systems was first demonstrated with direct codes, and chaos remains a thorny problem with them.

Direct methods are the method of choice for particle numbers under 1000. They are good for open star clusters. Aarseth's code is a finely tuned tool for that problem, where close encounters and binary formation dominate the dynamics. Direct codes have been cobbled to suppress two-body relaxation by softening forces for close encounters (thus desecrating all the beautiful features that make Aarseth's codes elegant) so they might be used to simulate galaxies, in which two-body relaxation times are very long. Not only is it painful to witness, but the physical justification is weak. Careful workers use other tools.

Recent extensions started with the introduction of a tree structure by Appel (1985). This brings in a new data structure. Mutually closest particles are paired. Particles are leaves and centroids of particle groups are nodes in a binary tree. The tree is extended to a node that represents the whole cluster. It has the advantage that particles that are close together in space are easily singled out for careful treatment, while a batch of particles that are pretty far away from the one you're looking at can be treated as a single blob without need to dissect it further. This speeds up force calculations. The force acting on any one particle due to all the others requires $O(\log n)$ operations (it's normally $O(n)$, proportional to the number of particles). But the tree soon becomes tangled as the stars move on their individual orbits, so tree maintenance becomes the principal programming problem. This line of development has been carried out by Jernighan and Porter, by Hernquist, and by Barnes and Hut. It has been carried to its highest level of development by Rokhlin and Greengard at Yale Computer Science (Greengard's thesis won the ACM Distinguished Dissertation Prize for 1987; it is published in Greengard 1988, where lots of detail is given). Greengard carves up the configuration space differently, but in so doing, he is able to evaluate the total force acting on each particle due to *all* the others in the system in $O(n)$ operations instead of the usual $O(n^2)$. The tree is so easy to construct from scratch that maintenance is no problem at all. The number of computer cycles required is linear in particle number, albeit with a fairly large numeric coefficient. Greengard (1988) shows how the method may be made to achieve any specified accuracy in force values. Tree methods promise to permit direct calculations to be extended to as many as 10^5 particles, but it is not obvious how to marry the Greengard-Rokhlin algorithms with the individual time-steps that made Aarseth's codes so efficient. The gain from using individual time-steps is greater than the gain from the tree method.

This is a development that bears watching, one that is exciting from a computational point of view. It is too new to have yet been used for useful scientific results. Tree codes do not provide any method to deal with chaos. Two scientific questions involving dependences on particle number may be best addressed with large-n tree codes: Lyapunov characteristic numbers and binary formation rates. Aside from these, they represent a misguided effort from a physical point of view because they do not face the principal physical problem of dealing with chaos. Developments in integration methods are likely to prove more fruitful in addressing chaos. Symplectic (Deprit 1969) or energy conserving (Marciniak 1985) methods deserve attention.

4.2 POISSON METHODS

Relaxation times are very long in galaxies, and the most straightforward way to build that into a program is to use many particles. Binary formation is thought not to be important as well. Both of these features urge smoothing the mass distribution. The mass distribution no longer looks like an assembly of point particles. Forces can then be derived from tabulated potentials, which in turn may be determined by solving the Poisson equation. That frees the force calculation from any dependence on particle number.

Galaxy dynamics can be studied with far fewer particles than the 10^{11} in a real galaxy. Only enough are needed to make the relaxation time much longer than the dynamical time. The ratio of the two time scales goes like $n/\log(n)$ and they are equal at 400 particles, so 10^5 particles is a good compromise.

The first successful examples, and still the most versatile versions, run in cartesian coordinates with potentials tabulated on a regular cartesian grid. The state of the art has a million particles on a 256^3 grid. Fourier transform methods provide the most efficient way to solve the Poisson equation in cartesian coordinates.

The problem of coping with chaos first motivated this kind of program. The simple time-centered leapfrog integrator, coupled with forces derived from a potential, gives an area-preserving computation that has an exact Liouville theorem. The physics is mapped directly to the computer, rather than first being approximated by a set of differential equations which are then further approximated to solve them on a digital computer. Emphasis is on the large number of integrals provided by the Liouville theorem. Large particle numbers (and consequent suitability for problems in the dynamics of galaxies) came as a serendipitous by-product. Programs are also elegant in having a good match between the refinement of force values (determined by the number of grid points) and the refinement of the integration method. Neither is much better than the other. This kind of program is responsible for most of the discoveries and insights described in §2. Grid methods were first used in a planar geometry to describe disk galaxies (with the notable exception of one example that used a two-dimensional Poisson solver and got forces between rods, rather than forces between point particles), but most use in the past 15 years has been fully three-dimensional.

The same idea can be extended to other geometries. Sellwood used polar coordinates in the plane a very effectively to study growth rates and structures of unstable modes in disk galaxies. Either cylindrical or spherical geometries are the evident three-dimensional extension. Spherical coordinates in three dimensions (or polar on the plane) have the happy feature that they automatically provide higher resolution near the center, where galaxies typically become denser and seem to support structures with higher gradients. However, a trick that gets rid of sidelobes in the truncated eigenfunction expansions in cartesian geometries cannot be extended to spherical harmonics. "Pseudoparticles" have strange shapes, and some spurious results are possible. The presence of a singular point (or line) in the coordinate system opens the door to other spurious effects. The tendency of the nucleus of a galaxy to wander off that singular point (item 11 of §2) gives a special kind of trouble. Potential expansions in spherical coordinates are the familiar multipoles; careful workers extend to $l = 6$ with all admissible m values included. Others stop at as low an order as quadrupole. Occasionally multipoles with a few harmonics are mentioned as equivalent to cartesian codes. It is hard to see how a method with so few degrees of freedom can be like a method that admits many more harmonics than particles.

Some useful science has been obtained from these programs. In addition to Sellwood's studies of growing disturbances, the dynamical instability with strongly anisotropic velocity dispersions (item 10 of §2) was found with a multipolar code by Merritt and Barnes. Multipolar methods can be measure-preserving (Liouville theorem) if the leapfrog integration method is used. However, occasionally other methods such as Runge-Kutta or predictor-corrector, are used. Not only do these lose the Liouville theorem, one of the strongest virtues of Poisson methods, but they are also ugly because of a sharp imbalance between the integration scheme and the potential solver.

A variant of the multipole method has been used in which there are multipole expansions about two centers to study galaxy collisions (vanAlbada; Villumsen). Forces are calculated at each particle's position rather than tabulated on a grid in yet another variant.

4.3 OTHER METHODS

Several other methods have been developed for special problems.

The Toomres (1972) used a "restricted three-body" approximation in which a swarm of test particles moves in the gravitational potential field generated by a pair of particles for their famous study of bridges and tails following galaxy collisions. They needed a fast inexpensive method to explore a high-dimensional parameter space in order to search out plausible configurations to produce things that look like certain objects in the sky. Self-consistency was not a concern. Indeed, the problem could not have been done had the galaxies been self-consistent. The method is useful in problems where self-consistency is known not to be a concern; however, self-consistency is the essence of galaxy (or star cluster) dynamics. The method has been abused by others by being applied in situations where self-consistency is important.

An extension to the "restricted three-body" programs was introduced by Borne (1984). The swarm of test particles moves under the influence of the two mass points that represent the galaxies, but the galaxy centroids respond to the force field developed by the swarm of particles. This method was used by Lin and Tremaine to study the decay of a satellite in orbit around a galaxy due to dynamical friction within the galaxy.

Monte-Carlo and hybrid methods have been used to study developments in globular clusters. Relaxation effects are thought to be important near the cores of globulars, where the relaxation time is well under a Hubble time, but where it is still long compared to a dynamical time (crossing time, or orbital period for a typical star). Hybrid methods treat the core region by n-body methods and join it to an envelope that is treated by Monte Carlo (McMillan and Lightman 1984). Boundary conditions where the two techniques join can be very fussy.

Shapiro and Teukolsky (1985) developed an n-body method to study (general) relativistic star clusters. Many new problems arise in this kind of calculation since the metric must be evaluated along with the integration, and some simplifying concepts such as simultaneity are no longer applicable. This accomplishment is a tour-de-force, even though there may be few astrophysical applications that demand that generality or power.

Additional rapid developments are taking place as astronomers respond to newer architectures (connection machine, cosmic cube, butterfly, etc.) and even in some special cases have constructed special purpose machines (digital orrery). Many of these are reviewed in the book of Hut and McMillan (1986).

5. Graphics

Galaxies and star clusters are beautiful objects. Graphical displays help dynamical studies capture some of that beauty. Dynamical developments have a beauty all their own. Much of the charm of dynamical studies comes from graphical presentations, especially from motion pictures that show the changing galaxy model.

The most important responses of galaxies or of star clusters are the least likely to be noticeable in purely numerical summaries. The most important things about a protogalaxy or a galaxy are things pertaining to its shape: whether it stayed a single galaxy or broke up into several pieces, whether it has a single density maximum or several, whether it is round (spherical), barlike, ring shaped, or triangular, whether it has sharp corners or a hole in its center, and things like that. The danger of ignoring these possibilities in numerical experiments is self-evident. They are best found graphically. The motion picture of center motions in a galaxy shown with this talk is a good example of unexpected phenomena made evident through graphics.

Graphical displays provide additional checks that the physics is being handled correctly. Graphical displays are also the best way to cope with unanticipated or unexpected results. Simply seeing something is the best way nature has given us of recognizing whether there is something strange about it.

Good graphics is vital to doing physics in numerical experiments. The richness and complexity of dynamical responses in galaxies could hardly have been appreciated without graphical displays to make it evident. Few of the discoveries listed in §2 could have been made without graphics. Graphics is not simply a luxury.

6. Scientific Productivity

A fair number of discoveries, accomplishments, and useful results are listed in §2. A good ratio of discoveries to effort marks numerical computation as more productive, on balance, than either analytic theory or observation over the time adequate computing power has been available.

Most discoveries are serendipitous. Something was noticed in an experiment that seemed a little odd. It was found to be real under further checking, and it was finally recognized to be a new effect. New tools, such as a computational approach, made the discovery possible.

Granted that numerical experiments in galaxy dynamics and in the dynamics of star clusters have been scientifically productive in the past, what are the prospects? I think they are good. The list of scientific problems I want to work on is longer today than it has ever been. We have now explored some of the simple early questions, so we have some confidence to tackle the more complicated realistic problems that characterize astronomy. We know a little better how to formulate the questions and how to design experiments to study those questions. A reasonable amount of careful, tasteful effort can yield a rich harvest of important results and even some new discoveries. Many interesting observational questions require features or details we have not yet learned how to include in our experiments. The real research opportunities lie in designing extensions to present experimental techniques that will permit new scientific questions to be addressed. There is no limit to that.

It is a pleasure to thank the many people with whom I have discussed the gravitational n-body problem and science done with n-body programs over many years. Kevin Prendergast, Bruce Smith, and Peter Vandervoort have been especially helpful. NASA-Ames Research Center is to be thanked for providing access to the necessary computing power. Recent research has been partially supported by funds allocated by the NASA-Ames Research Center under Cooperative Agreement NCC2-265 with the University of Chicago.

References

Aarseth, S. J. 1985. in *Multiple Time Scales*, Ed. J .U .Brackbill and B .I. Cohen (New York: Academic Press) pp. 377-418.

Appel, A. W. 1985. *SIAM J.Sci.&Stat.Comput.*, **6**, 85.

Deprit, A. 1969. *Celestial Mechanics*, **1**, 12.

Eckert, W. J., Brouwer, D., and Clemence, G. M. 1951. *Astronomical Papers Prepared for the Use of the American Ephemeris and Nautical Almanac*, Vol. XII, (Washington D.C.: U. S. Government Printing Office).

Greengard, L. F. 1988. *The Rapid Evaluation of Potential Fields in Particle Systems*, (Cambridge, MA: The MIT Press).

Holmberg, E. 1941. *Astrophys.J.* **94**, 385.

Hut, P., and McMillan, S. L. W. 1986. *The Use of Supercomputers in Stellar Dynamics, Proceedings, Princeton USA 1986*, (New York: Springer-Verlag).

Lindblad, P. O. 1960. *Stockholms Obs.Ann* **21**, No. 4.

Marciniak, A. 1985. *Numerical Solutions of the N-Body Problem*, (Boston: Reidel)

McMillan, S. L. W., and Lightman, A. P. 1984. *Astrophys.J.* **283**, 801, *Fiche* 95-F3.

Miller, R. H. 1985. *Celestial Mechanics* **37**, 307.

Miller, R. H. 1988, "Stellar Dynamics," in *Numerical Methods in Astrophysics*, ed. P. R. Woodward, (Boston: Academic Press) to be published.

Pasta, J., and Ulam, S. M. 1953. "Heuristic Studies in Problems of Mathematical Physics on High Speed Computing Machines." Los Alamos Report LA-1557.

Sellwood, J. A. 1987. *Ann.Rev.Astron.&Astrophys.*, **25**, 151.

Shapiro, S. L., and Teukolsky, S. A. 1985. *Astrophys.J.* **298**, 58, *Fiche* 113-E8.

Toomre, A., and Toomre, J. 1972. *Astrophys.J.* **178**, 623.

von Hoerner, S. 1960. *Zeitschr.für Astrophys.* **50**, 184.

APPLICATION OF COMPUTER IN QUALITATIVE STUDIES OF CELESTIAL MECHANICS

LIN LIU and YI-SUI SUN
Department of Astronomy
Nanjing University
Nanjing
China

1. INTRODUCTION

The ordinary differential equations describing the motion of celestial bodies or other problems and the corresponding mappings are almost all non-integrable, very complicated nonlinear systems. For this reason, even in studying qualitatively some behaviours of simple dynamical models such as instabilities, chaos and asympototic regime we will encounter theoretical difficulties, as V.I.Arnold pointed out: " The non-integrable problems of dynamics appear inaccessible to the tools of modern mathematics ". Therefore, it is important to explore the behaviour of the above problems with a computer. In other words, a computer can be a laboratory for qualitative studies in celestial mechanics. This paper will present some examples on the applications of a computer in studies of dynamical systems related to celestial mechanics.It will be divided into two parts.

2.THE PROBLEM OF EVOLUTION ON ORBITAL RESONANCE

The phenomenon of the orbital resonance exists universally in space. What role the mechanical characteristics really play in the dynamical evolution of celestial bodies is an open problem. In fact, it is very difficult mathematically to solve it even for the simplest three-body system of Sun, Jupiter and asteroid. The corresponding ordinary differential equations of orbital resonance are a very complicated nonlinear systems. But as one may handle an ordinary two-body problem with perturbations, the system can be separated into two parts:

$$F = F_{ir} + \Delta F \qquad (1)$$

Celestial Mechanics 45: 27–30, 1989.

in which

$$F_{1\gamma} = F_0(\tilde{L}) + F_1(\tilde{L})\cos\tilde{l} \qquad (2)$$

$$|F_1| / |F_0| = O(\varepsilon) \qquad (3)$$

where $\tilde{L}$ and $\tilde{l}$ are the resonant parameters, ε is a small parameter.

The Hamiltonian defined by (2) corresponds to an autonomous system with one degree of freedom $(\tilde{L},\tilde{l})$. $F_0(\tilde{L})$ and $F_1(\tilde{L})$ are still complicated but integrable. The distribution of its equilibrium points (they can only be centers and saddle points) and the trajectories near the equilibrium points corresponding to various values of p/q in the phase plane $(\tilde{L}-\tilde{l})$ is very clear. There are many close trajectories near the equilibrium points in the phase plane $(\tilde{L}-\tilde{l})$. The variations of their orbital semi-major axes a (or $\tilde{L}$) and resonant argument $\tilde{l}$ (it is a kind of synodic longitude) are restricted, and this is a fundamental characteristic of orbital resonances: it is called libration similar to a simple pendulum. The above conclusion can only express the asteroidal concentrations for every resonant belt and can not distinguish the difference from the resonant states corresponding to various values of p/q, namely, whether the structures of the distributions of trajectories are stable or not. Therefore, it is a key of the problem to investigate the variations of distribution of the above trajectories by considering the term ΔF.

Up till now, because the tools of modern mathematics are not sufficient, it has been very difficult to settle strictly the structural stability of the resonant belts. But, it will be possible to provide a way to resolve the above problem by a computer. For that, the orbital evolution and the corresponding Lyapunov Characteristic Numbers of 70 asteroids have been computed. Some significant results have been obtained and important information for further investigating the formation of the Kirkwood gaps has been found.

3. BEHAVIOUR OF THE MEASURE-PRESERVING MAPPINGS WITH ODD DIMENSION

The real problems in celestial mechanics are in general formalized as Hamiltonian systems. Under some conditions, they can be reduced to measure-preserving mappings by means of Poincare's surface of section. The mapping will present two advantages:

(1) The numerical computation is much faster (typically by a factor 1000).

(2) The essential properties are represented more clearly, for example, periodic and quasi-periodic orbits will be reduced to a fixed point, and an invariant curve on the surface of section of Poincare.

Many studies of measure-preserving mappings with even dimension have been made, because Hamiltonian systems always possess an even dimension and we can use the KAM theorem. But measure-preserving mappings with an odd dimension do not seem to have been considered, we do not know any theoretical conclusions yet. Just as we do in the case of ordinary diferential equations as discussed last section, because of the complexity of non-linear systems, we use the computer to explore numerically its behaviour. The numerical results can help us to formulate theoretical conjectures. We consider a special class of measure-preserving mappings with three dimensions, i.e. it is an extension of two-dimensional measure-preserving mappings to three-dimensions, which are often related to problems in dynamics. We have studied the following three dimensional mappings:

$$T: \begin{cases} x_{n+1} = f(x_n, y_n) + P(z_n), \\ y_{n+1} = g(x_n, y_n) + Q(z_n), \\ z_{n+1} = z_n + R(y_{n+1}) + D, \qquad (\bmod 2\pi) \end{cases} \tag{4}$$

where P,Q,R are small periodic perturbations, D is a small perturbation parameter and f,g are analytic functions at the origin or in an annulus surrounding the origin. When P=Q=R=0 and D=0 the mapping T is reduced to a two-dimensional measure-preserving mapping $\overline{T}$

$$\overline{T}: \begin{cases} x_{n+1} = f(x_n, y_n), \\ y_{n+1} = g(x_n, y_n), \end{cases} \tag{5}$$

where (0,0) is the fixed point with elliptic, parabolic or hyperbolic type respectively. We have used two methods: LCN's (Lyapunov Characteristic Numbers) and slice-cutting method. It is well known that the LCN's provide a more precise quantitative definition of chaos. As the LCN's are limit values, we need a large number of iterations to exhibit the limit behaviour. In general the number of iterations is taken to be $n=10^5$, sometimes to be $n=10^6$.

We have found a criterion for the existence of invariant manifolds and concluded that:

(1) The one- and two-dimensional invariant manifolds do not exist in the vicinity of the fixed point.

(2) When the one-dimensional invariant manifold exists, in a sufficiently small neighbourhood of it, there exist in general two-dimensional invariant manifolds.

(3) The invariant manifolds of the mapping T can only be generated in the subset of the invariant manifolds of the mapping $\bar{T}$.

(4) For mapping $\bar{T}$ with a parabolic or hyperbolic fixed point, the ordered region near the fixed point is destroyed by the perturbed extension to more distant regions. This result is opposite to the case in which the mapping $\bar{T}$ has an elliptic fixed point.

In qualitative studies of celestial mechanics, the importance of computers is increasing.

REFERENCES

C.L. Siegel and J.K.Moser, Lectures on Celestial Mechanics,Springer-Verlag Berlin Heidelberg New York,1971. pp 151-155.

L.Liu, K.A.Innanen and S.P.Zhang, Astron.J.,90, pp877, 1985.

L.Liu and K.A.Innanen, Astron.J., 90, pp887, 1985.

X.H.Liao and L.Liu, Acta Astronomica Sinica, 28, pp11, 1987.

L.Liu and X.H.Liao, Scientia Sinica (A),5, pp585, 1988.

L.Liu and X.H.Liao, In The Few Body Problem, ed. by M.J.Valtonen, pp135, 1988.

X.H.Liao and L.Liu, Acta Astronomica Sinica, 1, pp29, 1988.

T.L.Zhang and Y.S.Sun, Scientia Sinica (A), 31, pp 87-97, 1988.

Y.S.Sun and Z.M.Yan, To be appeared in Celestial Mechanics, 1988

Y.S.Sun, Celes.Mech., 30, pp7-19,1983.

Y.S.Sun, Celes.Mech., 33, pp111-125,1984.

Y.S.Sun, Scientia Sinica (A), 27, pp174-183,1984.

Y.S.Sun and C.Froeschle, Scientia Sinica (A), 25, pp750-757, 1982.

Y.S.Sun, Acta Astronomica Sinica, 24, pp128-135, 1983.

LEADING SPIRAL ARMS, RETROGRADE GALAXY ENCOUNTERS AND MASSIVE HALOS

Gene G. Byrd, University of Alabama, U.S.A.
M. Thomasson, Onsala Space Observatory, Sweden
K.J. Donner, Nordita, Denmark
B. Sundelius, Chalmers University of Technology, Sweden
T.Y. Huang, University of Nanjing, China
M.J. Valtonen, University of Turku, Finland

ABSTRACT. We study, theoretically and with N-body simulations, the formation of spiral patterns in retrograde galaxy encounters. A one-armed leading spiral dominates in a disk if the tidal perturbation from the companion is large and the disk is surrounded by a massive halo. Otherwise, a trailing pattern forms. The leading arm is made up of particles in slightly elongated orbits whose turning points outline the arm. The arm rotates opposite to the disk rotation. We have found one spiral galaxy, NGC4622, with a leading arm near its nucleus. From the literature, we find that very few spirals, if any, in a sample of strongly perturbed galaxies have leading arms. A possible reason for this is that few spiral galaxies have a halo with larger mass than the disk within the visible disk.

1. INTRODUCTION

Kormendy and Norman (1979) found in an observational survey that well defined spiral arm (grand design) patterns were preferentially found in disk galaxies that had rising rotation curves, bars or nearby companions. Galaxies with rising rotation curves will favor grand design patterns because differential rotation (which tends to erase arms) is weak in them. Barred galaxies favor arms because the bar can gravitationally generate or renew trailing arms. Disk galaxies with neither of these two favorable attributes which have grand design patterns were found to have companions. Elmegreen and Elmegreen (1982,1983,1987) observationally confirm this picture. There thus seems to be a strong connection between spiral patterns and tidal perturbation.

Much theoretical work has suggested a connection between tidal perturbation and spiral patterns (Lindblad 1960, Toomre 1969, Goldreich and Tremaine 1978, and Donner 1979). The pattern is thought to be produced by the orbiting perturber acting resonantly on perturbed disk orbits that precess with the angular orbital rate of the disturber. Two-armed spiral patterns are generated in direct

Celestial Mechanics **45**: 31–36, 1989.

encounters at the inner Lindblad resonance where the perturbed orbits close after two oscillations. Kalnajs(1975) and Athanassoula(1978) discuss how for retrograde companions the main resonance will be for perturbed orbits which close after one radial oscillation. In contrast to the direct encounters, these retrograde companions will produce one-armed leading spiral arm patterns.

2. DIRECT AND RETROGRADE TIDAL SPIRAL ARMS: SIMULATION AND THEORY

In previous computer studies (Sundelius, Thomasson, Valtonen, and Byrd 1987), we studied trailing spiral arm patterns created by direct encounters finding that the tidal arms are initially looser, winding up with time. Arm pitch in the Hubble classification system is thus indicated not to be immutable. We also found very long lived tidally-induced arms. In the present paper, we use computer simulations and theory to study the tidal generation of leading spiral arms. We also evaluate observations of the existence and frequency of leading spiral arms in disk galaxies. We then combine the results of these two studies to reach conclusions about the existence of massive halos of spiral galaxies. Full details of this investigation will be published in Thomasson, Donner, Sundelius, Byrd, Huang, and Valtonen (1989).

In our computer simulations, we use a modified version of the Miller polar program (Miller 1976). The modifications are described in Byrd, Saarinen, and Valtonen (1986), Byrd, Valtonen, Sundelius and Valtaoja (1986) and Sundelius et al. (1987). The program models a 60,000 particle self-gravitating disk inside an inert halo component. Matching the situation for many galaxies, the disk and halo each create a constant rotation velocity with distance from the disk center. The ratio of the halo and disk components can be varied.

The illustrated example (Fig. 1) is a system where the halo is 2.5 times the disk and a retrograde perturber 0.23 times the galaxy mass approached within 1.25 of the disk radius in a zero energy orbit. The tidally generated arm winds counterclockwise (the same as the disk rotation), but the arm pattern turns clockwise. The behavior of the arm matches that expected analytically.

We made many simulations of retrograde encounters varying the halo to disk ratio and also varying the level of tidal perturbation. Perturbation was in terms of the (satellite mass/galaxy mass)/(close approach distance in disk radii) to the 4th power. We examined the results of the encounter to see if leading or trailing arms materialized. We found that not all retrograde encounters produced leading arms. Unambiguous leading arm production required both rather strong levels of perturbation (greater than 0.1) and also a halo to disk ratio greater than 1.0. For both strong and weak retrograde encounters with halo/disk less than 1.0, trailing arms dominated the resulting pattern.

The variety in the above results can be understood in terms of disk self-gravity and swing amplification which can generate trailing spiral arms from small perturbations (Toomre 1981, Athanassoula 1984). For the particular case of our constant orbital velocity disk, two-armed spiral patterns will be amplified from even small perturbations for disk galaxies with no halo up to galaxies where the halo is comparable to the disk mass. This amplification of trailing patterns happens for both direct <u>and</u> retrograde encounters. If the halo mass is greater than the disk mass, no significant amplification occurs. On the other hand, leading arms are not amplified by swing amplification. These arms are swamped even in retrograde encounters by quickly amplified trailing arms if the halo is not too massive. Leading arms are dominant when a retrograde companion strongly perturbs a high halo disk galaxy.

3. OBSERVED FREQUENCY OF LEADING SPIRAL ARMS

Since strong retrograde encounters should not be uncommon, we can examine the observed frequency of leading spiral arms to find out how frequent massive halos of spiral galaxies might be. In a sample of 17 spirals, de Vaucouleurs (1958) found <u>all</u> to have trailing arms. Pasha (1985) examined 189 galaxies indicating that 4 had leading arms. Sharp and Keel (1985) examined two of these and found one to be a two-armed trailing spiral and the other not to be a spiral galaxy. It is not clear that the remaining two are leading arm spirals (Blackman 1982).

At this point, one might wonder whether <u>any</u> leading arm spirals exist. Kalnajs (1975) and Athanassoula (1978) have suggested that M32 in a retrograde orbit has created a leading arm in the inner parts of M31. However, M31's nearly edge-on orientation makes its inner spiral structure unclear. Moreover, Byrd (1976,1977,1978) estimates that a trailing pattern is best for M31's outer parts and that arm distortions and radial velocity irregularities favor a direct orbit for M32. While M31 is rather uncertain, we have examined photographs of NGC 4622 (see Fig. 2). It seems clear to us that this nearly face-on galaxy definitely has a single leading arm in its <u>inner</u> portions. The fraction of spirals with leading arms is evidently rather low although it is clear that they do exist. Incidently, NGC 4622 does have a nearby companion. It also has two (presumably) trailing arms away from the nuclear bulge where disk self-gravity is evidently more important.

Of course, our computer survey indicates we should be interested in the subset of galaxies which are strongly perturbed as candidates for leading arms. Examining Pasha's list again for galaxies which are clearly interacting or have a substantial companion (Arp 1966, Arp and Madore 1987, de Vaucouleurs, de Vaucouleurs and Corwin 1976, Nilson 1973, Vorontsov-Velyaminov 1959), we find 28 which conform to these criteria. An additional 53 galaxies from Pasha's list belong to pairs that might be physical or interacting. Therefore, even being generous and taking 2 galaxies in Pasha's sample to have leading arms, we estimate that at most only a few percent of the strongly perturbed galaxies in Pasha's sample have leading arms. We estimate that for orbits of random inclination, about 25% of the sample should have strong retrograde perturbation.

4. CONCLUSIONS

While single leading arms do exist in spiral galaxies, they are evidently underabundant or even quite rare in strongly perturbed disk galaxies. Combining this observational result with the results of our computer surveys, the most straightforward conclusion seems to be that massive halos do not appear to be the norm within the disk radii of spiral galaxies. The rarity of clear leading arms indicates that the disk is at least comparable to the halo mass within the disk radius. Beyond the disk radius, one must use other diagnostics such as distances and velocities of companions to indicate whether more extended massive halos exist. However, it is interesting that the simple observation of whether a galaxy's arms are single or double (leading or trailing) can tell us about its mysterious hidden halo.

REFERENCES

Arp, H.: 1966, *Astrophys. J. Suppl.* **14**, 1.

Arp. H., Madore, B.F.: 1987, *A Catalogue of Southern Peculiar Galaxies and Associations*, Cambridge Univ. Press, Cambridge.

Athanassoula, E.: 1978, *Astron. Astrophys.* **69**, 395.

Athanassoula, E.: 1984, *Physics Reports* **114**, 319.

Blackman, C.P.: 1982, *Monthly Notices Roy. Astron Soc.* **200**, 407.

Byrd, G.G.: 1976, *Astrophys. J.* **208**, 688.

Byrd, G.G.: 1977, *Astrophys. J.* **218**, 86.

Byrd, G.G.: 1978, *Astrophys. J.* **226**, 70.

Byrd, G.G., Valtonen, M.J., Sundelius, B., Valtaoja, L.: 1986, *Astron. Astrophys.* **166**, 75.

Byrd, G.G., Saarinen, S., Valtonen, M.J.: 1986, *Monthly Notices Roy. Astron. Soc.* **220**, 619.

Donner, K.J.: 1979, Ph.D. Thesis, University of Cambridge.

Elmegreen, D.M., Elmegreen, B.G.: 1982, *Monthly Notices Roy. Astron. Soc.* **201**, 1021.

Elmegreen, B.G., Elmegreen, D.M.: 1983, *Astrophys. J* **267**, 31.

Elmegreen, D.M., Elmegreen, B.G.: 1987, *Astrophys. J.* **314**, 3.

Goldreich, P., Tremaine, S.: 1978, *Astrophys. J* **222**, 850.

Kalnajs, A.J.: 1975, *La dynamique des galaxies spirales*, ed. L. Weliachew, p. 103, CNRS, Paris.

Kormendy, J., Norman, C.A.: 1979, *Astrophys. J.* **233**, 539.

Lindblad, P.O.: 1960, *Stockholms Obs. Ann.* **21**, No. 3.

Miller, R.H.: 1976, *J. Comp. Phys.* **21**, 400.

Nilson, P.: 1973, *Uppsala General Catalogue of Galaxies*, Uppsala Astron. Obs. Ann., vol. 6.

Pasha, I.I.: 1985, *Soviet Astron. Lett.* **11**, 1.

Sharp, N.A., Keel, W.C.: 1985, *Astron. J.* **90**, 469.

Sundelius, B., Thomasson, M., Valtonen, M.J., Byrd, G.G.: 1987, *Astron. Astrophys.* **174**, 67.

Thomasson, M., Donner, K.J., Sundelius, B., Byrd, G.G., Huang, T.Y. and Valtonen, M.J.: 1989, *Astron. Astrophys.* (in press).

Toomre, A.: 1969, *Astrophys. J.* **158**, 899.

Toomre, A.: 1981, *The Structure and Evolution of Normal Galaxies*, eds. S.M. Fall, D. Lynden-Bell, p. 111, Cambridge University Press, Cambridge.
de Vaucouleurs, G.: 1958, *Astrophys. J.* **127**, 487.
de Vaucouleurs, G., de Vaucoleurs, A., Corwin, H.G.: 1976, *Second Reference Catalogue of Bright Galaxies*, University of Texas Press, Austin, London.
Vorontsov-Velyaminov, B.A.: 1959, *Atlas and Catalogue of Interacting Galaxies*, Part I., Sternberg Institute, Moscow.

Figure 1. Leading arm formed in a retrograde encounter simulation.

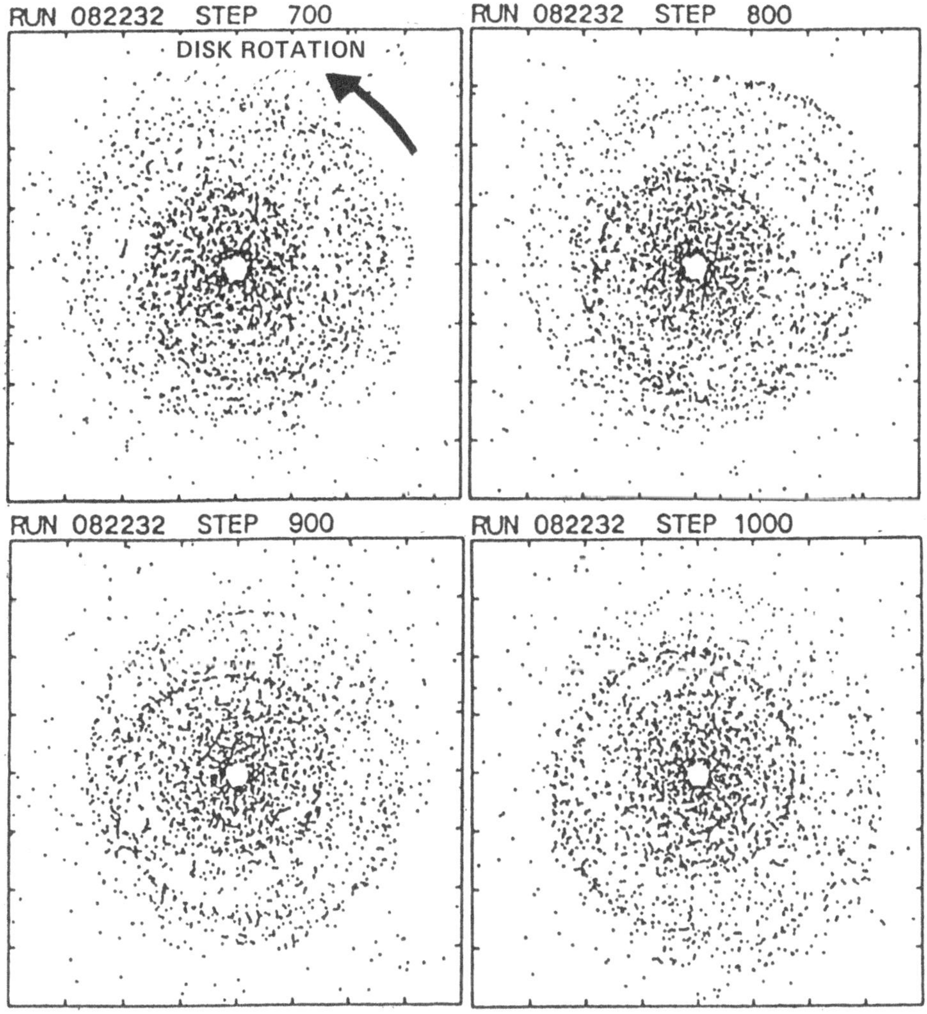

Figure 2. NGC 4622. Note candidate for a leading spiral arm inside ring near the nucleus. Compare to last frame of Figure 1. Photo - National Optical Astronomy Observatories.

G. Byrd acknowledges support by the National Science Foundation through EPSCoR grant RII 8610669.

ON CONSTRUCTION OF ANALYTIC SOLUTIONS TO THE THREE BODY PROBLEM BY USE OF COMPUTER EXPERIMENTS

M.J. Valtonen, S. Mikkola
Turku University Observatory, University of Turku SF 20500 Turku, Finland
and
T.Y. Huang
Department of Astronomy, University of Nanjing, China

Abstract: New analytic approximations of the general three-body problem are given. They are obtained by making use of approximate analytic theory as well as 26000 computed orbits. These solutions give the energy of the binary after the three-body system has broken up, or alternatively the terminal velocity of the escaping body. It is shown that the hard binary scattering and the break-up of bound three-body systems are described by the same analytic expression.

1. Introduction

In the general three-body problem few exact solutions are known. In practical applications it is usually adequate to construct approximate solutions by calculating a necessary number of three-body orbits and summarizing them in the form of analytic formulae. Even though the procedure is straightforward, it requires large amounts of computer time to cover the many dimensional orbit space satisfactorily. In fact, only a small part of the phase-space is covered to-date.

The existing analytic solutions may be classified in two categories:
(1) solutions containing all relevant initial parameters, and
(2) solutions in statistical the sense, which apply to ensembles of three-body orbits.

Examples of the first kind are given by Saslaw, Valtonen and Aarseth (1974), and Valtonen (1975), while the solutions of the second kind are found e.g. in Heggie (1975), Valtonen and Heggie (1979), Hut and Bahcall (1983) and Hut (1984). The construction of analytic formulae is greatly facilitated by analytic theory, even if approximate, such as the theories of Heggie (1975) and Monaghan (1976 a,b).

Recently we have derived new analytic expressions for three different kinds of three-body systems:

(a) interaction of a circular binary and a parabolic third body (solution of category 1),

Celestial Mechanics **45**: 37–43, 1989.

(b) break-up of a bound three-body system with different values of total angular momenta (solution of category 2), and

(c) scattering of a mildly hyperbolic third body by a binary (solution of category 2).

2. Circular binary

Consider a binary with masses m_1 and m_2 in a circular orbit (eccentricity $e = 0$) and a third body with mass m_3 passing by it in a parabolic orbit. The unperturbed pericenter distance q is greater than the binary semi-major axis a by a factor which is large enough for the binary to survive the encounter, i.e. no exchanges are allowed. We study arbitrary values of inclination i between the binary orbit and the parabolic orbit, as well as the whole range of parameters ω and Ω (argument of pericenter and longitude of ascending node, respectively).

Similarly to Heggie (1975), an expression for the change of the binding energy B_0 of the binary is derived. In the first order perturbation theory we obtain

$$\left(\frac{\Delta B}{B_0}\right)_1 = m_3(q/a)^{-\frac{3}{2}}(m_1+m_2)^{-\frac{1}{2}}[f_1 \sin 2(\Omega+\omega)+f_2 \sin 2(\Omega-\omega)+f_3 \sin(2\Omega)] \quad (1)$$

where f_1, f_2 and f_3 are functions of i and q. This functional form agrees with numerical experiments by Saslaw *et al.* (1974) and Valtonen (1975) except that there should be an additional term which does not average to zero. We suggest that this term should take the form

$$\left(\frac{\Delta B}{B_0}\right)_2 = {m_3}^2/(m_1+m_2)(q/a)^{-3} f_4, \quad (2)$$

where f_4 is also a function of i and q. The total change of the binary binding energy is then

$$\frac{\Delta B}{B_0} = \left(\frac{\Delta B}{B_0}\right)_1 + \left(\frac{\Delta B}{B_0}\right)_2 . \quad (3)$$

The functions f_1,..,f_4 are obtained by calculating 18000 numerical orbits with different values of the orbital parameters and by fitting results usin the least-squares method to suitable functions. The case studied had $m_1 = m_2 = \frac{1}{2}m_3$. It was subsequently checked that the results are valid for arbitrary mass ratios m_1/m_2 while the fit is good only around $m_3 \sim m_1 + m_2$. Figure 1 illustrates the agreement between the fitting function and the experiments for one set of initial parameters.

The functions f_1, .., f_4 as well as other details of the work are found in Huang and Valtonen (1987).

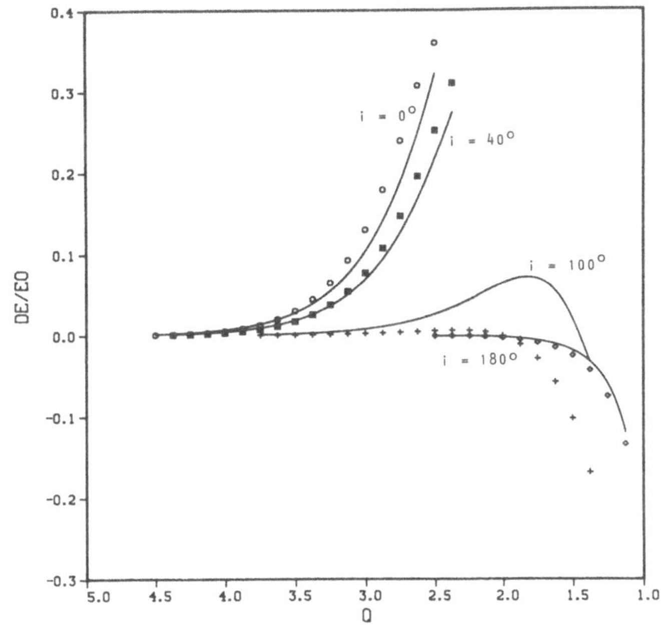

Figure 1: Binary energy change $\Delta E/E_0$ as a function of q.
Continuous curves are from the fitting equations. Data points are from numerical experiments. The used inclinations are shown in the figure and the other angles were $\Omega = 36^0$ and $\omega = 97^0$.

3. Bound systems

Here we study a binary of masses m_1 and m_2 interacting with a third body of mass m_3. The results are given in a system of units where the gravitational constant $G = 1$ and the total mass equals 3 units.

The initial orbit of the third body is elliptic and weakly bound to the binary. Orientations of the initial two-body orbits and the squares of eccentricities were uniformly randomized. However, the quantity

$$w = -c^2 E \tag{4}$$

which is the product of the square of the total angular momentum c and the total energy E was kept constant within each set of 400 experiments. Five different values of the parameter w were studied. As is well known, this quantity is the only (known) integral independent of scale (e.g. Donnison 1983). The values of w range from zero to a maximum, at which a strong three-body interaction is still possible. (For large w the outer orbit passes the binary at a typical distance $q \sim 0.75aw$ if the masses and gravitational constant are put equal to one, and a is the semi-major axis of the binary.)

The following mass families were studied: equal masses, as well as three families with masses $0.75, 1.0$ and 1.25, with each of the masses being the initial outer mass m_3 in turn. The total number of experiments was 8000. Among these experiments 145 turned out to be so long lived that the integration was halted before they had broken apart, but as a rule the three-body systems break-up at some point in time by ejecting one of the bodies (however, there are exceptions (Henon,1976)).

The most relevant outcome distribution is that of the binding energy B of the surviving binary. However, it is more convenient to use the parameter

$$z = |E|/B, \tag{5}$$

which is confined to the interval $(0, 1)$.

From previous work we know that, in case of naturally distributed angular momentum, the distribution of z is well represented by the power law

$$f(z) = (g + 1)z^g \tag{6}$$

where $g = 2.5$ (Saslaw *et al.* 1974, Heggie 1975). However, we find that the distribution depends strongly on w. A maximum likelihood fit to the data gives new power laws depending on values of w. These are $g = 1.0, 3.3, 5.5, 10.1$ and 15.9 corresponding to $w = 0, 1, 2, 3$, and 4. However, a power law of the form (6) is not a very satisfactory way to represent the data at $w > 0$, since it assumes that z may always obtain values down to $z \sim 0$. Actually there should be a lower limit to $z = z_i$ as explained by Mikkola and Valtonen (1986). On physical grounds a plausible form for the minimum z_i was deduced to be

$$z_i = 1 - \sqrt{1 - w/w_x}. \tag{7}$$

The corresponding distribution function of z, replacing Eq.6, is

$$f(z, w) = (g + 1)(z - z_i)^g/(1 - z_i)^{g+1}, \tag{8}$$

where also g varies with w. A maximum likelihood fit to the experimental data gives the approximations

$$w_x \approx 6.45$$

$$g \approx 1.0 + 1.73w \tag{9}$$

We note that the maximum value of w in Eq.7 is essentially the same as the 'no exchange' boundary for equal masses (this value $= 6.25$ is actually the one belonging to the rectilinear Lagrangian configuration (Donnison 1983, Marchal, Yoshida and Yi-Sui 1984)). This coincidence is probably explained by noting that around this value the interaction becomes very weak. Fig.2 illustrates the agreement between the analytic distribution function (Eq.8) and the experiments. Further results and other details of this work are found in Mikkola and Valtonen (1986).

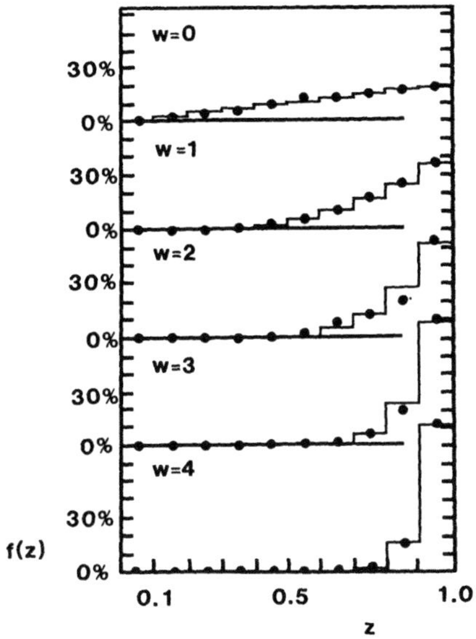

Figure 2: A comparison of the distributions $f(z,w)$ calculated from equation (8) with the empirical values (dots) from numerical experiments.

4. Scattering

The above results were derived for bound systems. Here we test the hypothesis that the distribution is valid also when the binary and the third body meet in an initially hyperbolic orbit of relative velocity V (at infinity). Hut (1984) has carried out scattering experiments between a binary and a third body. His results are given in the form of cross-section σ_Δ for the relative change Δ of the binding energy of the binary. On the other hand, we can calculate the cross-section

$$\sigma_z = \int f(z,w)\sigma_w \, dw, \tag{11}$$

where σ_w is the differential cross-section for w and $f(z,w)$ is given by Eqs.8 and 9. It is possible to evaluate Eq.11 in a scattering situation where σ_w is known. If the incoming velocity V is given in units of the critical velocity at which the total energy becomes zero, then we may write $z = (1 - V^2)/(1 + \Delta)$. Using this it is possible to carry out a conversion between σ_z and σ_Δ. A physically more significant quantity is the rate at which random encounters feed energy between the binding of the binary and its surrounding star field. This is proportional to $\Delta\sigma_\Delta$. It is convenient to further convert to the variable $\eta = 1 - z = \Delta/(1+\Delta)$ and remove the gravitational focusing and binary size factors. Thus we calculate $k(\eta) = \Delta(1+\Delta)^2 V^2 \sigma_\Delta/(\pi a^2)$. The result is shown in Fig.3 (solid line) when $V \approx 0.1$. We compare it with the experimental data from Hut (1984)(dots with error bars). The good agreement breaks down when V is increased to much greater values than used above. Further details of this work are found in Mikkola (1986).

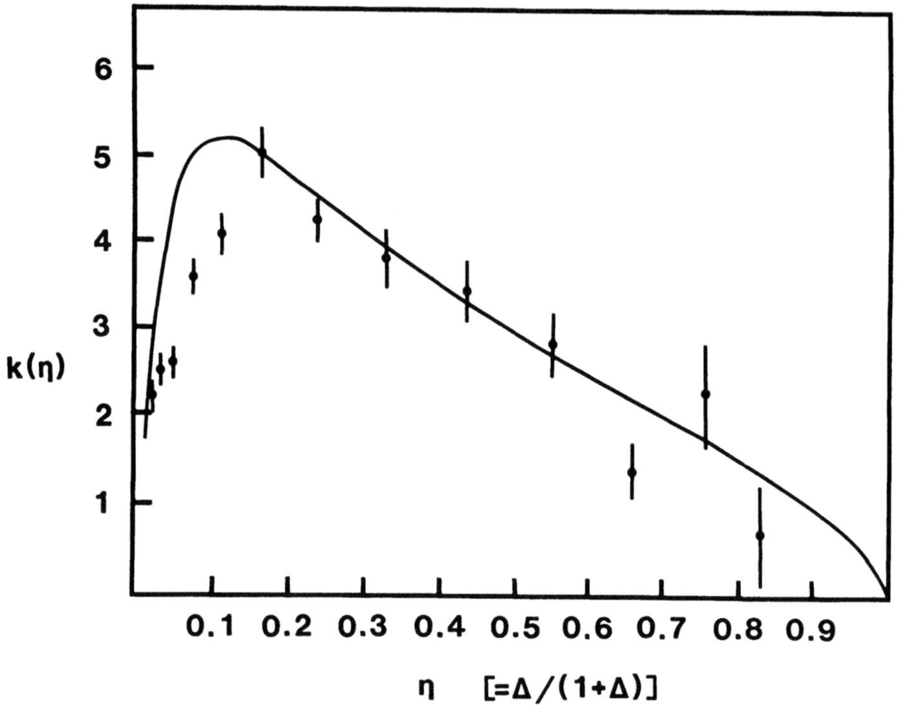

Figure 3: A comparison of the energy exchange cross-section (Hut 1984) with the semi-analytic model (full drawn curve). The ordinate is the function $k(\eta)$ which is explained in the text.

We conclude that the total energy and angular momentum are the decisive quantities determining the statistical behaviour of a three-body system. This is true as long as the system is either initially bound or the incoming energy of the third body is small compared to the binding of the binary; in other words if there is enough mixing in the orbits. However if the third body has a high incoming velocity then this quantity becomes an additional important parameter (in this case many of the interactions are of short duration and thus there is not enough orbital mixing to allow a purely statistical description).

References

Donnison,J.R. :(1983), *Cel. Mech.*, **32**, 145.
Heggie, D.C. :(1975), *Monthly Notices Roy. Astron. Soc.*, **173**, 729.
Henon, M. :(1976), *Cel. Mech.*, **13**, 267.
Huang, T.-Y. and Valtonen, M.J. :(1987), *Monthly Notices Roy. Astron. Soc.*, **229**, 333
Hut, P. :(1984), *Astrophys. J. Suppl. Ser.*, **55**, 301.
Hut, P. and Bahcall, J.N. :(1983), *Astrophys. J.*, **268**,319.
Marchal, C., Yoshida, J. and Yi-Sui, S. :(1984), *Cel. Mech.*, **34**,65.
Mikkola, S. :(1986), *Monthly Notices Roy. Astron. Soc.*, **223**, 757.
Mikkola, S. and Valtonen, M.J. :(1986), *Monthly Notices Roy. Astron. Soc.*, **223**, 269.
Monaghan, J.J. :(1976a), *Monthly Notices Roy. Astron. Soc.*, **176**, 63.
Monaghan, J.J. :(1976b), *Monthly Notices Roy. Astron. Soc.*, **177**, 583.
Saslaw, W.C., Valtonen, M.J. and Aarseth, S.J. :(1974), *Astrophys. J.*, **190**, 253.
Valtonen, M.J. :(1975), *Memoirs Roy. Astron. Soc.*, **80**, 61.
Valtonen, M.J. and Heggie, D.C. :(1979), *Cel. Mech.*, **19**, 53.

THE USE OF COMPUTER ALGORITHMS IN THE CONSTRUCTION OF A THEORY OF THE LONG-PERIOD PERTURBATIONS OF SATURN'S SATELLITE HYPERION

P.J. Message
Department of Applied Mathematics and Theoretical Physics,
The University,
Liverpool, L69 3BX,
U.K.

ABSTRACT. This paper briefly indicates some computer-based techniques which have been developed in the construction of a theory of the long-period perturbations of the orbit of Saturn's satellite Hyperion, in the method used to expand the disturbing function and its derivatives, in the fitting into the theory of observationally-derived values of the co-efficients of the main long-period terms, and in the construction of a differential correction system to find a dynamically consistent set of values of these co-efficients.

1. INTRODUCTION

The perturbations of the orbit of Hyperion are dominated by the effect of Titan, and in particular by the consequences of the 4:3 near-commensurability of their orbital periods, very much closer than for any other pair of bodies in the solar system, with the exception of the Trojan-type cases. The difficulties in constructing a theory are increased by the large and variable eccentricity of Hyperion (mean value about 0.10481, the largest periodic term having amplitude 0.0245) and the large ratio, 0.8251, of the major semi-axes of the two orbits. This paper concentrates on a brief indication of some of the computer-based techniques which have been developed to meet the particular problems encountered in the construction of a theory of the long-period perturbations of the satellite, describing briefly the development of the disturbing function, the calculation of the second-order terms, the fitting of the solution, and the construction of a differential correction process used in the comparison with observational data. Another paper will give a more complete account of the theory and its development, and of the results achieved.

A test of the fit of the theory to observation is given by a comparison of the estimates of the mass of Titan derived from each of the main long-period terms and from the mean motion of the apse. The previous theory, by Woltjer (1928), introduced special methods, but left these estimates more discrepant than the uncertainties of the observational data explain. That theory was restricted, from the nature of the

Celestial Mechanics **45**: 45–53, 1989.

method used, to the first power only of Titan's eccentricity. The present work includes effects of up to the third power of Titan's eccentricity. In fact it has proved necessary to include second-order perturbations to produce agreement between the estimates of the mass of Titan, using the observations between 1875 and 1922.

A further objective of the present work was to develop a system of differential correction of the co-efficients of the long-period terms, to find a set of values of these co-efficients consistent with dynamical theory.

2. METHOD OF DEVELOPMENT OF THE DISTURBING FUNCTION AND ITS DERIVATIVES

Of the disturbing function for the perturbation of Hyperion by Titan, that part relating to motion in the orbit plane may be expressed as a triple Fourier series in the angular arguments

$$\phi = \lambda - \lambda', \quad \theta = 4\lambda - 3\lambda' - \varpi, \text{ and } \sigma = \varpi - \varpi', \tag{1}$$

(where λ and ϖ refer to Hyperion's orbit, and λ' and ϖ' to Titan's) of which θ is the critical argument associated with the very close commensurability of orbital periods, ϕ is the short-period argument, and σ that of long period. Now convergence of the critical and long-period terms is very slow, because of the large ratio of the major semi-axes, and the large eccentricity, but convergence is improved by using, in place of θ, the quantity ω, defined by

$$\theta = \pi + q \sin \omega, \tag{2}$$

for a suitably chosen constant q. This change of variable was used by Woltjer in his solution of a simplified form of the equation governing θ, but it is used here simply to gain the advantage of rapid convergence. Use of $q = 54^o$, just larger than the greatest departure of θ from 180° in the motion, ensures that only real values of ω occur in the motion, so that the fall-off of the co-efficients in a Fourier series over ω gives a true indication of how well the series represents the function. We note that

$$\sum_{j=0}^{\infty} a_j \cos j\theta = c_0 + \sum_{m=1}^{\infty} c_m \cos 2m\omega ,$$

and

$$\sum_{j=1}^{\infty} b_j \sin j\theta = \sum_{m=0}^{\infty} d_m \sin (2m+1)\omega , \tag{3}$$

where

$$c_0 = \sum_{j=0}^{\infty} (-1)^j a_j J_0(jq),$$

$$c_m = 2 \sum_{j=1}^{\infty} (-1)^j a_j J_{2m}(jq) \qquad \text{(for } m \neq 0\text{)},$$

$$\text{and} \quad d_m = 2 \sum_{j=1}^{\infty} (-1)^j \; b_j \; J_{2m+1}(jq).$$

The rapid convergence is helped by the rapid fall-of with increasing p of the Bessel function $J_p(x)$, even for quite large x.

Dependence of the disturbing function and its derivatives on a and e is given by Taylor series in powers of $a - a_o$ and $e - e_o$, where a_o and e_o are constant values, chosen close to the mean values taken in the motion. Powers up to $(e - e_o)^6$ were found to give adequate accuracy to match the observational data.

To provide the means to develop the long-period theory, a programme was written in Algol to calculate the means over ϕ of the disturbing function, and its derivatives with respect to a and e up to the required orders, evaluated for $a = a_o$ and $e = e_o$, and for a set of values of ω and of σ. Fourier analysis over σ and ω was carried out, and it was found that use of eight values of σ, and five of ω, was adequate. (For example, the term independent of σ in the series for $\frac{1}{Gm'}\frac{\partial \bar{R}}{\partial e}$ has the expansion over ω:

$$\begin{aligned}
&-1.307\ 7519 \\
&-0.016\ 5765 \cos 2\omega \\
&+0.001\ 5534 \cos 4\omega \\
&-0.000\ 0584 \cos 6\omega \\
&+0.000\ 0018 \cos 8\omega \ .)
\end{aligned}$$

Since the Fourier series over σ gave terms up to those in 3σ, the contributions corresponding to the third power of Titan's eccentricity were included. Then $\bar{R}$, the mean over ϕ of the disturbing function, R, is represented by a series

$$\bar{R} = \sum_{m=0}^{\infty} \sum_{k=0}^{\infty} \left\{ L_{m,k} \cos 2m\omega \cos k\sigma + M_{m,k} \sin(2m+1)\omega \sin k\sigma \right\} ,$$

with similar series for the derivatives of $\bar{R}$. Derivatives with respect to θ are given by

$$\frac{\partial \bar{R}}{\partial \theta} = \sum_{m=0}^{\infty} \sum_{k=0}^{\infty} \left\{ L^*_{m,k} \sin (2m+1)\omega \cos k\sigma + M^*_{m,k} \cos 2m\omega \sin k\sigma \right\},$$

where

$$L^*_{m,k} = \frac{4}{q} \sum_{s=1}^{\infty} (-1)^s \, (m + s) \, L_{m+s,k} ,$$

$$M^*_{0,k} = \frac{1}{q} \sum_{s=0}^{\infty} (-1)^s \, (2s + 1) \, M_{s,k} ,$$

$$\text{and} \quad M^*_{m,k} = \frac{2}{q} \sum_{s=1}^{\infty} (-1)^s \, (2m + 2s + 1) \, M_{m+s,k} .$$

3. CALCULATION OF THE SECOND-ORDER TERMS IN THE LONG-PERIOD THEORY

A Lie series transformation is used to form the long-period system, in which ϕ is ignorable. To first order, the disturbing function, R, is replaced by $\bar{R}$, its mean over ϕ, and, to second order, to it must be added the mean over ϕ of the Poisson bracket expression $\frac{1}{2}\{R, S_1\}$, where S_1 is that solution of

$$n \frac{\partial S_1}{\partial \lambda} + n' \frac{\partial S_1}{\partial \lambda'} = \bar{R} - R, \qquad (4)$$

which is periodic in ϕ with no constant term. To compute this, we note that (4) may be written

$$(n - n') \frac{\partial S_1}{\partial \phi} + (4n - 3n') \frac{\partial S_1}{\partial \theta} = \bar{R} - R, \qquad (4')$$

and that $4n - 3n'$ is very close to zero, because of the very close commensurability, so that

$$S_1(\phi_0, \theta) \simeq \frac{1}{2\pi(n-n')} \int_{\phi_0}^{2\pi+\phi_0} \phi(\bar{R} - R)\, d\phi, \qquad (5)$$

which gives a starting value for S_1, for a given θ, from which S_1 itself can be computed for any ϕ, using

$$S_1(\phi, \theta) \simeq S_1(\phi_0, \theta) + \frac{1}{(n - n')} \int_{\phi_0}^{\phi} (\bar{R} - R)\, d\phi \qquad (6)$$

so that S_1 can be computed as a function of ϕ and θ. Its derivatives with respect to the orbital elements are found correspondingly, using equations found by differentiation of (4').

The equations for the Lie series transformation show that the means over ϕ of λ, ϖ, a, and e in the actual motion are equal, to second order, to their counterparts in the long-period system of equations, plus the means over ϕ of, respectively, the Poisson bracket expressions

$$\left\{ \frac{2}{na}\frac{\partial S_1}{\partial a} - \frac{Y}{na^2}\frac{\partial S_1}{\partial e},\ S_1 \right\}, \quad \left\{ -\frac{X}{na^2}\frac{\partial S_1}{\partial e},\ S_1 \right\},$$

$$\left\{ \frac{2}{na}\frac{\partial S_1}{\partial \lambda},\ S_1 \right\}, \text{ and } \left\{ -\frac{Y}{na^2}\frac{\partial S_1}{\partial \lambda} - \frac{X}{na^2}\frac{\partial S_1}{\partial \varpi},\ S_1 \right\}, \qquad (7)$$

where $X = \sqrt{(1 - e^2)}/e$, and $Y = \sqrt{(1 - e^2)}(1/e - X)$.

4. THE FITTING OF THE SOLUTIONS TO THE EQUATIONS

The form of the solution of the long-period system of equations which corresponds to libration of the critical argument θ is based on double Fourier series in two linear functions of time, $\tau = \nu t + \tau_0$, with the period, about 21 months, of the libration, and $\zeta = \chi t + \zeta_0$, the linear part of the difference of the apse longitudes, with period about 18¾ years.

The solution takes the form

$$\theta = \pi + \sum \theta_{ij} \sin (i\tau + j\zeta),$$

$$\sigma = \zeta + \sum \varpi_{ij} \sin (i\tau + j\zeta),$$

$$a = a_{00} + \sum a_{ij} \cos (i\tau + j\zeta),$$

and $$e = e_{00} + \sum e_{ij} \cos (i\tau + j\zeta), \qquad (8)$$

the summations being over all integer pairs (i,j) with $i \geq 0$. These expressions must be substituted into the Lagrange equations, giving, on equating the co-efficients of each Fourier term, a series of equations relating the θ_{ij}, ϖ_{ij}, a_{ij}, e_{ij}, ν, and χ to the mass of Titan, leaving just two free quantities amongst them. The phase constants τ_0 and ζ_0 provide the two other constants of integration for the motion in the orbit plane. To carry out this substitution into the expressions for the disturbing function and its derivatives, we need double Fourier series expressions for $\sin (2m+1)\omega$ and $\cos 2m\omega$ as well as $\sin k\sigma$ and $\cos k\sigma$. We have

$$\sin \omega = \frac{1}{q} \sum \theta_{ij} \sin (i\tau + j\zeta),$$

from which we can express each $\sin (2m+1)\omega$ in the form $\sum B_{ij} \sin(i\tau+j\zeta)$ and each $\cos 2m\omega$ in the form $\sum C_{ij} \cos (i\tau+ j\zeta)$. In the course of these and later developments, it is frequently necessary to form the product of two double Fourier series of either of these two forms, and an Algol procedure has been written to do this, building up each co-efficient in the product series in turn. We also need expressions in these two forms for the quantities $\cos k\sigma$ and $\sin k\sigma$. To form them, we note that ϖ_{01} $(=-\beta$, say) is the largest co-efficient numerically in the expression for σ, so we put $\sigma_1 = \zeta - \beta \sin\zeta$, which is the dominant part of σ, and then

$$\cos k\sigma_1 = J_k(k\beta) + \sum_{m=1}^{\infty} \left\{ J_{k-m}(k\beta) + J_{k+m}(k\beta) \right\} \cos m\zeta ,$$

and $$\sin k\sigma_1 = \sum_{m=1}^{\infty} \left\{ J_{m-k}(k\beta) - J_{m+k}(k\beta) \right\} \sin m\zeta . \qquad (9)$$

We then use the Taylor series

$$\cos k\sigma = \cos k\sigma_1 - k(\sigma - \sigma_1) \sin k\sigma_1 - \tfrac{1}{2}k^2(\sigma - \sigma_1)^2 \cos k\sigma_1 \text{ \&ce.},$$

and $$\sin k\sigma = \sin k\sigma_1 + k(\sigma - \sigma_1) \cos k\sigma_1 - \tfrac{1}{2}k^2(\sigma - \sigma_1)^2 \sin k\sigma_1 \text{ \&ce.}, \qquad (10)$$

We also need the double Fourier series expansion of the quantity $X = \sqrt{(1 - e^2)}/e$, which appears on the right-hand side of the Lagrange equations, and in the Poisson bracket expressions which we need.

We note that $e_1 = e_{00} + e_{01}\cos\zeta$ is the dominant part of e, and use Taylor series in powers of $e - e_1$. The values of X, and its derivatives with respect to e, evaluated at $e = e_1$, are given by the expression

$$(-1)^m\, m!\, \kappa^{-m-1} \left\{ C_0^{(-m-1)}(\alpha) + 2\sum_{n=1}^{\infty} C_n^{(-m-1)}(\alpha)\cos n\zeta \right\}$$

$$+ \sum_{k=m}^{\infty} v_k \frac{(2k-1)!}{(2k-m-1)!}\, \kappa^{2k-m-1} \left\{ C_0^{(2k-m-1)}(\alpha) + 2\sum_{n=1}^{\infty} C_n^{(2k-m-1)}(\alpha)\cos n\zeta \right\}$$

$$+ d_m\, m!\, v_{(m-1)/2}, \qquad (11)$$

which gives X if $m=0$, and $\frac{d^mX}{de^m}$ if $m > 0$. The quantities arising in this expression are κ and α, defined by $\kappa(1 + \alpha^2) = e_{00}$ and $2\kappa\alpha = e_{01}$, and also $C_n^{(s)}(\alpha) = \frac{s(s-1)(s-2)\ldots(s-n+1)}{1.2.3.\ldots\ldots\ldots.n}\,\alpha^2 \sum_{m=0}^{\infty} u_{s,n,m}$

in which $u_{s,n,m} = \frac{(s-m+1)(s-n-m+1)}{m(m+n)}\,\alpha^2\, u_{s,n,m-1}$

with $u_{s,n,0} = 1$,

$v_k = (2k-3)\, v_{k-1}/(2k), \quad v_0 = 1,$

and d_m is 1 if m is odd, and zero if m is even.

An Algol programme has been set up to carry out these successive developments on being supplied with a set of values for the largest co-efficients in the expressions (8). This was run using values obtained from a new analysis of the opposition mean elements obtained by Woltjer in his reduction of the observations of Hyperion from 1875 to 1922. At this stage the Taylor series in powers of $e - e_0$ and $a - a_0$ were summed. (To indicate the convergence, successive contributions to that part of $\frac{1}{Gm'}\frac{\partial R}{\partial e}$ which is independent of τ and ζ are:

with $a = a_0$ and $e = e_0$:	-1.318 756 0
$e - e_0$ term:	+0.023 082 9
$a - a_0$ term:	-0.000 208 6
$(e - e_0)^2$ term:	-0.010 925 5
$(e - e_0)^3$ term:	+0.000 072 8
$(e - e_0)^4$ term:	-0.000 042 9
$(e - e_0)^5$ term:	-0.000 001 9
Sum:	-1.306 779 1

)

Then equating co-efficients of each Fourier term in each of the Lagrange

We note that $e_1 = e_{00} + e_{01} \cos \zeta$ is the dominant part of e, and use Taylor series in powers of $e - e_1$. The values of X and its derivatives with respect to e, evaluated at $e = e_1$, are given by the expression

$$(-1)^m \, m! \, \kappa^{-m-1} \left\{ C_0^{(-m-1)}(\alpha) + 2\sum_{n=1}^{\infty} C_n^{(-m-1)}(\alpha) \cos n\zeta \right\}$$

$$+ \sum_{k=m^*}^{\infty} v_k \frac{(2k-1)!}{(2k-m-1)!} \kappa^{2k-m-1} \left\{ C_0^{(2k-m-1)}(\alpha) + 2\sum_{n=1}^{\infty} C_n^{(2k-m-1)}(\alpha) \cos n\zeta \right\}$$

$$+ d_m \, m! \, v_{(m+1)/2} , \qquad (11)$$

(where m^* is the integral part of $\frac{1}{2}(m+3)$), which gives X if $m = 0$, and $\frac{d^m X}{de^m}$ if $m > 0$. The quantities arising in this expression are and , defined by $\kappa(1 + \alpha^2) = e_{00}$ and $2\kappa\alpha = e_{01}$, and also

$$C_n^{(s)}(\alpha) = \frac{s(s-1)(s-2)\ldots(s-n+1)}{1.2.3.\ldots\ldots\ldots.n} \alpha^2 \sum_{m=0}^{\infty} u_{s,n,m} ,$$

(in which $u_{s,n,m} = \frac{(s-m+1)(s-n-m+1)}{m(m+n)} \alpha^2 \, u_{s,n,m-1}$, with $u_{s,n,0} = 1$),

$$v_k = (2k-3)\, v_{k-1}/(2k), \text{ with } v_0 = 1,$$

and d_m is 1 if m is odd, and zero if m is even.

An Algol programme has been set up to carry out these successive developments on being supplied with a set of values for the largest co-efficients in the expressions (8). This was run using values obtained from a new analysis of the opposition mean elements obtained by Woltjer in his reduction of the observations of Hyperion from 1875 to 1922. At this stage the Taylor series in powers of $e - e_0$ and $a - a_0$ were summed. (To indicate the convergence, successive contributions to that part of $\frac{1}{Gm'} \frac{\partial \bar{R}}{\partial e}$ which is independent of τ and ζ are:

with $a = a_0$ and $e = e_0$:	-1.318 756 0
$e - e_0$ term:	+0.023 082 9
$a - a_0$ term:	-0.000 208 6
$(e - e_0)^2$ term:	-0.010 925 5
$(e - e_0)^3$ term:	+0.000 072 8
$(e - e_0)^4$ term:	-0.000 042 9
$(e - e_0)^5$ term:	-0.000 001 9
Sum:	-1.306 779 1

)

Then equating co-efficients of each Fourier term in each of the Lagrange equations gives a series of estimates of the mass of Titan, of which the best determined are given here, in units of 10^{-4} times the mass of Saturn:

from the mean motion of ϖ	:	2.346 ± 0.012
from the 21-month term in	θ:	2.361 ± 0.026
from the 18¾-year term in	e:	2.285 ± 0.053
from the 18¾-year term in	$\dot{\varpi}$:	2.420 ± 0.098
from the 21-month term in	e:	2.43 ± 0.21

These are consistent within the uncertainties, which follow from the standard errors of the co-efficients derived from the observational data. The weighted mean is 2.347 ± 0.016. (The value derived from the Voyager I flyby is 2.3664 ± 0.0008 (Tyler et al., 1981).)

5. DIFFERENTIAL CORRECTIONS

A programme was written to set up a differential correctection scheme, which was used both in calculating the standard errors of the estimates of the mass of Titan just given, and also in order to seek a dynamically consistent set of values of the co-efficients. Thus, for example, suppose that the Lagrange equation for ϖ, after substituting the Fourier series in τ and ζ, becomes

$$\dot{\varpi} = \frac{X}{na^2}\frac{\partial R}{\partial e} = \sum P_{ij} \cos(i\tau + j\zeta).$$

Then we need the partial derivatives of the co-efficients P_{ij} with respect to the most significant of the co-efficients θ_{ij}, ϖ_{ij}, e_{ij}, and a_{ij}. Then, from the matrix of partial derivatives, there were derived expressions for the changes in the most significant of these co-efficients, linear in changes in θ_{10}, (the amplitude of the free libration in θ), in n^* (the constant part of $\dot{\lambda}$), and in the mass of Titan. Then the values of the co-efficients derived from the observations give equations of condition for this differential correction process. Two solutions have been carried out, one using the co-efficients derived from the 1875 to 1922 observations, as described above, and the other using values of the co-efficients obtained in an analysis of observations from 1967 to 1983 (Taylor et al., 1987). Each of these two independent solutions leads to a set of values for the co-efficients of the main long-period terms, each set consistent with dynamical theory. Initial results show close agreement between the two sets, indicating good representation by the theory of the long-period perturbations of the satellite. A fuller account of the work will be given elsewhere.

6. CONCLUSIONS

The special computer-based procedures, whose construction is indicated here, to meet the particular difficulties in the case of Hyperion, do enable the development of expressions for the long-period perturbations, which fit well to the observational data, and also enable the construction of a differential correction scheme able to find dynamically consistent sets of co-efficients of the main long-period perturbations from a fit to observationally derived values.

REFERENCES

Taylor, D.B., Sinclair, A.T., & Message, P.J., 1987: in Astronomy and Astrophysics, 181, 383-390.

Tyler, G.L., Eshleman, V.R., Anderson, J.D., Levy, G.S., Lindal, G.F., Wood, G.E., & Croft, T.A., 1981: in Science, 212, 201-206.

Woltjer, J., 1928: in Annalen van der Sterrewacht Leiden, XVI, part 3.

SOME NUMERICAL ASPECTS IN A SEARCH FOR l=1 PERIODIIC ORBITS FOR HECUBA ASTEROIDS

T. Kiang *Dunsink Observatory, Dublin, Ireland*

I have searched for periodic orbits with periods equal to one librational period in the critical resonance angle (l=1 orbits) in the simple circular model of Sun-Jupiter-asteroid, and this paper reports some numerical aspect of the work. I integrated the usual equations of motion in rotating coordinates and velocities x, y, u, v, considered as a system of four first-order equations. The Jacobi constant C was kept strictly constant by adjusting the larger of u and v at the end of each integration step. The computer used was a VAX 11/780 and double precision arithmetic was used, which had a precision of 16 decimal digits, largely equivalent to 16 decimal places in the four variables used.

Hamming's (1959) fourth-order, predictor-modifier-corrector was used with a variable step h for the integration. Before each run, the value of a parameter labelled TOL is fixed, and h is to be halved (or doubled) whenever the absolute difference between the predicted and corrected values, $|p-c|$, given by Hamming's scheme, for any one of the variables exceeds 2 TOL (or falls below 0.01 TOL). For a typical orbit of interest with a period of some 425 yr, the following empirical relation between the pre-set value of TOL and the number of steps N taken to complete the integration was obtained:

$$\lg N = 5.355 - 0.21\ (\lg \mathrm{TOL} + 13) \qquad (1)$$

Generally speaking, the accuracy of numerical integration of differential equations is ultimately conditioned by the propagation

Celestial Mechanics **45**: 55–60, 1989.

and accumulation of (i) truncation error e_{tr} and (ii) round-off error e_{ro}, operating at each step of the integration. For a first-order equation, $dy/dt=f(t,y)$, Hamming's method is known to be stable in the sense that, if $\partial f/\partial y < 0$, then e_{tr} has an exponentially decreasing after-effect, so that the error due to truncation accumulated at the end of a large number N of steps is not very different from that at a single step. On the other hand, it is known that, if $\partial f/\partial y > 0$, then all methods will give an exponentially increasing after-affect. In the present case, we are dealing with a fourth-order system and the natural analogue of $\partial f/\partial y$ is the determinant $||\partial f_i/\partial y_k||$, and this, in our case, has alternating signs, so neither of the above conclusions apply. Obviously, the average truncation error at a single step, $\langle e_{tr}\rangle$, must be regarded as the absolute lower limit for the error accumulated after N steps. A reasonable guess for the latter might be the half-power law, $\langle e_{tr}\rangle\sqrt{N}$.

In Hamming's method, the expected value of e_{tr} can be estimated from the known value of $p-c$, $e_{tr}=-(9/121)(p-c)$. Assuming the values of $|p-c|$ to be uniformly distributed between 2 TOL and 0.01 TOL (reasonable, since $p-c$ can be either positive or negative), we have

$$\langle e_{tr}\rangle = 0.075 \text{ TOL}. \tag{2}$$

From (1) and (2), we have

$$\lg (\sqrt{N}\langle e_{tr}\rangle) = -11.45 + 0.895 (\lg \text{TOL} + 13) \tag{3}$$

To find something about the propagation of the round-off error, e_{ro}, in the present case, I adopted an experimental approach: I chose a fixed value of h so small that e_{tr} can be regarded as negligible compared to e_{ro} and integrated the equations for a precessing Keplerian ellipse (setting the mass of Jupiter to 0), for which the precise result is known analytically. Then the march of the difference between the numerical analytical results is none other than the evolution of e_{ro}. I took h=2.0(-5), and an ellipse with e=0.15 and precessing by 1/36 of the circle per conjunction. Starting at a

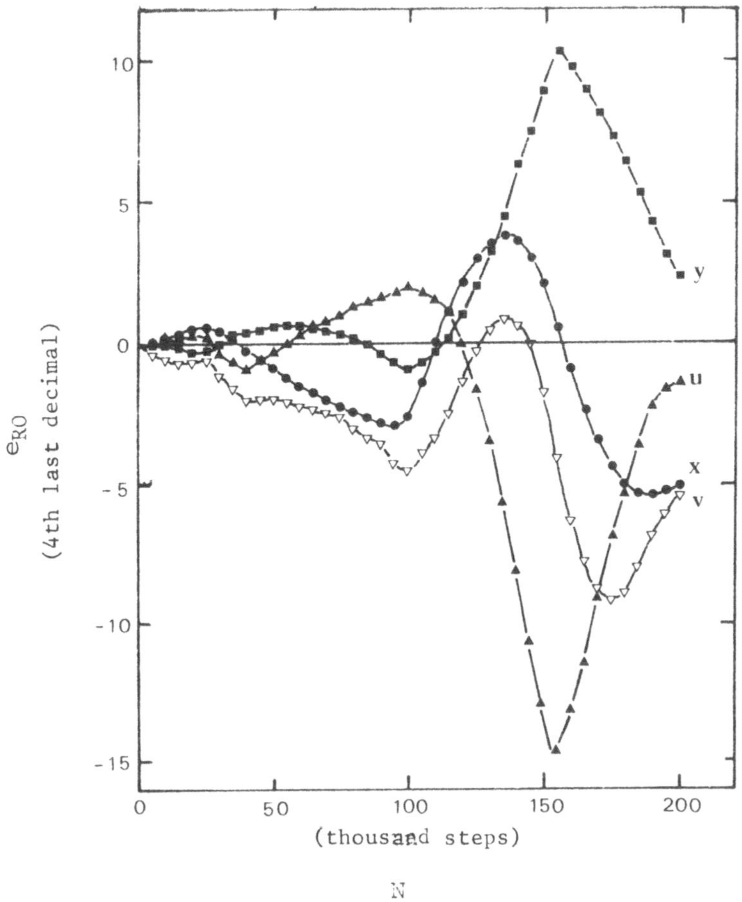

Fig. 1 Empirically determined round-off error in each of the 4 variables *x*, *y*, *u*, *v*, as function of the number of steps of integration. The ordinate is in units of the fourth last decimal place. The evolution of the round-off error takes the form of oscillation with increasing amplitude and wavelength.

perihelion conjunction, the integration was followed for 400 000 steps. Results for the first half of the run are shown in Fig.1. We see that the evolution of e_{ro} takes the form of oscillations of

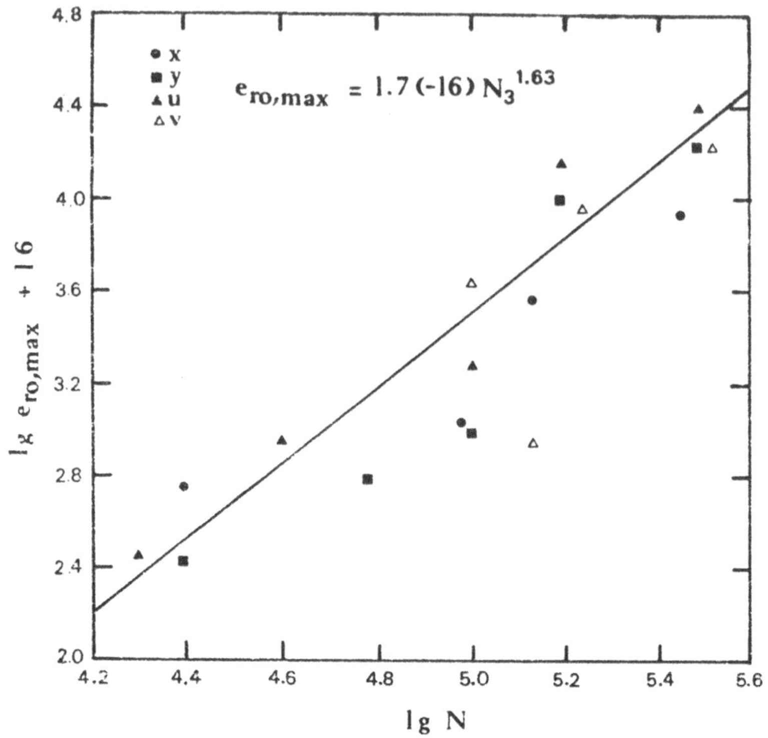

Fig. 2 Peak amplitude - location plot for 4 round-off error curves in a numerical integration using double precision arithmetic on a VAX 11/780 of the equations of the circular model of three bodies. N is the number of integration steps. The mean relation is indicated by the solid line and is shown.

growing amplitudes and wavelengths. (The additional downward trend in the first part of the *v*-curve is probably due to the fact that, for the first 10 000 steps covering the first 1/8 of a synodic period, *v* was the "adjusted" variable mentioned earlier, and so it had a "bad start"). Fig.2 is the maximum amplitude-location plot; the mean relation shown is

$$\lg e_{ro,max} + 16 = -4.67 + 1.63 \lg N. \tag{4}$$

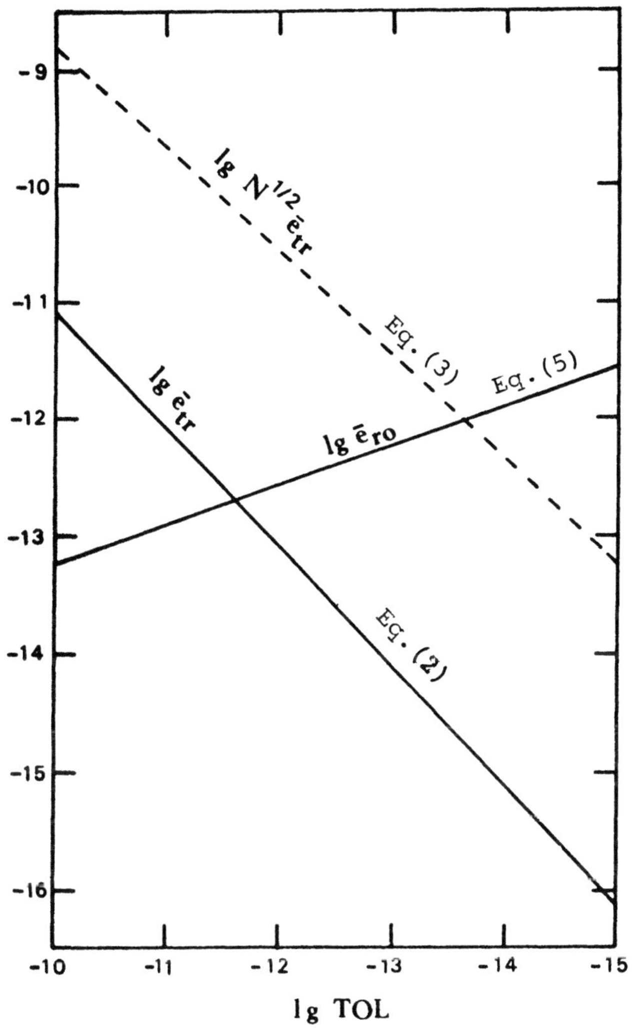

Fig. 3 Comparisons of Eqs.(2) and (3) with Eq.(5) show that the accuracy of numerical integration of a periodic orbit of period of around 400 years in the circular model of three bodies using double precision arithmetic on a VAX 11/780 Computer will be limited by *truncation error* for lg TOL > -11.5 and by *roundoff error* for lg TOL < -13.5. The actual errors in integrations with lg TOL = -13.0 (corresponding to about 250 000 integration steps) were found to be quite close to the value given by Eq.(5) for the roundoff error.

We can reasonably take half the maximum error as the average error. Using (4) and (1), we then have

$$\lg \langle e_{ro} \rangle = -12.24 - 0.34\ (\lg \mathrm{TOL} + 13) \qquad (5)$$

The three relations (2), (3), and (5) are graphed in Fig.3. It may be concluded that the accuracy of the present integration will be truncation error limited if lg TOL > -11,6, and round-off error limited if lg TOL < -13.6.

Using TOL = 1.0(-13), I calculated a number of l=1 periodic orbits with periods T between 179 and 471 yr (containing between 11 and 40 synodic periods). In all cases, the initial coordinates and velocities were reproduced to 12 decimal places after T. This degree of precision was confirmed by integration in the reversed direction (Szebehely 1967): in all cases, the misclosure at the end of the round trip was in the 13th decimal place. Referring to Fig.3, we see that this result is in good agreement with the expected accumulated round-off error. Thus, at lg TOL = -13, the accuracy of the presnt integration is round-off error limited and there is an indication that the behaviour of the truncation error is even better than the "mild" half-power law when integrating the equations of the circular model by Hamming's method.

ACKNOWLEDGEMENT That $||\partial f/\partial y||$ in the present case has changing signs so that the results when it has consistently one sign do not apply was pointed out to me by a Referee of this paper.

REFERENCES

Hamming R.W., (1959), *J.Ass.Comput.Mach.*6, 37-47.

Szebehely V., (1967), "Theory of Orbits", p.518 Academic Press.

SHORDE I Program System and Applications

He Miaofu, Zhu Wenyao, Feng Chugang, Huang Cheng and Chang Hua
Shanghai Observatory, Academia Sinica
Shanghai, China

Abstract. As one of the Analysis Centers of SLR global data, a program system named SHORDE I has been developed at Shanghai Observatory. This software package is used for processing observational data of satellites and estimating satellite orbit and dynamical and geodetic parameters. The reference systems and the force model required for constructing the package were first realized at the cm-level. In order to save running time and the storage on the computer, a multistage-multiarc procedure has been especially designed.

The technique has been applied to the SLR data from LAGEOS and earth rotation parameters and precise station coordinates were derived using the data from Nov. 1984 to Jan. 1988.

1. SHORDE I Program System

Since early 1970s, when the Doppler and Laser tracking systems for satellites were set up and operated, the Shanghai Observatory has become involved in satellite orbit determination. Nowadays, as one of the global analysis centers of SLR data, we have been developing the program system named SHORDE I, to process the range data and to estimate satellite orbit, dynamical and geodetic parameters

The SHORDE I program system consists of three parts. The first part includes preprocessing the full-rate data, generating short-arc/long-arc trajectories for satellites to remove some of the effects of the unmodeled perturbations, obtaining normal point data and identifying bad observations.

The second part is for predicting the satellite orbit, and is based on the numerical integration of the perturbation equations, using the Adams-Cowell integrator, the Adams-Bashforth integrator, or some other integrators.

The third part is for the determination of the satellite orbit, dynamical and geodetic parameters (ERP and station coordinates) and some geophysical parameters. Different estimation methods have been provided for different purposes. The program is written in Fortran 77 language and runs on IBM 4341 or 3031 computer. For constructing a more accurate force model, a detailed analysis of perturbations on artificial satellites has been done (*He Miaofu*, 1984; *Huang Cheng & He Miaofu*, 1986,1988).

2. The Multistage-Multiarc Method

In the usual long arc solution of the satellite motion, the precise satellite state, earth rotation parameters, station geodetic parameters and dynamical parameters are all estimated simultaneously. These adjustable prameters can be devided into three groups. The first group, the global parameters, include the station coordinates and certain geophysical constants and these are adjusted in the whole of the long arc (several months) in order to obtain stable and reliable results. The second group, the first-stage local parameters, such

Celestial Mechanics **45**: 61–64, 1989.

as the orbit elements, are independently adjusted in the different subarcs (10 or 15 days) so as to reduce the effect of some unmodeled perturbations and obtain stable solutions. The third group, the second-stage local parameters, such as ERPs, are independently adjusted from one sub-subarcs (1-5 days) to the next so as to model properly their variation.

For this big, time-consuming job, we have designed a software package embodying a " multistage-multiarc " procedure in order to save running time and storage on the computer (*He Miaofu et al,* 1986).

Let X(i,j) be the state correction vector of the second-stage local parameters in the ith sub-subarc of the jth subarc;

Y(j) be the state correction vector of the first-stage local parameters in the jth subarc;

Z be the state correction vector of the global parameters.

where i=1,2,...,m. n,m are the number of sub-subarcs in each subarc and the number of subarcs in the whole long arc, respectively.

The adjustment for whole long arc is to be accomplished step by step from one sub-subarc to the next.

After applying a series of Givens-Gentleman orthogonal transformation to the data of a sub-subarc, we obtain the following normal equation

$$D_{ij}^{\frac{1}{2}}\begin{bmatrix} U_{X(i,j)} & R_{X(i,j)Y(i,j)} & R_{X(i,j)Z(i,j)} \\ & U_{Y(i,j)} & R_{Y(i,j)Z(i,j)} \\ & & U_{Z(i,j)} \end{bmatrix}\begin{bmatrix} X(i,j) \\ Y(i,j) \\ Z(i,j) \end{bmatrix} = D_{ij}^{\frac{1}{2}}\begin{bmatrix} B_{X(i,j)} \\ B_{Y(i,j)} \\ B_{Z(i,j)} \end{bmatrix}$$

with

$$D_{ij}^{\frac{1}{2}} = \begin{bmatrix} D_{X(i,j)}^{\frac{1}{2}} & & \\ & D_{Y(i,j)}^{\frac{1}{2}} & \\ & & D_{Z(i,j)}^{\frac{1}{2}} \end{bmatrix}$$

where $D_{X(i,j)}^{\frac{1}{2}}$, $D_{Y(i,j)}^{\frac{1}{2}}$, $D_{Z(i,j)}^{\frac{1}{2}}$ are the nonsingular diagonal matrices and $U_{X(i,j)}$, $U_{Y(i,j)}$, $U_{Z(i,j)}$ are the unit upper triangular matrices.

After finishing the calculation for the ith sub-subarc, the values obtained, $D_{X(i,j)}^{\frac{1}{2}}$, $U_{X(i,j)}$, $R_{X(i,j)Y(i,j)}$, $R_{X(i,j)Z(i,j)}$ and $B_{X(i,j)}$ related to X(i,j) are stored to be used later, and these will have no effect on the next sub-subarc.

While the following matrix and vector which have been calculated for the ith sub-subarc

$$\begin{bmatrix} D^{\frac{1}{2}}_{Y(i,j)} & \\ & D^{\frac{1}{2}}_{Z(i,j)} \end{bmatrix} \begin{bmatrix} U_{Y(i,j)} & R_{Y(i,j)Z(i,j)} \\ & U_{Z(i,j)} \end{bmatrix} \text{ and } \begin{bmatrix} D^{\frac{1}{2}}_{Y(i,j)} & \\ & D^{\frac{1}{2}}_{Z(i,j)} \end{bmatrix} \begin{bmatrix} B_{Y(i,j)} \\ B_{Z(i,j)} \end{bmatrix}$$

are taken as the apriori information of Y(j) and Z in the process of next sub-subarc.

After the processing of the jth subarc, i.e., finishing the nth sub-subarc of jth subarc, the information $D^{\frac{1}{2}}_{Y(n,j)}U_{Y(n,j)}$, $D^{\frac{1}{2}}_{Y(n,j)}R_{Y(n,j)Z(n,j)}$ and $D^{\frac{1}{2}}_{Y(n,j)}B_{Y(n,j)}$ related to Y(j) is stored and these will have no effect on the next subarc.

The matrix $D^{\frac{1}{2}}_{Z(n,j)}U_{Z(n,j)}$ and the vector $D^{\frac{1}{2}}_{Z(n,j)}B_{Z(n,j)}$ are then taken as the apriori information of the global parameters Z in the process of next subarc.

The process runs through the whole long arc from the 1st sub-subarc of the 1st subarc to the nth sub-subarc of the mth subarc. After the processing of the whole long arc, the best linear, unbiased, minimum variance estimation of $\hat{X}(i,j)$, $\hat{Y}(j)$, $\hat{Z}$ will be determined by the following expressions

$$\begin{aligned}
\hat{Z} &= U^{-1}_{Z(n,m)}\, B_{Z(n,m)} \\
\hat{Y}(j) &= U^{-1}_{Y(n,j)}\,(B_{Y(n,j)} - R_{Y(n,j)Z(n,j)}\hat{Z}) \\
\hat{X}(i,j) &= U^{-1}_{X(i,j)}\,(B_{X(i,j)} - R_{X(i,j)Y(i,j)}\hat{Y}(j) - R_{X(i,j)Z(i,j)}\hat{Z})
\end{aligned}$$

and the corresponding covariance matrices can be expressed as:

$$\begin{aligned}
P_{X(i,j)} &= (U^T_{X(i,j)}\, D_{X(i,j)}\, U_{X(i,j)})^{-1} \\
P_{Y(j)} &= (U^T_{Y(n,j)}\, D_{Y(n,j)}\, U_{Y(n,j)} + \sum_{i=1}^{n} R^T_{X(i,j)Y(i,j)}\, D_{X(i,j)}\, R_{X(i,j)Y(i,j)})^{-1} \\
P_Z &= (U^T_{Z(n,m)}\, D_{Z(n,m)}\, U_{Z(n,m)} + \sum_{j=1}^{m} R^T_{Y(n,j)Z(n,j)}\, D_{Y(n,j)}\, R_{Y(n,j)Z(n,j)} \\
&\quad + \sum_{j=1}^{m}\sum_{i=1}^{n} R^T_{X(i,j)Z(i,j)}\, D_{X(i,j)}\, R_{X(i,j)Z(i,j)})^{-1}
\end{aligned}$$

It can be strictly proven that the results of using this technique are the same as the those from the overall adjustment of the whole long arc used in the usual method.

3. Applications

Applications are mainly to SLR data of LAGEOS. To determine a series of the earth rotation parameters ERP(SHA) 87L01, the three year SLR data of LAGEOS from Nov. 1984 to Jan. 1988 have been processed with the SHORDE I program system. Each set of epoch states for LAGEOS was estimated simultaneously with 10-days orbit element corrections and 5-days ERP (three earth rotation parameters X_P, Y_P, D_R, as well as the rate of the polar motion $\dot{X}_p$, $\dot{Y}_p$) (*Zhu Wenyao, et al,* 1988).

The standard errors σ_{X_P}, σ_{Y_P} and σ_{D_R} of X_P, Y_P and D_R are

$\sigma_{X_P} = 1.3$ mas, $\sigma_{Y_P} = 1.2$ mas , $\sigma_{D_R} = 0.07$ ms/day .

The ability to detect the rate of polar motion with 5-day arc solution has been investigated and found to be better than 1 mas/day.

In addition, the overall RMS fit of the laser observations to the orbits in 10-day arcs is generally about 10 cm.

A comparison of SHORDE I results with UTOPIA (CSR of Texas University at Austin) ones for ERP solution series indicates the external precisions to be in X_P, 1.6 mas; in Y_P, 1.7 mas; in D_R, 0.16 ms/day. These values which are rather close to the standard errors given above have provided a good check for our program. We conclude that the SHORDE I program works well.

4. Acknowledgements

We are grateful to Professors B.D. Tapley, B.E. Schutz, and Drs. C.K. Shum and R.J. Eanes of CSR, Texas University at Austin, for their constant kind help and fruitful suggestions.

5. References

He Miaofu, Chin. Astro. Astrophys., 8 (1984), 113-118.

He Miaofu, Feng Chugang & Huang Cheng, Ibid, 10, (1986), 336-341.

Huang Cheng & He Miaofu, Ibid, 12, (1988), 34-37.

Huang Cheng & He Miaofu, Space Dynamical and Celestial Mechanics, 273-279, D.Reidel Publishing Company (1986).

Zhu Wenyao, Feng Chugang, Chang Hua & Teng Zhanming, BIH Annual Report for 1987 (in press).

A SEMI-NUMERICAL EXPANSION OF THE AVERAGED DISTURBING FUNCTION FOR SOME VERY-HIGH-ECCENTRICITY ORBITS

S.FERRAZ-MELLO
Instituto Astronômico e Geofísico
Universidade de São Paulo
04301 - São Paulo, SP
Brazil

ABSTRACT. A note on the construction of expansions of the disturbing function valid for any eccentricity and on some applications.

1. Introduction

We consider in this paper the development of an idea first used by Woltjer (1928) in his studies of the motion of Hyperion. The basis of Woltjer's idea is the fact that the disturbing function is an analytical function in some domain of the complex plane defined by the polar coordinates (e,σ), where e is the eccentricity of the disturbed orbit and σ is a suitable angle related to its pericenter, *e.g.*, the longitude of the pericenter itself. In this sense the classical Laplacian expansion is a Taylor expansion about the origin. The choice of the origin for this expansion is certainly the most suitable one for the study of general problems but it is not mandatory. For many problems it may be convenient to have another choice.

An example is the case of the motion of librating asteroids in the resonance (p+q):p with Jupiter. In this case the most suitable σ is defined by

$$q\sigma = (p+q)\lambda_1 - p\lambda - q\varpi$$

(Schubart, 1968) where λ,λ_1 are the mean longitudes of the asteroid and Jupiter, respectively, and ϖ is the longitude of the asteroid's perihelion. This angle just has small variations and the motion, in the complex plane, lies in the neighborhood of a point (e_0,σ_0) generally far away from the origin. In this case it is obvious that a Taylor expansion about (e_0,σ_0) will represent the disturbing function better than the classical expansion. For instance, in the study of the motion of Hyperion, Woltjer has been able to obtain the relation between the period of the libration of Hyperion and the mass of Titan with good precision, just using a quadratic expansion about the libration center e_0=0.104, $\sigma_0=\pi$.

Celestial Mechanics 45: 65–68, 1989.

However, in order to obtain a similar result using the symmetric Laplacian expansion more than 30 harmonics of the critical argument σ are necessary (Duriez, 1988).

2. The averaged asymmetric expansion

Consider the planar problem of the motion of an asteroid in resonance (p+q):p with Jupiter. The disturbing function may be written as a function of three angles: σ, σ_1, Q, where

$$q\sigma_1 = (p+q)\lambda_1 - p\lambda - q\varpi_1$$

and

$$qQ = \lambda - \lambda_1$$

(ϖ_1 is the longitude of Jupiter's perihelion).

The expansion of the disturbing function involves three main steps:

(a) Taylor expansion about $e_1=0$ (e_1 is the eccentricity of Jupiter's orbit). The result is a series in the powers of $e_1.\cos\sigma_1$ and $e_1.\sin\sigma_1$. Every coefficient in this expansion is a function of e, σ, Q and of the semi-major axes a, a_1.

(b) Taylor expansion of every coefficient about (e_0, σ_0), say

$$\Gamma = \Gamma_0 + \Gamma'_k(k-k_0) + \Gamma'_h(h-h_0) + \tfrac{1}{2}\Gamma''_{kk}(k-k_0)^2 + \Gamma''_{kh}(h-h_0)(k-k_0) + \tfrac{1}{2}\Gamma''_{hh}(h-h_0)^2 + \ldots$$

where

$$k + ih = e.\exp i\sigma, \qquad (i=\sqrt{-1})$$
$$k_0 + ih_0 = e_0.\exp i\sigma_0.$$

The coefficients Γ_* are functions of Q, a, a_1 and of the parameters k_0, h_0.

(c) Averaging over Q performed numerically for fixed values of a, a_1, k_0, h_0.

3. High-eccentricity (HE) and very-high-eccentricity (VHE) expansions

While the original work of Woltjer was completely numerical we preferred, for the sake of having an expansion easy to use in many situations, to perform steps (a) and (b) above, analytically. The coefficients Γ_* obtained in this way depend on Q and a through the radius-vector r and the true longitude Φ. In order to introduce a and λ we need to introduce the solutions of the two-body problem. One easy way to do it is through the use of the series solutions. Since only a few relations are needed (r, Φ and their derivatives with respect to k and h), we may, for r and Φ, use high-order expansions:

$$\sum_{0}^{N} P_n(e^2)\, e^n \exp in\,[\sigma + (p+q)Q]$$

where $\sigma + (p+q)Q$ is the mean anomaly. High-eccentricity (HE) expansions have thus been obtained in the particular case $q=1$, $h_0=0$ (Ferraz-Mello, 1987). They allowed us to explore the known first-order librations and to show that not only does the eccentricity have an important variation in a libration period but also does the semi-major axis (up to 3% for some asteroids); these variations are executed about a center whose eccentricity and semi-major axis are not independent but related by a *law of structure* (Ferraz-Mello, 1988). However, even if a good representation of the disturbing function was possible for eccentricities as high as 0.55, we must be aware that the radius of convergence of the series solutions of the two-body problem is 0.6627 (see Hagihara, 1970, sec. 5.14).

Anyway, those series are only algorithms used to compute r,Φ and their derivatives, and can be substituted by close equations. The resulting very-high-eccentricity (VHE) expansion has been obtained for the general case (any k_0, h_0 and q) (Ferraz-Mello and Sato, 1989). The VHE expansion allowed us to explore the problem of corotation orbits and to obtain solutions for resonant orbits with any eccentricity (see next section). One practical question considered is related to the fact that the variables involved in Kepler's equation are singular for $e=0$ and special considerations are necessary to obtain a formulary valid for any value of e, without singularity at the origin.

4. Applications of the VHE expansion

The VHE expansion was used to construct solutions of the averaged elliptical asteroidal problem near first-order resonances in order to study high-eccentricity librations and corotations. Figure 1 shows the locus of the stable (dotted) and unstable (non dotted) corotation solutions

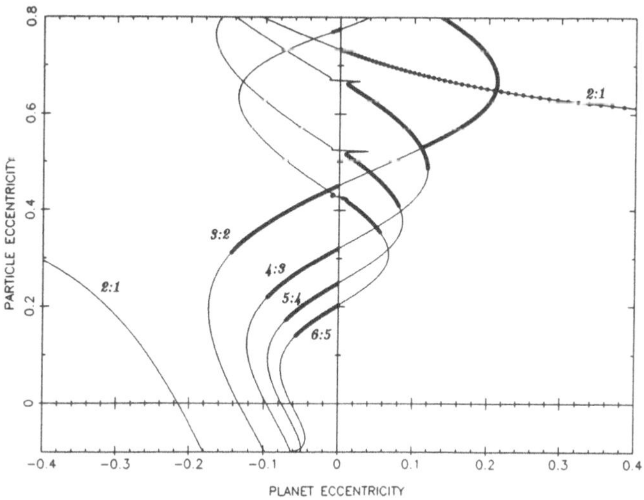

Figure 1. Locus of exact corotation orbits.

obtained for the resonances 2:1, 3:2, 4:3, 5:4 and 6:5 in the Sun-Jupiter-asteroid problem (the eccentricity of Jupiter is not fixed). The solutions are such that the orbits of the particle and Jupiter share the same apsidal line and have conjunctions on it. We adopted algebraic values for the eccentricities which are positive or negative following the body is at the perihelion or at the aphelion at the moment of a conjunction. We note that the intersections with the vertical axis at e=0.73, for the resonance 2:1, and e=0.45, for the resonance 3:2, are in full agreement with known results for the bifurcation points of families of periodic orbits of the restricted problem (see Hadjidemetriou, 1988). For the other resonances two intersections appear in the figure. However only the lower one corresponds to a true bifurcation; the higher one has a complex structure still not studied in detail.

The averaged asymmetric expansion is being used in the construction of formal theories for the long period perturbations of Pluto and Hyperion; in the case of Pluto, Nitto (1988) derived an expansion considering also the resonance involving the space variables i,Ω.

5. Conclusion

The averaged asymmetric expansion of the disturbing function allowed us to analytically study solutions of the elliptic restricted problem averaged near resonances in situations not allowed with the classical Laplacian expansion. In particular it allowed us to establish the law of structure, relating the semi-major axis and the eccentricity of the centers of libration, and the law relating the eccentricities of the disturbing and disturbed bodies in exact corotation. In both cases algebraic equations were found, relating coefficients of the averaged asymmetric expansion and some orbital elements.

References

Duriez, L.: 1988, *Astron. Astrophys.* **194**, 309-318.
Ferraz-Mello, S.: 1987, *Astron. Astrophys.* **183**, 397-402.
Ferraz-Mello, S.: 1988, *Astron. J.* **96**, 400-408.
Ferraz-Mello, S. and Sato. M.: 1989, *Astron. Astrophys.* (submitted).
Hadjidemetriou, J.D.: 1988, *Celest. Mech.* (in press).
Hagihara, Y.: 1970: *Celestial Mechanics*, vol.I, MIT Press, Cambridge.
Nitto, M.A.: 1988, *comm. pres. annual meeting Soc. Astron. Brasil.* (july 1988).
Schubart, J.: 1968, *Astron. J.* **73**, 99-103.
Woltjer Jr., J.: 1928, *Ann. Sterrew. Leiden* **16**(3).

SYMBOLIC ALGEBRA SOFTWARE FOR PLANETARY THEORIES

F. VÁRADI
Department of Mathematics and
Institute of Geophysics and Planetary Physics
University of California
Los Angeles, CA 90024
USA

ABSTRACT. A symbolic algebra software package under development in MACSYMA is described briefly. Its capabilities are illustrated with a simple application.

1. Introduction

In the last few years there has been considerable progress in planetary theories but there are still discrepancies. The results of long-term numerical integrations (Applegate *et al.*, 1987, and Carpino *et al.*, 1987) are in good agreement but the same level of consistency has not been reached by the most recent analytical (Duriez, 1979) and semi-analytical (Laskar, 1987) theories. Our work is toward a symbolic planetary theory based on a Hamiltonian Lie-series formalism; at present we intend to use classical D'Alembert-type expansions. Since this work is still in progress, we only report solutions to some technical problems and a simple application to the 6:1 resonance in the circular restricted three body problem. The software performance estimates refer to this type of application.

2. Software

For its flexibility we prefer MACSYMA to other algebraic manipulation packages, like Maple. Our software runs under UNIX on a Sun-3 workstation. MACSYMA expressions are very large, we store most parts of the expansions in disk files and process them separately. UNIX C-shell scripts support file management, file size is anywhere between 10K and 100K, they are also compressed to about one tenth of their original size. Files are arranged according to powers of eccentricity-type variables, e.g. the expansion of the disturbing function is stored in 28 files when the expansion is carried out up to the sixth power of eccentricity.

In each file the expressions are represented as lists, with an indexing according to the Fourier harmonics in mean longitudes. They are handled with functions operating on lists, e.g. addition consists of appending and adding the coefficients of common terms. Actually, trigonometric functions or MACSYMA's Poisson series representation are never used.

Celestial Mechanics **45**: 69–72, 1989.

The Fourier coefficients are rational expressions in which Laplace coefficients are present as symbolic functions. The necessary operations can be carried out in a straightforward manner, the calculation of Poisson brackets is the most time consuming part, a typical running time is 5 hours.

In our Hamiltonian formalism we use complex variables to represent eccentricity-type variables and complex exponentials of mean longitudes. Expansions of elliptic motion in terms of Poincaré variables has been determined. Lie series turned out to be very efficient, the expansions can be calculated up to 8th power of eccentricity in less than an hour.

We use brute force differentiation to determine expansions of perturbing functions. First, an expansion in powers of eccentricity-type variables is calculated using expansions of elliptic motion. Further expansion is carried out in terms of Laplace coefficients, at this point simplifications and representation conversions are made. The whole operation takes about 4 hours.

The generator and the normalized Hamiltonian are determined using the well-known Lie series calculus, removing the mean longitudes (and the longitude of pericenter in the application below). Of course, we keep resonant terms whenever we wish.

Other programs are also available, e.g. a numerical evaluator for MACSYMA expressions, written in C, Birkhoff normalization, operations on quasi-periodic functions, etc.

3. An Application

In order to gain some experience we calculated the 5th order resonant system in the planar circular restricted three body problem in a rotating coordinate frame. We follow Henrard and Lemaître (1983) and Lemaître (1984) in a slightly different formalism.

The unperturbed system is

$$H_0 = -\frac{1}{2\Lambda^2} - \Lambda + \frac{1}{2}(z\bar{z})$$

where $z = \xi + i\eta$ and $(\lambda, \eta, \Lambda, \xi)$ are Poincare variables. We introduce

$$\omega(\Lambda) = \frac{\partial H_0}{\partial \Lambda} = \frac{1}{\Lambda^3} - 1$$

The equation for a general term W in the Lie series generator is

$$\{H_0, W\} = B(\Lambda) z^k \bar{z}^l \exp im\lambda$$

and it is easily solved

$$W = \frac{B(\Lambda)}{i(-m\omega(\Lambda) + l - k)} z^k \bar{z}^l \exp im\lambda$$

Resonance occurs when the denominator is zero (or very small), i.e. $\omega(\Lambda) = (l-k)/m = p/q$, where p and q are fixed integers (and they are relative primes). The order of resonance is p, this corresponds to a $p+q:q$ resonance in non-rotating frame.

In order to keep $z = 0$ a regular point of the phase space, we introduce the resonant variables $w = u + iv$ and $\bar{w}$ as $w = z\exp(-i\lambda q/p)$. All resonant terms are products of w and $\bar{w}$ and we also have $z\bar{z} = w\bar{w}$. With

$$F = \Lambda + \frac{q}{2p} z\bar{z}$$

and $f = \lambda$ the transformation

$$(\lambda, \Lambda, \eta, \xi) \to (f, F, u, v)$$

is canonical. Since f is not present in the averaged Hamiltonian, F is a first integral.

In the new variables the unperturbed Hamiltonian is

$$H_0 = -\frac{1}{2(F - \frac{q}{2p}w\bar{w})^2} - F + \frac{q}{2p}w\bar{w} + \frac{1}{2}w\bar{w}$$

In the averaged perturbing function $\hat{R}_\mu$ we have the resonant terms $w^k\bar{w}^l$, $l - k = np$, $n = 0, -1, 1, -2, 2, \ldots$, i.e.

$$\hat{R}_\mu = \mu \sum_{l-k=np} A_{k,l} w^k \bar{w}^l$$

and $\hat{R}_\mu$ includes secular terms (n=0). Expanding H_0 around $w = 0$ we obtain

$$H_0 = -\frac{1}{2F^2} - F + \frac{q}{2p}\left(\frac{p}{q} - \frac{1}{F^3} + 1\right)w\bar{w} - \frac{3}{2F^4}\left(\frac{q}{2p}\right)^2 w^2\bar{w}^2 - \frac{2}{F^5}\left(\frac{q}{2p}\right)^3 w^2\bar{w}^3$$

F is constant and near the resonance the quantity

$$\frac{p}{q} - \frac{1}{F^3} + 1 = \frac{p}{q} - \omega(F)$$

is small because the difference between F and Λ is of order $\|w\|^2$ (approximately the square of eccentricity). The averaged perturbing function should also be expanded but for the sake of brevity we ignored most of these additional quantities.

The full averaged Hamiltonian can be written as

$$H = \alpha w\bar{w} + \beta w^2\bar{w}^2 + \gamma w^3\bar{w}^3 + \mu \sum_{k-l=np} A_{k,l} w^k \bar{w}^l$$

where α, β and γ are known functions. In this Hamiltonian α and μ are small and β and γ are of order unity. We used our software to obtain numerical values for $A_{i,j}$ taking into account the first five Laplace coefficients.

Birkhoff normalization is not possible for this Hamiltonian since α becomes zero at a certain value of F, i.e. we have a codimension 1 bifurcation of an elliptic equilibrium point of a Hamiltonian vector field. A picture of the bifurcation is shown on Fig. 1.

Acknowledgements. This work is supported by NSF Grant ATM86-15424.

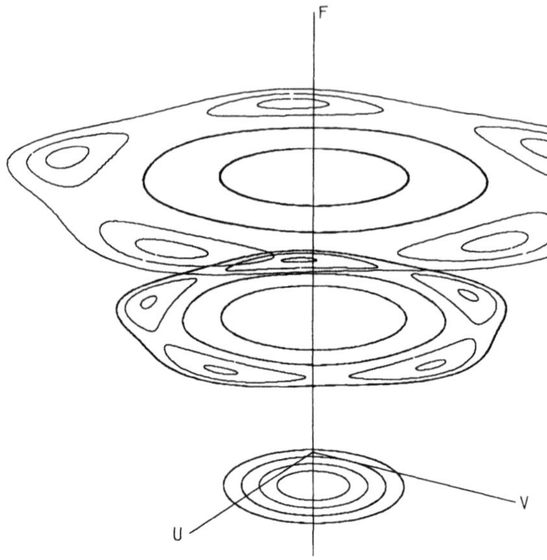

Fig. 1. 6/1 Resonance.

References

Applegate, J. H., Douglas, M. R., Gursel, Y., Sussman, G. J. and Wisdom, J. (1986) 'The outer Solar System for 200 million years', *Astron. J.* **92**(1), 176-194.

Carpino, M., Milani, A., and Nobili, A. M. (1987) 'Long-term numerical integrations and synthetic theories for the motion of the outer planets', *Astron. Astrophys.* **181**, 182-194.

Duriez, L. (1979) 'Approche d'une Theorie Générale Planétaire en variables elliptiques héliocentriques', *Théses, Lille*

Henrard, J. and Lemaître, A. (1983) 'A second fundamental model for resonance', *Celest. Mech.* **30**, 197-218.

Laskar, J. (1987) 'Secular evolution of the Solar System over 20 million years', submitted to *Astron. Astrophys.*

Lemaître, A. (1984) 'High order resonances in the restricted three body problem', *Celest. Mech.* **32**, 109-126.

TESTS OF THE EXTRAPOLATION METHOD FOR THE NUMERICAL INTEGRATION OF THE KEPLERIAN MOTION

Toshio Fukushima
Satellite Geodesy Office
Hydrographic Department
5-3-1, Tsukiji, Chuo-ku, Tokyo 104
Japan

ABSTRACT. Two extrapolation methods (Gragg's method and GBS method) were tested for the numerical integration of the Keplerian motion. The tests cleared that the Gragg's method is better for the orbit computation and is 1) easy to control, 2) highly accurate, 3) tough, 4) economic, and 5) very flexible against the close encounter.

1. Introduction

The extrapolation method (for the numerical integration of an ordinary differential equation) can be classified as a one-step method, although it is much different from the Runge-Kutta type methods (Gear, 1971). The extrapolation method is the extension of Richardson's extrapolation to h = 0 for a primitive numerical integration method such as the Euler method. In fact this method works as follows:

Consider to integrate a differential equation dx/dt = f(t,x) from t = 0 to H. If one uses a certain integration rule and performs numerically an integration with a steplength h, then the resulting value is a function of h as X(h) and is generally different from the true solution x(t=H). In principle, the true solution is equal to X(0), which we cannot obtain directly. Then the extrapolation method estimates it by an extrapolation based on the knowledge on multiple trial integration results $X(h_i)$ (i=1,2,...), where trial stepsizes h_i are submultiples of H as H/2, H/3, and so on.

Thus there are extrapolation methods as many as the number of combinations of the trial integration rule and the extrapolation scheme. However, there are two powerful combinations using the modified mid-point rule: the Gragg's method using the Aitken (polynomial) extrapolation scheme (Gragg, 1965) and the GBS (Gragg-Bulirsch-Stoer) method using a rational extrapolation scheme (Bulirsch and Stoer, 1966).

This brief report describes some characteristics of the extrapolation method for the orbit computation of celestial bodies. In the followings, we test the Gragg's method for the integration of Keplerian orbit starting from the pericenter using 18 decimals arithmatics by evaluating the relative error in the positional vector

Celestial Mechanics **45**: 73–76, 1989.

unless we mention otherwise.

2. Characteristics of The Extrapolation Method

First we seeked for the highest accuracy which the extrapolation method can attain. Our test in 36decimals arithmatics shows that the limitting number of effective digits for the circular 2-body motion after one revolution of orbit is 35 for the extrapolation method and 28 for the Adams-Moulton method of order 16, respectively. In both cases we could not obtain better results because of the numerical instability in the former and of the saturation due to the round-off error for the latter.

Second we cleared the controllability of the extrapolation method. There are two control parameters in the extrapolation method. The one is the (major) steplength H and the other is the required accuracy d. The extrapolation method is much different from other methods in the sense that H and d can be chosen almost independently with each other. This is true as long as 1) d is not so small that the round-off error does not exceed the truncation one, and 2) H is not so large that the numerical instability does not occur.

The former assertion is proved by Figure 1 where the obtained accuracy after one revolution of a Keplerian orbit (e = 0.6) depends not on H but on d as long as $d > 10^{-30}$ (36decimals arithmatics is used). The latter is understood by Figure 2 showing the obtained error of a circular orbit as a function of H and the number of steps integrated. This figure reveals that the extrapolation method is so stable that H can be chosen as long as about 1radian.

Third we examined the extrapolation scheme. Figure 3 indicates that both the rational and polynomial extrapolation schemes produce the same result for a smooth problem (Keplerian orbit with e = 0.01), however,

Figure 1. Accuracy Controllability of the Extrapolation Method for a Keplerian Orbit of e=0.6

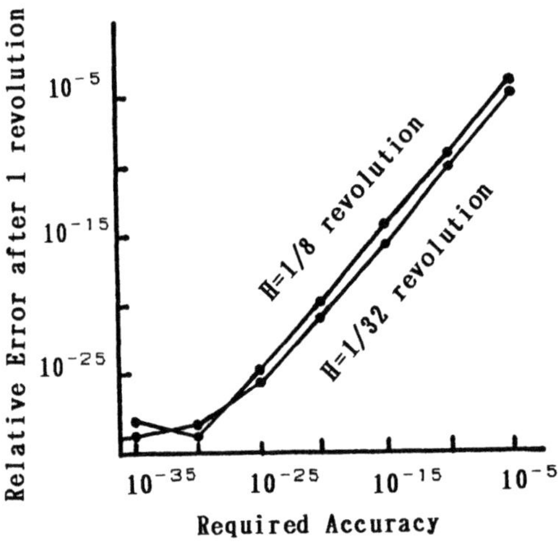

Figure 2. Stability of the Extrapolation Method for a Circular Orbit

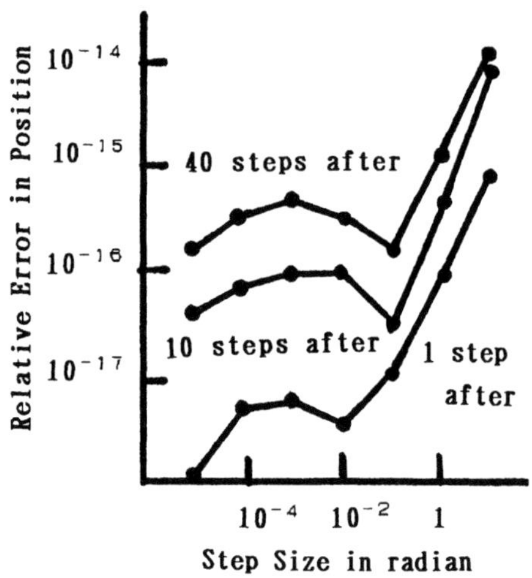

Figure 3. Effect of Extrapolation Scheme : Rational (solid line) vs Polynomial (broken line)

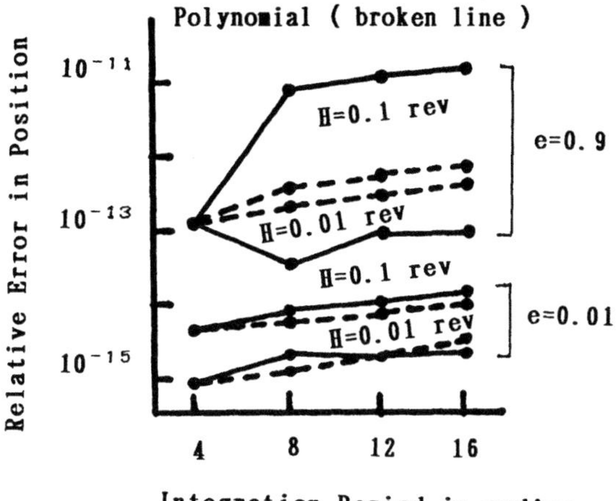

Figure 4. Sensitivity of Cost Performance with Step Size for a Keplerian Motion of e=0.6

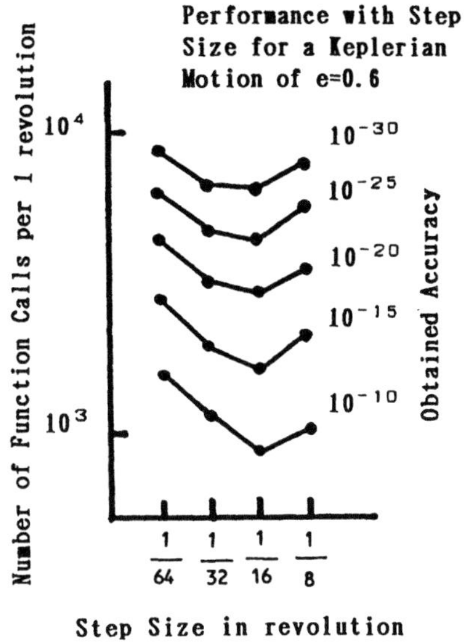

the polynomial one is more reliable for a rapidly varying problem (Keplerian orbit with e = 0.9). As for the sequence of the trial steplengths, our experience assured that the most economical one, i.e. $h_i = H/(i+1)$ for i = 1, 2,... is sufficiently stable for the Keplerian orbit even when e = 0.9.

Fourth we tested the cost performance of the extrapolation method. Usually the cost performance of an integration method is measured by the the ratio of number of function calls to the obtained accuracy. Figure 4 shows that the choice of H makes not so much difference in the cost performance although the highest cost performance is roughly achieved by H ≅ 20degree for the case of orbit with e = 0.6 from our experience.

Finally we compared the cost performance of the extrapolation method with those of other methods for some Keplerian orbit integrations after a period of about 35 revolutions as 1) with DVDQ and RKF7(8) for e = 0.02, 2) with DVDQ and Adams-Moulton for e = 0.1, and 3) with Adams-Moulton for e = 0.9 in Figure 5, respectively. Here the results of DVDQ and RKF7(8) are taught by Prof. Kinoshita (1980, private communication).

3.Conclusion

From these results, we conclude that the extrapolation method is 1) easy to control, 2) highly accurate, 3) tough, 4) economic, and 5) very flexible against the close encounter so as to be suitable for the problems a) requiring the highest accuracy, b) varying rapidly, or c) whose characteristics is unknown. Especially we recommend the Gragg's method of the trial stepsize sequence H/(i+1) with H of about 10 to 40degree for the computation of nearly Keplerian orbit.

References

Bulirsch, R. and Stoer, J. (1966) Numerische Mathematik **8**, 1.
Gear, C.W. (1971) Numerical Initial Value Problems in Ordinary Differential Equations, Prentice-Hall, Inc., New Jersy.
Gragg, W.B. (1965) SIAM J. Numer. Anal. Ser.B, **2**, 384.

Figure 5. Comparison of Cost Performance of the Extrapolation Method with Other Methods for the 35 revolutions Integration of Keplerian Orbits of e=0.02 , 0.1 and 0.9 .

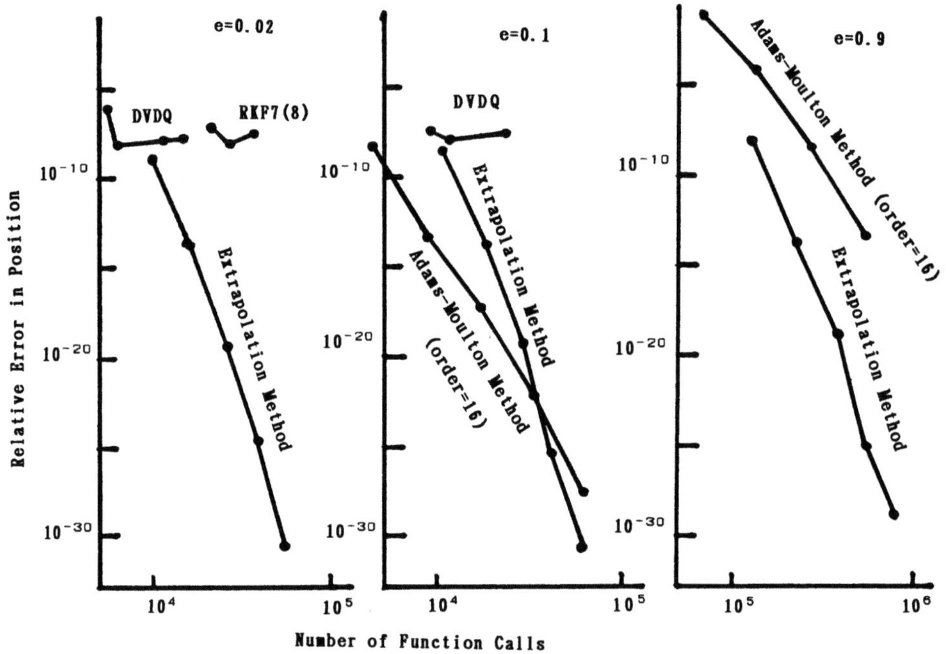

APPLICATION OF A MASSIVELY PARALLEL COMPUTER TO THE N-BODY PROBLEM

PAUL HERTZ
E.O. Hulburt Center for Space Research
Naval Research Laboratory
Washington, DC 20375 USA

STEPHEN L.W. McMILLAN
Department of Physics
Drexel University
Philadelphia, PA 19104 USA

Abstract. Parallel processor computers represent a new technology that has recently become available for astronomical applications. We have implemented an N-body code on a TMC Connection Machine CM-2 in order to investigate the advantages of a massively parallel computer over serial machines, including conventional supercomputers. For collisionless problems following N stars, a direct integration code scales as $O(N^2)$ on serial machines and on the CM-2 as $O(\log(N))$ for small N and $O(N \log(N))$ for large N. The CM-2 outperforms workstations for $N > 50$ and supercomputers for $N > 4000$.

1. Introduction: Collisionless N-body Problems

A class of intrinsically three dimensional stellar dynamics problems, generally termed "collisionless" N-body problems, may be treated using simple algorithms. Examples of collisionless N-body problems are galaxy simulations, where binary encounters are explicitly turned off by the use of a softened gravitational potential, and violent relaxation, where the system is not followed long enough for binaries to play an important role and energy is distributed within a star cluster by collective, rather than two-body, effects. Stellar timesteps can be no longer than a crossing time, and the cost of a single timestep is governed by the cost of determining the force on a star.

In the simplest, direct summation codes, this is an $O(N)$ operation; since there are N stars in the system, the time to perform a crossing time simulation is proportional to N^2. On a small machine (e.g. VAX 11/780, SUN-3) a 100-body simulation requires ~2 minutes of CPU time, while a vectorized supercomputer (e.g. Cray-2, Cyber 205) runs ~100 times faster but still scales as $O(N^2)$.

Celestial Mechanics **45**: 77–80, 1989.

2. The Connection Machine: A Massively Parallel Computer

One solution is the use of a general purpose, parallel computer. By performing many operations simultaneously, rather than sequentially, large speedups can be achieved. A rule of thumb is that an O(N) loop becomes O(log(N)) in a parallel architecture (using a binary tree algorithm), so that parallel direct summation schemes will code as O(N log(N)) or even O(log(N)).

We have implemented a simple direct integration N-body code on the NRL Connection Machine CM-2 (Hillis 1985). The CM-2 is a massively parallel SIMD (single instruction, multiple data) computer; in it a very large number of processors are imbedded directly in memory, where the data reside. Parallel operations are performed on all data simultaneously, and programming is effected by selectively activating processors and by interprocessor communication. Code is written, compiled, and executed on a front end machine (e.g. VAX 8800); instructions and data are sent to the CM-2 in real time through a NEXUS microcontroller.

The full CM-2 contains up to 64k (1k = 1024) bit serial custom processors, a Weitek fpa for every 16 processors, and a general router for interprocessor communication. For completely parallel operations, aggregate sustained processing speeds of 2500 MIPS (32-bit integer addition) and 500 Mflops (single precision) can be attained. Each processor is associated with 64k bits of memory, for an aggregate memory of 512 Mbytes. Memory can be written into or read from at 300 Gbits/sec, and data exchange between processors can occur at an aggregate rate of up to 8 Gbits/sec. The NRL CM-2 is a quarter machine, containing 16k processors and 128 Mbytes of memory.

The CM-2 was programmed using C*, an extension of C containing parallel constructs. The extensions provide for (i) I/O between the front-end machine and CM-2, (ii) interprocessor communication, (iii) arithmetic and logical operations on parallel data, (iv) selection of active processors, and (v) reduction of parallel data to scalar values.

3. CMBODY: A Parallel N-body Code

We have converted the standard Aarseth (1963) NBODY0/1 FORTRAN code into a serial C code on the VAX 8800; we then parallelized it. The parallel program, CMBODY, has two options for calculating the total force acting on each body by the other N-1 bodies (DIRECT, SIMULTANEOUS) which use two different mappings of data onto the CM processors. A third mapping, TREE_CODE, has been used by Makino and Hut (1988).

DIRECT: Each processor represents one star, and all properties of a star are stored in the memory associated with that processor. The predictor and corrector steps are done simultaneously for all stars, and are thus O(1). The force calculation determines the force on a single star due to all other stars simultaneously, but individual stars are treated serially. It takes log(N) steps to sum forces, so determining the forces on N stars scales as O(N log(N)).

SIMULTANEOUS: During the force calculation, each processor represents the interaction between two stars. Since the forces between all pairs of stars are determined simultaneously, and it takes log(N) steps to sum the forces on a star, this code scales as O(log(N)). On a finite machine, this code quickly runs out of memory. The NRL CM-2 has enough memory to handle N = 1024, or more than one million virtual processors.

The most important code modifications that must be made in order to utilize efficiently the power of the CM-2 include: declaring parallel variables for each star, configuring the CM-2 to have the correct number of processors, coding efficient interprocessor transfer of data in the force calculations, and coding global reductions (e.g. summing over forces, choosing smallest timestep) to run in log(N) time. All of these changes are straightforward in C*, and are supported by the parallel operators and data types available. The low level language Paris (PARallel Instruction Set) was used to reconfigure the CM-2 processors from N (physical) processors to N^2 (virtual) processors when the SIMULTANEOUS force calculation was used.

4. Timing Comparisons: Serial vs. Parallel

We have run comparison timings on a large number of serial computers. The same code was used throughout, and the $O(N^2)$ scaling was verified. In the table we show timings of selected serial machines in CPU minutes for N-body simulations run at various N which use Aarseth's standard NBODY1 code. We also show timings for the parallel code CMBODY with both force calculation algorithms; CMBODY used Aarseth's NBODY0 time step algorithm. The timings are for a one crossing time integration; the codes were written in FORTRAN, except where indicated. From the timings and scalings, we estimate that the CM-2 can perform an N-body simulation faster than a SUN-3 for N > 50, faster than a Cray X/MP for N > 4000, and faster than an ETA-10 for N > 10,000.

5. Conclusions

Implementation of a simple N-body code in a parallel environment results in performance superior to the fastest supercomputers currently available for modest values of N -- only a few thousand. Implementation of a tree code, rather than direct summation, may replace the N^2 processors by a few times N log(N) processors. Thus an O(log(N)) scaling may be possible for N up to several thousand, the same scaling as the SIMULTANEOUS force calculation. Due to the large overhead associated with tree codes, TREE_CODE will be less efficient than SIMULTANEOUS for small N; however TREE_CODE is not as memory intensive as SIMULTANEOUS and can be used at large N.

Parallelization of existing codes follow similar rules to vectorization -- simple algorithms, with few special cases, benefit most from the process. Conceptually, the conversion itself is not significantly more difficult than porting a code from a VAX to a Cray. Improvements in the CM-2 operating system promise to make its use easier and faster still -- indirect addressing is now possible and FORTRAN will soon be available. Problems involving the simulation of up to ~10^4 bodies for tens of crossing times can now easily be performed overnight, at a much lower cost than on state-of-the-art supercomputers.

We thank the staff of the NRL Connection Machine Facility, especially Robert Whaley, for programming assistance. This work is supported by the Office of Naval Research.

6. References

Aarseth, S.J. 1963, Mon.Not.Roy.astr.Soc., **126**, 223.
Hillis, W.D. 1985, The Connection Machine, (MIT Press: Cambridge, MA).
Makino, J., and Hut, P. 1988, these proceedings.

CPU Time in Minutes for One Crossing Time

	N = 25	N = 100	N = 400
MicroVAX II (fpa)	0.320	4.847	80.95
VAX 11/750 (fpa)	0.279	4.007	62.72
VAX 11/780 (fpa)	0.187	2.864	46.67
SUN-3 (fpa)	0.153	1.781	25.25
Convex C-1 (unvectorized)	0.188	1.489	21.65
VAX 8600 (fpa)	0.046	0.757	
VAX 8800 (fpa)	0.037	0.581	7.500
Convex C-1 (vectorized)	0.097	0.484	5.988
Cray X/MP (vectorized, cft)	0.0041	0.0218	0.256
Cray-2 (vectorized, cft77)	0.0030	0.0181	0.226
Cyber 205 (vectorized)	0.0058	0.0347	0.224
ETA-10 (vectorized)	0.0015	0.0106	0.0894
VAX 8800 (C, fpa, float)	0.044	0.049	6.143
CM-2 (C*, fpa, DIRECT)	0.321	0.793	3.140
CM-2 (C*, fpa, SIMULTANEOUS)	0.230	0.256	0.440

HYDRODYNAMICAL MODELS OF ELLIPTICAL GALAXIES

E. Moreno and P. Pişmiş
Instituto de Astronomía, UNAM
Apartado Postal 70-264
04510 México, D. F.

Hydrodynamical models of elliptical galaxies can be obtained by means of Jeans' equations (see for example Ortega and de Freitas Pacheco, 1986). Following this approach, we consider a mass distribution in steady state which approximates de Vaucouleurs' $r^{\frac{1}{4}}$ law, and assume the principal axes of the velocity distribution to be parallel to unitary vectors in cylindrical and spherical coordinates, with the ratio between dispersions in different directions a constant or a function of position.

Jeans' equations are solved numerically along their characteristics. In this scheme one needs up to the second derivatives of the potential function. We have been able to obtain the needed expressions in analytic form for oblate, prolate, and triaxial systems.

We have used Runge-Kutta and Adams-Moulton predictor-corrector algorithms to solve the resulting system of differential equations along the characteristics. As the proposed mass distribution has a singularity at the origin of coordinates, a combination of both methods has been found particularly useful in solving the characteristics that reach the central region.

Many runs have been computed with Fehlberg's Runge-Kutta algorithms of second, fourth, and seventh order (Fehlberg 1968, 1969). The precision is similar with any of them, but the computing time is strongly reduced with the second order formula.

Given the position-dependent ratios between velocity dispersions along the coordinate axes, the solution may not be a physically acceptable one; for example the non-zero centroid rotational velocity at the center of the system shown in Figure 1(c)-(d). A right way to follow is to start with just enough information to generate a partial solution, the rest of it being determined by the appropriate constraints. The examples shown in Figure 1(a)-(b) are obtained in this way.

Figure 1 shows σ_z, the velocity dispersion in the z direction (the axis of rotation of the system), and $\overline{\Theta}$, the centroid rotational velocity, as function of distance, R, from the center of some types of axisymmetric systems. The motion of centroids is assumed to be purely

Celestial Mechanics **45**: 81–82, 1989.

circular. The ratio between velocity dispersions is a constant in all cases. At present we are generating solutions with position-dependent ratios between velocity dispersions in oblate, prolate, and (slightly) triaxial systems.

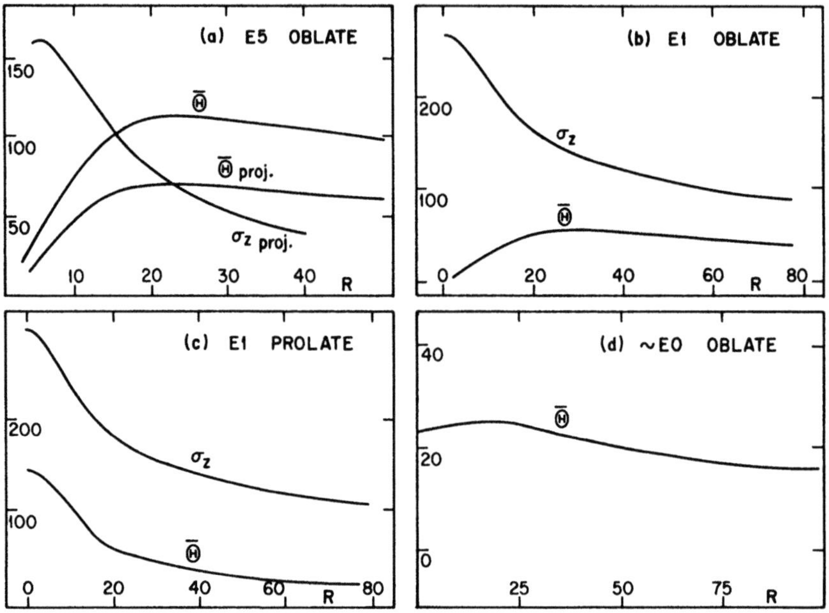

Figure 1. Some examples of the variation of σ_z and $\overline{\Theta}$ with distance R from the center of the system. We give the variation in the z=0 (symmetry) plane. An ellipticity type is assigned in each case for a view along this plane. In (a) the (integrated) projection at an inclination of 45 degrees is also shown. In all cases, vertical axis units are km sec^{-1}, and kpc. for the horizontal.

References.

Fehlberg, E. (1968) "Classical fifth-, sixth-, seventh-, and eighth-order Runge-Kutta formulas with stepsize control", NASA TR R-287.

Fehlberg, E. (1969) "Low-order classical Runge-Kutta formulae with stepsize control and their application to some heat transfer problems", NASA TR R-315.

Ortega, V. G. and de Freitas Pacheco, J. A. (1986) "Hydrodynamical models of elliptical galaxies", Rev. Mexicana Astron. Astrof., 12, 97-100.

A MULTIFLUID CODE FOR PROBLEMS OF ELLIPTICAL GALAXY FORMATION AND EVOLUTION

E. Athanassoula[1], S. Morin[1], K.H. Prendergast[2]

1 Observatoire de Marseille, 2 Place Le Verrier,
13248 Marseille Cedex 04, France
2 Columbia University, Dept. of Astronomy, 538 W. 120th St.,
1414 Pupin, New York, NY 10027

The general purpose of this ongoing investigation is to explore the consequences of various assumptions concerning the ways in which galaxies form and evolve. This is clearly a formidable undertaking, and to make any progress we kept the formulation of the problem as simple as possible. At present we use a multifluid code to follow the motions of stars, gas and dark matter in spherical geometry. In addition to self-gravitation we consider the following:

i)	star formation	governed by $\alpha(p,\rho)$
ii)	mass loss from stars	governed by $\beta(T)$
iii)	radiative cooling of gas	governed by $\kappa(p,\rho)$
iv)	heating of gas due to stars	governed by $\eta(T)$
v)	stellar evolution	governed by t_{evol}

In the above expressions, p and ρ are the pressure and density of the gas, and T is the (local) mean age of the stars.

The underlying hydrodynamic code is an Eulerian, first order, explicit, conservative, flux-splitting scheme based on gas-kinetic theory, and therefore bears some resemblance to the beam scheme. In the current code fluxes of mass, momentum and energy are computed from exact (error function) integrals over the Maxwellian velocity distribution. The complete multifluid code solves rate equations for the interconversion of stars and gas, for the heating of gas due to the presence of stars and for the radiative cooling of the gas at each hydrodynamic time step. The gravitational potential is computed at the same time, and is represented by a step function. In the absence of specific effects which would alter the energy (such as gas cooling) the scheme conserves total energy, including gravitational energy, to within machine accuracy. A more stringent test is that it preserves a bounded self-gravitating isothermal gas sphere for at least 50 crossing times. For initial conditions we take a sphere of cold gas of uniform density ρ and radius R_0. The program is simple and fast enough to enable us to explore parameter space in some detail (so far we have made over 200 runs). This is vitally necessary, because our knowledge of the rate coefficients is fragmentary at best. The "correct" values will ultimately be chosen to give the best match to observations.

Celestial Mechanics **45**: 83–84, 1989.

The figure shows the evolution of a number of models in a plot of central velocity dispersion in projection vs. total mass of stars. The numbers refer to the particular model and the points are equally spaced in time. The models have the same rate coefficients and initial density, but different initial radii and therefore masses. All models start with negligible velocity dispersion and stellar mass. The most rapid phase of evolution occurs in each case at the time of maximum contraction (approximately one initial free fall time). What happens thereafter depends on the parameters of the model and in particular its mass. Models 53, 52 and 51 have settled down to a slowly evolving phase in which the remaining gas is gradually consumed by star formation. Model 54 undergoes outgassing by bubble formation starting at the center and thereafter readjusts to a more distended configuration with declining central velocity dispersion and constant stellar mass. Model 55 outgasses so early that the entire configuration is disrupted. Qualitatively similar plots are obtained with other values of the rate coefficients.

Observations of central surface brightness, core radius, absolute magnitude and velocity dispersion (cf. Kormendy 1985, Ap. J. 295, 73) show that the dwarf and giant ellipticals form distinct sequences. It has been suggested that outgassing may be responsible for the very substantial differences between these two classes of objects. In every mass sequence of models we have run, outgassing occurs for intermediate and low mass models, and does not occur for high masses. The precise location of the break between those models which outgas and those which do not depends, of course, on the rate coefficients, but the qualitative results on outgassing appear to be robust.

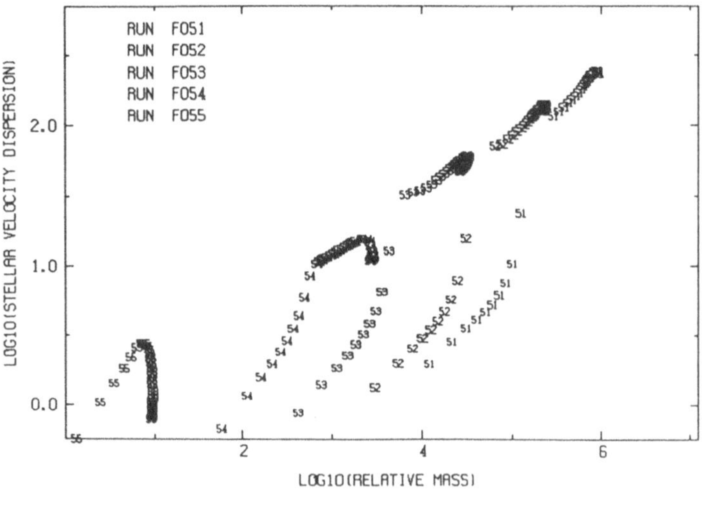

3D HYDRODYNAMICS AND RADIATIVE TRANSFER

A. P. BOSS
DTM, Carnegie Institution of Washington
5241 Broad Branch Road, N.W.
Washington, D.C. 20015
U.S.A.

ABSTRACT. Finite difference codes for the solution of the fully three dimensional equations of hydrodynamics, self-gravitation, and radiative transfer have been developed for use on relatively modest computers. Originally developed to study the collapse and fragmentation of protostellar clouds, this family of codes has been used to study a variety of astrophysical problems, with a particular emphasis on cosmogonical issues.

1. Equations of Motion

The time evolution of a nonmagnetic, inviscid, 3D gas cloud is determined by the continuity equation (in spherical coordinates r, θ, ϕ)

$$\frac{\partial \rho}{\partial t} + \nabla \cdot (\rho \mathbf{v}) = 0,$$

and the three momentum equations

$$\frac{\partial(\rho v_r)}{\partial t} + \nabla \cdot (\rho v_r \mathbf{v}) = -(\rho \frac{\partial \Phi}{\partial r} + \frac{\partial p}{\partial r}) + \frac{\rho}{r}(v_\theta^2 + v_\phi^2),$$

$$\frac{\partial(\rho v_\theta)}{\partial t} + \nabla \cdot (\rho v_\theta \mathbf{v}) = -\frac{1}{r}(\rho \frac{\partial \Phi}{\partial \theta} + \frac{\partial p}{\partial \theta}) - \frac{\rho}{r}(v_r v_\theta - v_\phi^2 cot\theta),$$

$$\frac{\partial(\rho A)}{\partial t} + \nabla \cdot (\rho A \mathbf{v}) = -(\rho \frac{\partial \Phi}{\partial \phi} + \frac{\partial p}{\partial \phi}),$$

where t is time, ρ is the mass density, $\mathbf{v} = (v_r, v_\theta, v_\phi)$ is the fluid velocity, $A = rsin\theta v_\phi$ is the specific angular momentum, and p is the pressure. The specific internal energy E is determined by

$$\frac{\partial(\rho E)}{\partial t} + \nabla \cdot (\rho E \mathbf{v}) = -p \nabla \cdot v + L,$$

where L is the time rate of change of energy per unit volume due to radiative transfer. These hydrodynamical equations are written in conservation law form and are solved using modified donor-cell differences and explicit time differencing.

Celestial Mechanics **45**: 85–88, 1989.

The gravitational potential Φ is determined by Poisson's equation

$$\nabla^2\Phi = 4\pi G\rho,$$

where G is the gravitational constant. Poisson's equation is solved by a spherical harmonic expansion and a tridiagonal matrix solver.
In the diffusion approximation for radiative transfer, L is given by

$$L = \nabla \cdot (\frac{4}{3\kappa\rho}\nabla(\sigma T^4)),$$

where κ is the Rosseland mean opacity, T is the temperature, and σ is the Stefan-Boltzmann constant. The diffusion approximation is strictly valid only for optically thick regions. In this approximation, the radiative transfer terms are included directly in the energy equation, which is then solved in the same manner as the hydrodynamical equations. In problems involving both optically thick and optically thin regions, the Eddington approximation provides a description of the radiation field superior to that of the diffusion approximation, at the expense of solving an added equation. In the Eddington approximation,

$$L = 4\pi\kappa\rho(J - B),$$

where J is the mean intensity, and $B = \sigma T^4/\pi$ is the Planck function. The mean intensity is the normalized angular integral of the specific intensity I, and is determined by the equation

$$\frac{1}{3}\frac{1}{\kappa\rho}\nabla \cdot (\frac{1}{\kappa\rho}\nabla J) = J - B.$$

The mean intensity equation is solved by an alternating directions implicit method. Three additional relations are needed to close the system of equations, namely the dependences of the specific internal energy, pressure, and opacity on the density and temperature of the gas. More details of the solution methods may be found in Boss (1980, 1984).

2. Applications

The code has been applied to the study of a wide range of astrophysical problems, in 1D, 2D, and 3D, and with thermodynamical assumptions ranging from isothermality to Eddington approximation radiative transfer. The problems which have been successfully studied to date include: the collapse and fragmentation of interstellar clouds and the formation of single or binary protostellar objects (reviewed in Boss 1988b); collapse of pre-supernova, rotating stellar cores (Tohline *et al.* 1980); solar nebula formation (Boss 1988c); explosive expansion of unstable polytropes (Barnes and Boss 1984); tidal disruption of dissipative planetesimals (Mizuno and Boss 1985); dynamic fission instability in both rapidly rotating, inviscid polytropes (Durisen *et al.* 1986) and dissipative planets (Boss 1986); structure and evolution of the envelopes of accreting protostellar cores (Boss 1987); the Papaloizou-Pringle instability in accretion toroids (Boss 1988a); and the dynamical instability of protostellar cores (Boss and Tscharnuter 1989).

The code was developed originally to study the problem of the collapse and possible fragmentation of interstellar clouds into single and multiple protostars. Protostellar collapse calculations are extremely demanding, because of the wide range in densities (increases of up to 20 orders of magnitude), temperatures (10K to 10^6K), and length scales (pc to $R_\odot$) typically encountered. Two representative examples of this type of calculation are presented in Figures 1 and 2.

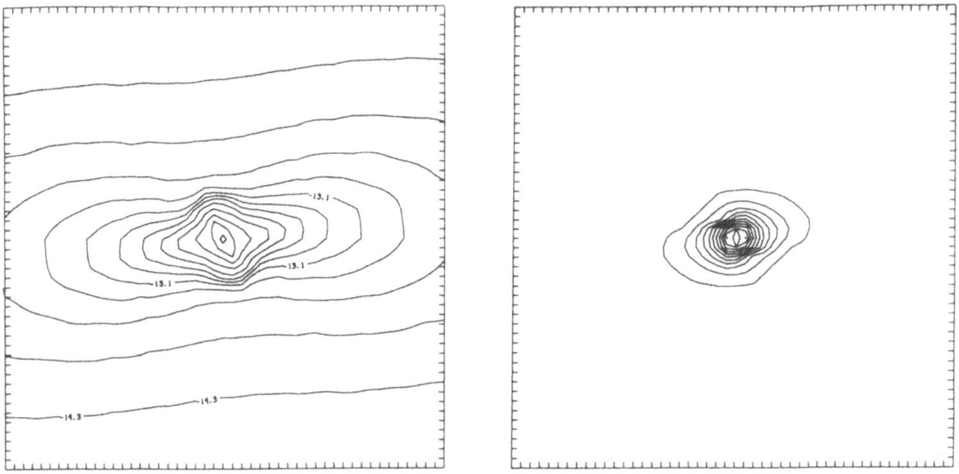

Figure 1. Density (left) and temperature (right) contours in the midplane of a collapsing, rotating protostellar cloud (Boss 1988d). A $1 M_{\odot}$ cloud has formed a bar surrounding a single, central protostar. Region shown is 1200 AU in radius. Maximum density is 10^{-11} g cm^{-3}; each contour corresponds to a change in density by a factor of two. Maximum temperature is 120 K; each contour corresponds to a change of 10 K. Central regions are optically thick, while the outer regions are optically thin; 3D Eddington approximation radiative transfer was used.

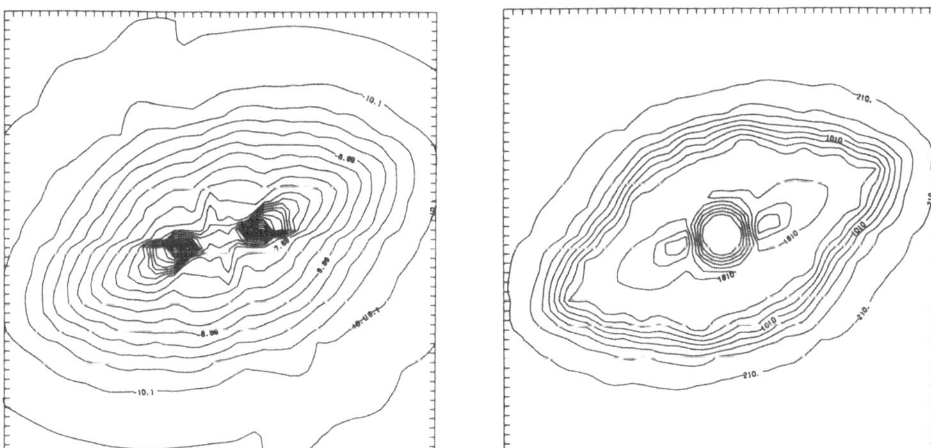

Figure 2. Equatorial density (left) and temperature (right) contours in a model of solar nebula formation (Boss 1988c). The $1 M_{\odot}$ nebula has collapsed and fragmented into a transient binary system that will undergo orbital decay. Region shown is 4 AU in radius. Maximum density is 10^{-4} g cm^{-3}; each contour corresponds to a change in density by a factor of two. Maximum temperature is 2210 K; each contour corresponds to a change of 200 K. The entire region shown is optically thick; 3D diffusion approximation radiative transfer was used.

3. Computational Demands

3D hydrodynamical calculations involve large numbers of grid points, and consequently require enormous amounts of computer time. When radiative transfer is included as well, the computational demands can soar. For example, because of the iterative solution required for the mean intensity equation, in optically thick regions the Eddington approximation code requires roughly 10 times more computer time than the diffusion approximation code. Run times for a typical 3D model of protostellar collapse are on the order of a week of Numerix NMX-432 array processor time, even with fairly coarse spatial resolution (an active grid of about $20^3 = 8000$ points). Ideally, one would like perhaps 1000 grid points in each spatial direction, but this would require about $50^4 = 6.25 \times 10^6$ as many computations (for an explicit code, the number of time steps increases when the grid is refined). Hence for the foreseeable future, 3D calculations will always benefit from increased computational power.

Production runs are currently being computed on the NMX-432 array processor and VAX computers of the Carnegie Institution. The code runs about 5 times faster on the NMX-432 array processor than on a VAX-11/780 FPA computer. Because of the nature of much of the coding, present vector machines do not offer a great advantage over scalar machines; e.g., on a CDC Cyber 205, the code runs only about a factor of 25 times faster than on the VAX-11/780. By running the NMX-432 continuously, then, one can achieve about 1/5 the speed of a supercomputer, at a small fraction (about 1/50) of the purchase price. Partly because of advanced compiler features such as dataflow analysis, the new 'personal supercomputers' (e.g., the Apollo 10000 series) promise speeds nearly equal to supercomputers, at an even lower fraction of the price, leading to a price/performance ratio close to *100 times lower* than present supercomputers.

Acknowledgments

This research is partially supported by the NSF Stars and Stellar Evolution Program (AST-8515644) and by the NASA Planetary Geology and Geophysics Program (NAGW-1410).

References

Barnes, A., and Boss, A. P. 1984, *Ap. J.*, **280**, 819.
Boss, A. P. 1980, *Ap. J.*, **236**, 619.
—— 1984, *Ap. J.*, **277**, 768.
—— 1986, *Icarus*, **66**, 330.
—— 1987, *Ap. J.*, **316**, 721.
—— 1988a, *M. N. R. A. S.*, **231**, 739.
—— 1988b, *Comments Ap.*, **12**, 169.
—— 1988c, *Science*, **241**, 565.
—— 1988d, *Ap. J.*, **331**, 370.
Boss, A. P., and Tscharnuter, W. M. 1989, *Ap. J. (Lett.)*, submitted.
Durisen, R. H., Gingold, R. A., Tohline, J. E., and Boss, A. P. 1986, *Ap. J.*, **305**, 281.
Mizuno, H., and Boss, A. P. 1985, *Icarus*, **63**, 109.
Tohline, J. E., Schombert, J. M., and Boss, A. P. 1980, *Space Sci. Rev.*, **27**, 555.

First-order and Second-order 3-D Hydrodynamics: A Comparison

Harold A. Williams
Department of Terrestrial Magnetism
Carnegie Institution of Washington
5241 Broad Branch Road, N.W.
Washington, D.C. 20015-1305, USA

Why, if you have a first-order hydrodynamic code, should you go to the trouble to make it second-order in space? The short answer is that *time scales* or *growth rates* in a first-order code will be *overestimated* or *underestimated*, respectively, when compared with a second-order code. A second-order scheme is not numerically viscous, or diffusive, like a first-order scheme.

When a hydrodynamic problem is solved on a computer, the continuous variables must be made discrete and derivatives must be approximated by differences. If, in going from continuous to discrete, the error terms are proportional to the grid size, the hydrocode is said to be first-order accurate in space. If the error terms are proportional to the grid size squared, the hydrocode is said to be second-order accurate.

The following two graphs of growth rates for the bar-mode instability in a first-order donor cell method (Figure 1) and a second-order Van Leer method (Figure 2) illustrate why second-order methods are better. To understand the astronomical motivation for such calculations, see Williams and Tohline (1987a, 1988). β is the ratio of the rotational energy to the absolute value of the gravitational potential energy. n is the polytropic index used in the adiabatic equation of state. D_m is a measure of the relative strength of the "$\cos m\theta$ mode," and it is derived from a Fourier analysis of the azimuthal density distribution at a particular radius in the equatorial plane of a model. D_2 measures the "bar-mode," which is predicted to be the fastest growing mode by the linear tensor virial equation, (Chandrasekhar 1969). The growth rate, the slope of $\ln D_2$ versus time, of the $\cos 2\theta$, bar-mode rotational-gravitational dynamic instability, which occurs when $\beta > \beta_d = 0.27$, no longer needs to be corrected for numerical viscosity in the second-order code. A second-order hydrodynamic code was written even though a well-tested, first-order hydrodynamic code was available principally because of the numerical diffusion in any first-order scheme. When studying nonaxisymmetric dynamic instabilities, this numerical diffusion manifests itself by being Fourier mode dependent; see equation

Celestial Mechanics **45**: 89–92, 1989.

(14), Williams and Tohline (1987a). Numerical diffusion damps the growth rate of higher-m structures. When β_i, the initial β for the model, is nearer β_d, a non-diffusive code is required to investigate what happens, since the bar-mode growth rate is gets smaller as β_i approaches β_d. A reliable time scale or growth rate in a hydrodynamic calculation requires a second-order calculation.

The solid line on Figure 1 has a slope equal to the growth rate predicted by a linear tensor virial equation analysis; this line is also coincident with the slope of the D_2 curve during its linear growth phase, corrected by numerical viscosity as detailed in equation (14) of Williams and Tohline (1987a). Figure 2 shows this simulation run in my second-order code with everything similarly defined. The slope of the solid line is again taken from predicted growth rates derived from a linear tensor virial equation analysis. The second-order code *performs as expected*, exhibiting growth rates very close to the predictions of the linear tensor virial theory. In the nondiffusive second-order code, no correction for growth rates is permitted or needed!

In the second-order hydrocode, a delicate and complicated interplay of power going from one Fourier mode to another is revealed. In Figure 2, during the linear growth regime, notice that when D_2 increases, D_4 increases as well. Contrast this to the decrease (damping) of the higher mode amplitudes D_4 and D_6 in Figure 1 (the first-order code) during the time period t =1000–2000 in polytropic units.

When written in conservative form, the continuity equation, the equation of motion, and the energy equation have a flux or advection term and source terms. It is fairly trivial to write the source terms, which are simple derivatives of variables such as pressure and gravitational potential, as second-order differences. It is not so trivial to write the flux or advection term, which contains the nonlinear terms and the coupling between the partial differential equations, in a second-order accurate form that is numerically stable. Norman and Winkler (1986) describe a method using Van Leer monotonic interpolation that is second-order accurate in the grid size. Van Albada et al. (1982) compare different computational methods in cosmic gas dynamics. The Van Leer monotonic interpolation method is described in Van Leer (1977a, 1977b, 1979). Williams (1988) describes rather explicitly how one actually goes about solving the hydrodynamic equations in Chapter 2 (including correction of a typographical error in Norman and Winkler (1986) in Van Leer's interpolation equation.

Appendix C in Williams (1988) has a program listing. Discussion of vectorization and the use of conditional vector merge to calculate the Van Leer slopes and Van Leer interpolation in a way that does not disrupt machines with a pipelined architecture is described in Chapter 3. Conditional vector merge functions, a Cray FORTRAN extension, can be created for the FPS M64/60 (264) and for many other machines by the use of statement functions (Williams and Tohline 1987b). In some machines, such as the IBM 3090 with vector facility and Cray machines,

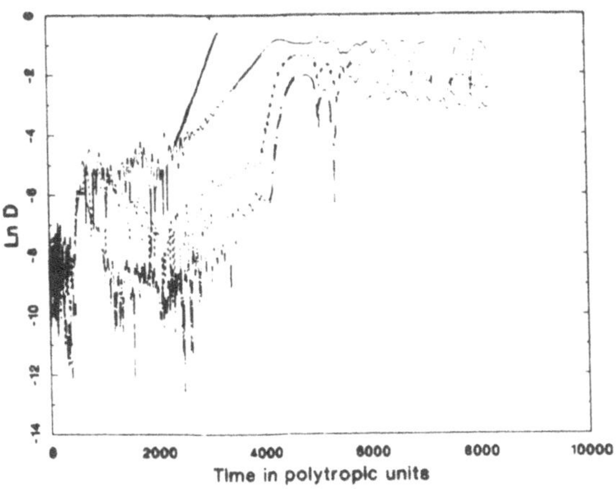

Figure 1.—The $n = 1.8$ and $\beta_i = 0.310$ model simulation run in Tohline's first-order code at $r/R_{eq} = 0.64$ on the equatorial plane. The solid curve is D_2, the dashed curve is D_4, and the chained-dashed curve is D_6, all versus time in polytropic units. The solid line has a slope equal to the growth rate predicted by the linear tensor virial equation analysis.

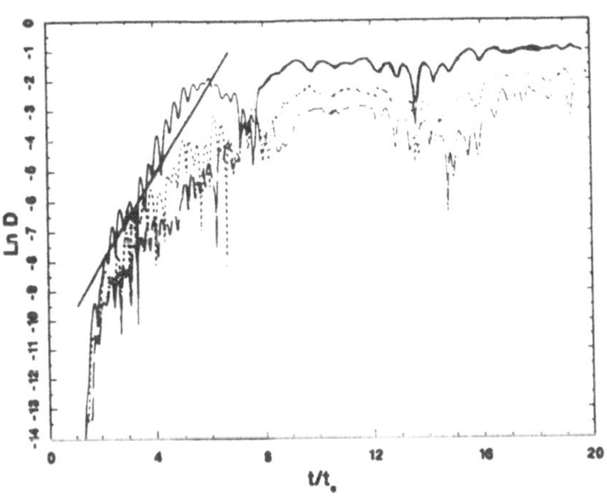

Figure 2.—The $n = 1.8$ and $\beta_i = 0.313$ model simulation run in my second-order code at $r/R_{eq} = 0.64$ on the equatorial plane, where the solid curve is D_2, the dashed curve is D_4, and the chained-dashed curve is D_6 and is versus time in central rotation periods. The solid line has a slope equal to the growth rate predicted by a linear tensor virial equation analysis for a $\beta_i = 0.310$ model instead of $\beta_i = 0.313$. The growth rate predicted using the linear tensor virial equation for a $\beta_i = 0.313$ should be a little larger than for a $\beta_i = 0.310$. The time units are given in terms of central rotation periods, where $t_c = 2\pi/\omega_c$ and ω_c is the angular velocity in the center of the model.

Van Leer slopes and Van Leer interpolation can be written using simple IF statements that will be automatically vectorized by the compiler by replacing the IF statements with conditional vector merge functionality. It is rather important in a second-order code to ensure that it vectorize while calculating the Van Leer slopes and Van Leer interpolation during the advection part of the calculation so that the code will compute rapidly. The second-order advection creates many more opportunities for the code to fail to vectorize because of branching, unless due caution is exercised.

My research into 3-D hydrodynamics to study star formation is partly supported by NASA grant NAGW-398 to the Carnegie Institution of Washington.

REFERENCES

Chandrasekhar, S. 1969, *Ellipsoidal Figures of Equilibrium* (New Haven: Yale University Press).

Norman, M.L., and Winkler, K.-H.A. 1986, 2-D Eulerian Hydrodynamics with Fluid Interfaces, Self-Gravity and Rotation. In *Astrophysical Radiation Hydrodynamics*, ed. K.-H. A. Winkler and M. L. Norman (Dordrecht: D. Reidel), p. 187.

Van Albada, G.D., van Leer, B., and Roberts, W.W., Jr. 1982, A Comparative Study of Computational Methods in Cosmic Gas Dynamics. *Astr. Ap.*, **108**, 76.

Van Leer, B. 1977a, Towards the Ultimate Conservative Difference Scheme III. Upstream-Centered Finite-Difference Schemes for Ideal Compressible Flow. *J. Comp. Phy.*, **23**, 263.

Van Leer, B. 1977b, Towards the Ultimate Conservative Difference Scheme IV. A New Approach to Numerical Convection. *J. Comp. Phy.*, **23**, 276.

Van Leer, B. 1979, Towards the Ultimate Conservative Difference Scheme V. A Second-Order Sequel to Godunov's Method. *J. Comp. Phy.*, **32**, 101.

Williams, H.A. 1988, Star Formation, Using 3-D Explicit Eulerian Hydrodynamics. (Ann Arbor: University Microfilm) Ph.D. thesis.

Williams, H.A. and Tohline J.E. 1987a, Linear and Nonlinear Dynamic Instability of Rotating Polytropes. *Ap. J.*, **315**.

Williams, H.A. and Tohline J.E. 1987b, A Second-Order, Three-Dimensional Hydrodynamic Computer Code (talk to the Working Group for Astronomical Software 170th meeting A.A.S.). *Bull. Am. Ast. Soc.*, **19**, 744.

Williams, H.A. and Tohline J.E. 1988, Circumstellar Ring Formation in Rapidly Rotating Protostars. *Ap. J.*, **334**, 000.

CHAOTIC DYNAMICS AND MONTE CARLO MODELLING

C. Froeschlé[1] and H. Rickman[2]

(1) Observatoire de Nice, B.P. 139, F-06003 Nice - France

(2) Astronomiska Observatoriet, Box 515, S-75120 Uppsala - Sweden

AbstractIn practice, i.e., as long as the initial conditions cannot be specified exactlty, the outcome of a chaotic dynamical system can only be specified in statistical terms. Evolution equations (e.g., the Fokker-Planck equation) for a distribution of test particles can then be formulated, and as an alternative to analytical, mostly approximate or idealized solutions one may simulate the problem using Monte Carlo techniques. Such simulations are a well-known tool in the study of completely chaotic many-body systems such as star clusters or planetary rings, where the sample of test particles can indeed be taken to represent a random set of true solutions according to Bowen's shadowing lemma. In this sense the Monte Carlo modelling plays a role analogous to that of averaging or mapping in regular dynamics, i.e.: the exact dynamical system is replaced by a model overlooking the details of the short-term motion but yielding a good approximation to the long-term behaviour. By a further discretization of the problem the stochastic system can be modelled as a Markov chain. Both Monte Carlo simulations and Markov models have been used in cometary dynamics, and we review some examples from this work to illustrate the success as well as limitations of these stochastic modelling techniques. Lyapunov characteristic exponents and Kolmogorov entropy appear to be suitable tools for estimating the underlying stochasticity to which Monte Carlo simulations refer.

Chaotic dynamics plays an important role for many problems of stellar dynamics (the classical N-body problem) and celestial mechanics (the dynamics of planetary ring particles and comets). Actually since the initial conditions cannot be specified exactly and also since the precision of integration is finite, the chaoticity of an orbit will cause exponential divergence of the computed orbit from the "true" orbit. Hence the outcome of a chaotic dynamical system can only be specified in statistical terms, and therefore a Monte Carlo approach becomes relevant. Indeed for the study of the dynamical evolution of globular clusters such an approach was undertaken already in the 1960's (see Hénon 1973).

Globular Cluster and Planetary Ring Dynamics

For both globular star clusters and planetary rings we are basically dealing with an N-body problem where N is very large. Thus the corresponding dynamical system has an underlying endogeneous stochasticity due to the high number of degrees of freedom. Indeed Froeschlé and Scheidecker (1975) and Benettin et al. (1979) have shown using as a model a one-dimensional self-gravitating system that both the size of the stochastic zone and the Kolmogorov entropy (see Froeschlé 1984) increases drastically with the number of degrees of freedom. The size of the regular component correspondingly drops very quickly to zero and indeed already for such an N-body problem with $N = 6$ it shrinks to zero (see Table 1).

An examination of N-body integration shows that even with the use of individual time steps the total computing time is for an interval t_R [relaxation time; see (Hénon 1973)] is proportional to N^3 Indeed the large amount of computing time thus involved for a globular cluster is not needed since we are not interested in the exact positions and velocities of individual stars but rather in the slow evolution of the system as a whole under a diffusion

Celestial Mechanics 45: 93–98, 1989.

process. Such a process is represented by the Fokker-Planck equation where both the statistical approach and symmetry arguments allow to reduce drastically the number of variables. Unfortunately this equation is still too complicated to be solved – it is a partial differential equation with four variables involving terms which are themselves quadruple integrals. But precisely Monte Carlo methods are efficient for the evaluation of multiple integrals as well as for solving partial differential equations: A random walk is devised in some phase space in such a way as to reproduce the equation to be solved. At this stage we are back to the original problem which was in fact a random walk due to the strong intrinsic stochasticity of the N-body problem, but as long as the resulting Fokker-Planck equation is the same we can alter the basic equation of the random walk. Instead of considering the effect of all the stars on a given star and of integrating all around the orbit one selects randomly just one point of the orbit and computes the perturbation from a randomly selected star. Finally this perturbation is multiplied by an appropriate factor in order to take into account all the time points and all the field stars.

The same approach has been used by Petit and Hénon (1987) to study a numerical model of planetary rings in which many particles interact through gravitation and inelastic collisions. Again only the essential variables are retained for each particle orbit, i.e., its semimajor axis and its mass. Other parameters are chosen at random from the appropriate distribution. The semimajor axis a evolves under the effect of encounters with other particles. Not all the interactions are computed but a pair of particles is drawn randomly according to a distribution which reflects the actual frequency of interactions between the particles. Such a precaution will also be discussed below for cometary dynamics in terms of "dynamical homogeneity". Then the two particles interact so that their semimajor axes change, and the process is repeated over and over again.

Table 1

N	μ_S	ρ	h
2	0.	0.	0.
3	0.04	0.02	0.0008
4	0.86	0.04	0.034
5	0.99	0.073	0.072
6	1	0.108	0.108
7	1	0.13	0.13
8	1	0.18	0.18

Relative measure μ_S of the stochastic region, entropy density ρ (the sum of all positive Lyapunov characteristic numbers) and Kolmogorov entropy h (the Lebesgue integral of ρ over the energy manifold computed by a Monte Carlo procedure). From Benettin *et al.* (1979).

It should be noted that in a chaotic region the maximum Lyapunov characteristic exponent (Froeschlé 1984) is strictly positive, which means that two orbits starting with very close initial conditions will diverge exponentially. Consequently, the validity of a Monte Carlo simulation cannot be checked by accurate numerical integration. It would be necessary to compute thousands of false orbits in order to compare the statistical distribution of the results with that of the Monte Carlo simulation, and this would in a sense make the simulation superfluous. Concerning the fact that computed orbits are "false" orbits we can gain some confidence from the "shadowing lemma" (Bowen 1978) which implies that while a computer may not calculate the orbit which was hoped for, what it does find is nonetheless an approximation of some true orbit of the system provided the orbit belongs to a hyperbolic invariant set, i.e., the system is stochastic. Indeed, like the averaging principle in regular dynamics, we can speak of a "shadowing principle" for chaotic dynamics (Froeschlé and Rickman 1988).

Cometary Dynamics

Contrary to the pronounced stochasticity of the above-mentioned problems, cometary dynamics involves both stochastic and regular components (see Rickman and Froeschlé 1988). Thus to judge the validity of the Monte Carlo simulation technique we have to investigate the underlying stochasticity of the model, which is different for stellar and planetary perturbations.

OORT CLOUD DYNAMICS

While it has recently been realized that an important part of the dynamics of Oort cloud comets arises from regular motion in the Galactic tidal field (Heisler and Tremaine 1986), a decisive role is nonetheless played by individual stellar encounters. Given the physical parameters of such an encounter, its effect on the cometary motion is fully determined. However, the parameters of individual stellar encounters are unpredictable so the stellar perturbations impose a stochastic variation on the cometary orbits.

In Monte Carlo simulations of stellar perturbations (see, e.g., Weissman 1982; Remy and Mignard 1985) the dynamical evolution of a cloud of comets is studied as follows. At a given starting epoch each comet is initialized by choosing a set of orbital elements. These are perturbed by the gravitational effect of passing stars. The geometrical parameters of the stellar encounters are chosen at random. During the passage of a star a comet receives a heliocentric impulse through the interactions of the star with the comet and the Sun. This induces a change in the cometary orbital elements, so these are updated and the comet moves along a new Keplerian ellipse until the next encounter with a random star. In Weissman's procedure the impulses are taken in an even more simplified way from a pre-determined distribution. Due to the perturbations caused by these "random stars" a stochastic process is imposed on the cometary orbital elements, which therefore undergo a random walk. The stochasticity is exogeneous since the stellar encounters occur at random, and the Monte Carlo simulation is justified being a more or less accurate representation of a real stochastic process.

As an Oort cloud comet enters the planetary region, the sequence of planetary perturbations at successive perihelion passages is uncorrelated, i.e., fully stochastic due to the long intervening periods. Just like in the above-mentioned case of a globular cluster it is then of no use to model this by following rigorously the motions of individual comets. The process can be directly simulated by picking random perturbations from pre-calculated samples (Everhart 1977).

SHORT-PERIOD COMET DYNAMICS

As comets are captured into short-period orbits, i.e., with orbital periods of the same order as those of the perturbing planets, the situation changes in a fundamental way. We now have to deal with only the intrinsic stochasticity of a dynamical system which can usually be approximated by a three-body problem: Sun-planet-comet (Rickman and Froeschlé 1988). This stochasticity derives mainly from close encounters with the planet, and during the intervals between such encounters the cometary motion is quasi-regular and predictable. In more precise terms this means that the motion is stochastic only over time scales longer than the typical interval between encounters with the perturbing planet. Thus the phase space domain of short-period comets presents chaotic and ordered regions in an intricate mixture.

Stochastic modelling of this motion is justified by the shadowing principle only if the time step is chosen long enough. Thus we may consider the following procedure (Rickman and Vaghi 1976; Froeschlé and Rickman 1980): we pick a large number of initial cometary orbits at random and integrate these with Jupiter as the perturbing planet over a time interval Δt, which can be different for different comets. From the resulting set of perturbations we construct stochastic orbital evolutions by picking independently a random perturbation for each successive interval Δt. For these evolutions to simulate real ones we would have a constraint on the choice of Δt so that a minimum time step for the random walk could in principle be defined.

However, there are practical problems of such a procedure which are not easy to solve. These are connected with the question of the definition of the perturbation samples: how should one choose the "boxes" of phase space where the initial conditions are to be picked? As already mentioned, the dimension of the boxes can be reduced by symmetry arguments and in the present case a critical point is that short-period comet orbits are known to be generally of low inclinations. One is therefore close to a planar problem with a four-dimensional phase space, and within the limits chosen for the inclination the choice of inclination and orientation of the nodal line is statistically immaterial. Furthermore, since we are dealing with Jupiter-crossing or Jupiter-tangent orbits and Jupiter's eccentricity is quite small, there is an approximate circular symmetry such that encounters with the planet can occur with equal probability independent of the orientation of the apsidal line. The choice of the latter is hence also immaterial, and obviously the time-related parameter expressing Jupiter's position at the time of the cometary perihelion passage is the real stochastic variable of the problem which should be taken at random with its true probability distribution. We are left with two orbital parameters which can be taken as, e.g., semi major axis and eccentricity (a, e), or aphelion and perihelion distances (Q, q).

Boxes in the (Q, q)-plane were considered in the Monte Carlo simulations cited above. Let us note that a further reduction of the dimension appears unfeasible. The Tisserand criterion for a co-planar cometary orbit :

$$T = \frac{2\,a_J}{Q+q} + 2\sqrt{\frac{2\,Q\,q}{a_J\,(Q+q)}} = constant \qquad (1)$$

might be used to restrict the random walk $(\Delta Q, \Delta q)$ to one-dimensional curves, and the Monte Carlo simulation would then consist of a set of independent, parallel simulations for the different values of T. In each of these one would consider a suitable orbital parameter varying along the curve, such as the inverse semimajor axis $z = 1/a$, and there would be a random walk with step size distributions $f_i(\Delta z)$ computed for intervals $[z_{i-1}, z_i]$ along the z axis. However, although small (Froeschlé and Rickman 1981), the perturbations ΔT in the elliptic restricted problem are indeed important for the outcome of the low-velocity encounters with Jupiter occurring in low-eccentricity planet-tangent orbits, and those encounters appear essential for the dynamical transfer of comets.

Quite obviously the number of (Q, q)-boxes is limited in practice by the requirement to compute a sample of perturbations large enough to give a fair representation of the dynamics of the region in question. E.g., pronounced non-Gaussian tails are known to exist in the perturbation distributions and the sample must extend far enough into these tails to cover their significant parts. This means that the number of orbits to be integrated in each box may be very large [for a possible way to reduce this problem, see Rickman and Froeschlé (1983)]. At present it thus appears difficult to go much beyond the dynamical resolution of the 30 boxes. On the other hand, this already guarantees a certain "dynamical homogeneity" in the sense that the perturbation distribution over a short interval of time (one orbital period) should not vary too much from one side of a box to the opposite one.

However, within all the boxes we can expect to find smaller regions corresponding to "resonant strips" with $a \approx a_{res}$, where the comets are close to a simple mean motion resonance with Jupiter. Within such a region there may be slow circulation of the critical argument, in which case it is extremely difficult to define a mean interval between encounters and to find an integration interval Δt which is everywhere sufficient. But actually the situation is even worse, because there also exist integrable regions of phase space corresponding to stable librations of the critical argument, where encounters with Jupiter never occur.

The main result of a Monte Carlo simulation of cometary dynamics is a picture of the distribution of comets over the various orbits connected by the stochastic transfer process in question. If the short-period comets are viewed in their most general framework, the dynamics includes perturbations by all the planets as well as nongravitational perturbations, and it would then be interesting to estimate the number of comets trapped in such quasi-stable resonant regions. How this goal would be achieved is not yet clear: the problem is to find an appropriate definition of the perturbation sample boxes and corresponding integration intervals for such a detailed investigation. If on the other hand one considers the dynamical transfer in the three-body problem (Sun-Jupiter-comet), the problem is instead to reach what may be called "topological homogeneity" (Froeschlé and Rickman 1988): Comets should then follow only chaotic routes and the relevant boxes from which the initial conditions are to be picked are in fact the intersections of the usual (Q, q)-boxes with the chaotic part of phase space. If this restriction is not taken into account, the rate of the transfer is artificially slowed down by the inclusion of irrelevant trappings or an overestimated probability of small perturbations.

Markov Chain Modelling

If indeed we consider the orbital distribution of comets as the principal result to be attained, we can again argue that the random walks by individual sample comets considered in the Monte Carlo simulation contain too much information: the only interesting quantity is the number of comets in each (Q, q)-box either as a function of time or in a steady state. The same perturbation samples used for the Monte Carlo simulation can then be used to calculate "jump probabilities" between the various (Q, q)-boxes over a common time interval Δt. Defining the state vector $\vec{n}$ of the cometary population to be the set of numbers of comets in the different boxes, and calling τ_{ij} the jump probability from box i to box j, $\vec{n}$ evolves according to a Markov chain:

$$\vec{n}(t + \Delta t) = \vec{n}(t) \cdot \tau \tag{2}$$

where $\tau = (\tau_{ij})$ is the transformation matrix. This method was first used in cometary dynamics by Rickman and Froeschlé (1979) for the same domain of the (Q, q)-plane as in the above-mentioned Monte Carlo simulation and later on for Oort cloud dynamics by Lago and Cazenave (1983).

This Markov method has the advantage of extreme efficiency, in particular for finding steady-state solutions where Eq. (2) reduces to:

$$\vec{n}_{ss} = \vec{n}_{ss} \cdot \tau, \tag{3}$$

i.e., a system of linear equations. Its main disadvantage is that, just like for the Monte Carlo simulation, the number of (Q, q)-boxes is limited by the large number of integrated orbits required to obtain accurate estimates of the jump probabilities. If e.g. the 30 boxes of Rickman and Froeschlé are considered, the information obtained on the (Q, q) distribution is strictly limited to the 30 sample points represented in the state vector.

Conclusions

This paper is intended to present some thoughts about the relations between the intrinsic chaos that is known to be present in non-integrable Hamiltonian systems and Monte Carlo modelling. Such Monte Carlo procedures appear to be straightforward for strongly chaotic systems with a large number of particles like globular clusters or even planetary rings. For systems with fewer degrees of freedom the stochasticity is less evident. However, since comets follow chaotic routes Monte Carlo simulations can be applied and have indeed been extensively used. In another paper (Froeschlé and Rickman 1988) we have presented a detailed discussion of Monte Carlo modelling of cometary dynamics, and here we try to penetrate deeper into some aspects of this discussion.

However, the present ideas are only a beginning and much work still remains especially concerning the dynamics of short-period comets in order to develop a Monte Carlo modelling procedure which is optimally adapted to the intrinsic stochasticity of this problem.

References

Benettin, G., Froeschlé, C., Scheidecker, J.-P. : 1979, "Kolmogorov Entropy of Dynamical Systems with Increasing Number of Degrees of Freedom",*Phys. Rev. A* **19**, 2454-2460

Bowen, R. : 1978, "On Axiom A Diffeomorphisms", CBMS Regional Conferences Series in *Mathematics*, Vol **35**, MS Publications : Providence

Everhart, E. : 1977, "The Evolution of Comet Orbits as Perturbed by Uranus and Neptune", in *Comets, Asteroids, Meteorites* (ed. A.H. Delsemme), Univ. Toledo Press, pp. 99-104

Froeschlé, C. : 1984, " The Lyapunov Characteristic Exponents - Applications to Celestial Mechanics", *Celest. Mech.* **34**, 95-115

Froeschlé, C., Rickman, H. : 1980, " New Monte Carlo Simulations of the Orbital Evolution of Short-Period Comets and Comparison with Observations", *Astron. Astrophys.* **82**, 183-194

Froeschlé, C., Rickman, H. : 1981, " A Monte Carlo Investigation of Jovian Perturbations on Short-Period Comet Orbits", *Icarus* **46**, 400-414

Froeschlé, C., Rickman, H. : 1988, " Monte Carlo Modelling of Cometary Dynamics", *Celest. Mech.*, in press

Froeschlé, C., Scheidecker, J.-P. : 1975, "Stochasticity of Dynamical Systems with Increasing Number of Degrees of Freedom", *Phys. Rev. A* **12**, 2137-2143

Heisler, J., Tremaine S. : 1986, "The Influence of the Galactic Tidal Field on the Oort Comet Cloud", *Icarus* **65**, 13-26

Hénon, M. : 1973, "Collisional Dynamics in Spherical Stellar Systems", in *Dynamical Structure and Evolution of Stellar Systems* (eds. L. Martinet and M. Mayor), Swiss Soc. of Astron. and Astrophys., pp. 183-260

Hénon, M., Heiles, C. : 1964, " The Applicability of the Third Integral of Motion; Some Numerical Experiments", *Astron. J.* **69**, 73-79

Lago, B., Casenave, A. : 1983, "Evolution of the Cometary Perihelion Distance in the Oort Cloud : Another Statistical Approach", *Icarus* **53**,68-83

Petit, J.M., Hénon, M. : 1987, "A Numerical Simulation of Planetary Rings. II. Monte Carlo Model ", *Astron. Astrophys.* **188**, 198-205

Remy, F., Mignard, F. : 1985, "Dynamical Evolution of the Oort Cloud. I. A Monte Carlo Simulation", *Icarus* **63**, 1-19

Rickman, H., Froeschlé, C. : 1979, "Orbital Evolution of Short-Period Comets Treated as a Markov Process", *Astron. J.* **84**, 1910-1917

Rickman, H., Froeschlé, C. : 1983, "A Keplerian Method to Estimate Perturbations in the Restricted Three-Body Problem", *Moon and Planets* **28**, 69-86

Rickman, H., Froeschlé, C. : 1988, "Cometary Dynamics", *Celest. Mech.*, in press

Rickman, H., Vaghi, S. : 1976, "A Monte Carlo Simulation of the Orbital Evolution of Comets in the Inner Planetary Region", *Astron. Astrophys.* **51**, 327-342

Weissman, P.R. : 1982, "Dynamical History of the Oort Cloud", In *Comets* (ed. L.L. Wilkening), Univ. Arizona Press, Tucson, pp. 637-658

Software Tools for Nonlinear Dynamics

A. Abad, A. Elipe and M.L. Sein-Echaluce
Grupo de Mecánica Espacial
Universidad de Zaragoza. 50009 Zaragoza. Spain

Abstract

An algebraic processor plus an interactive graphics system transform a desk computer into a very handy tool for research in non linear mechanics.

1 Introduction

We present an instance of algebraic calculation and interactive graphics on a desk computer (2 M bytes of memory, 70 Mb of disk storage) This sort of equipment hardware and software is capable of solving large class of dynamical systems in non-linear dynamics.

2 The dynamical problem

The dynamical problem is described by the following Hamiltonian:

$$\mathcal{H} = \frac{1}{2}(X^2 + Y^2) + \frac{1}{2}\omega^2(x^2 + y^2) + \omega^2 \alpha x^2 y + \eta\omega^2\beta^2 y^4 \tag{1}$$

The system is studied in the neighborhood of the stable equilibrium at the origin. Accordingly the Hamiltonian (1) the system presents a 1-1 resonance, therefore, in order to avoid zero-divisors and other complications, Lissajous variables are introduced (for details, see [1]).

3 Normalization by computer

The Hamiltonian in Lissajous variables is such that $\mathcal{H}_{0,0}$ reduces to $\omega\, L$, while $\mathcal{H}_{1,0}$, $\mathcal{H}_{2,0}$ are Fourier series in ℓ and g with coefficients being polynomials in σ and δ , which depend on L and G. Therefore we build the normalization - in other words we eliminate the angle ℓ - by a Lie transformation. The normalizing Lie transformation has been built automatically with a Processor of Poisson series. We received the package from Dr. Dasenbrock ([2]) He designed his program for a VAX 780. We took upon ourselves to transform Dasenbrock's processor from that minicomputer into our desk computer. Besides slight incompatibilities between the

Celestial Mechanics **45**: 99–102, 1989.

two FORTRANS, the problems we encountered the most were, of course, in the area of input/output. We checked our transposed version on classical elementary problems, when solutions we could easily compute by hand to a non-trivial order (solving Kepler's equation by Lie transformation, reconstructing Hansen's coefficients).

In the present case, the normalized Hamiltonian can out as the sum

$$\mathcal{H} = \mathcal{H}_{0,0} + \mathcal{H}_{0,1} + \frac{1}{2}\mathcal{H}_{0,2}$$

with

$$\begin{aligned}
\mathcal{H}_{0,0} &= \omega^2(\sigma^2 + \delta^2) \\
\mathcal{H}_{0,1} &= 0 \\
\mathcal{H}_{0,2} &= \left(A(\sigma^4 + \delta^4) + B\sigma^2\delta^2 + C\sigma\delta(\sigma^2 + \delta^2)\cos 2g + D\sigma^2\delta^2 \cos 4g\right)
\end{aligned}$$

The coefficients A, B, C and D are functions of the parameters α, β,η which were explicitly determined by the program. For instance, we found that

$$A = \frac{3}{4}\eta\beta^2 - \frac{1}{24}\alpha^2$$

4 From analysis to graphics, and back

After normalization, it turns out that the phase space of the system is a sphere. This is a very fortunate circumstance, and we intend to exploit it to the maximum. Indeed spheres and spherical curves are very easy to draw on the screen of a desk computer, provided of course the graphic resolution is sharp enough. In the orthographic projection we do not need sophisticated algorithms to deal with hidden lines.

With appropriate Cartesian coordinates (ξ_1, ξ_2, ξ_3), such that

$$(\xi_1, \xi_2) = \xi_3 \quad , \quad (\xi_2, \xi_3) = \xi_1 \quad , \quad (\xi_3, \xi_1) = \xi_2$$

the sphere that is the phase space on the integral manifold $L = ct.$ is represented by the equations

$$\xi_1^2 + \xi_2^2 + \xi_3^2 = \frac{1}{4}L^2$$

and dropping terms dependent exclusively on L, we get for the Hamiltonian function

$$\mathcal{H} = L + \varepsilon^2 \left[L^2 \left(\frac{3}{8}\eta\beta^2 - \frac{25}{48}\alpha^2\right) + \right.$$

$$L\xi_1\left(\frac{3}{2}\eta\beta^2+\frac{5}{12}\alpha^2\right)+\xi_1^2\left(\frac{3}{2}\eta\beta^2+\frac{5}{4}\alpha^2\right)+2\alpha^2\xi_3^2\Big]$$

Hence the equations of averaged motion

$$\begin{aligned}\dot{\xi}_1 &= -\quad 8\alpha^2\xi_2\xi_3\\ \dot{\xi}_2 &= -\quad L(3\eta\beta^2+\frac{5}{6}\alpha^2)\xi_3-(6\eta\beta^2-3\alpha^2)\xi_1\xi_3\\ \dot{\xi}_3 &= \quad L(3\eta\beta^2+\frac{5}{6}\alpha^2)\xi_2+(6\eta\beta^2+5\alpha^2)\xi_1\xi_2\end{aligned}$$

In order to understand how, qualitatively speaking, the phase flow varies with the parameters α, β, η, one should first determine the equilibria of the averaged system. This kind of analysis is often very awkward. But we are greatly assisted if we can generate at will portraits of the phase flow for any values of those parameters. As an example, we reproduce two portraits, one when $\alpha = \frac{1}{9}, \beta = \frac{1}{10}, \eta = 1$, (Fig. 1) and the other when $\alpha = \frac{1}{9}, \beta = \frac{1}{10}, \eta = -1$, (Fig. 2). The curves were obtained integrating numerically the averaged equations - a 4th order Runge-Kutta algorithm is sufficient at the resolution level of the screen we had available.

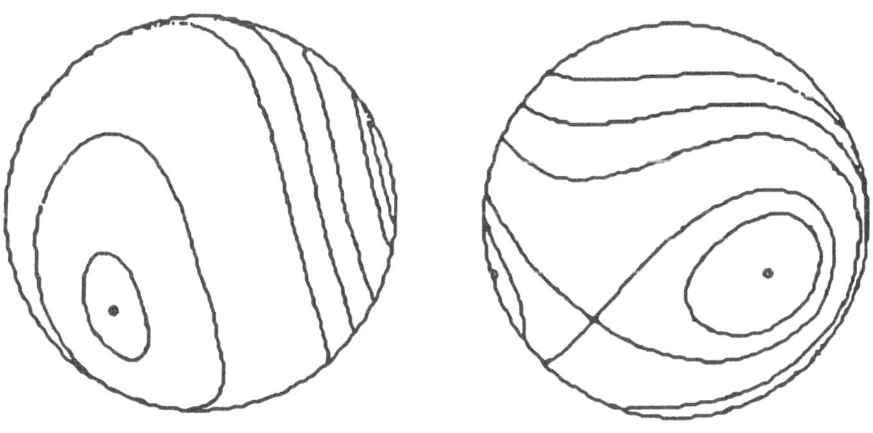

Fig. 1. Trajectories on the sphere when $\alpha = \frac{1}{9}, \beta = \frac{1}{10}, \eta = 1$.

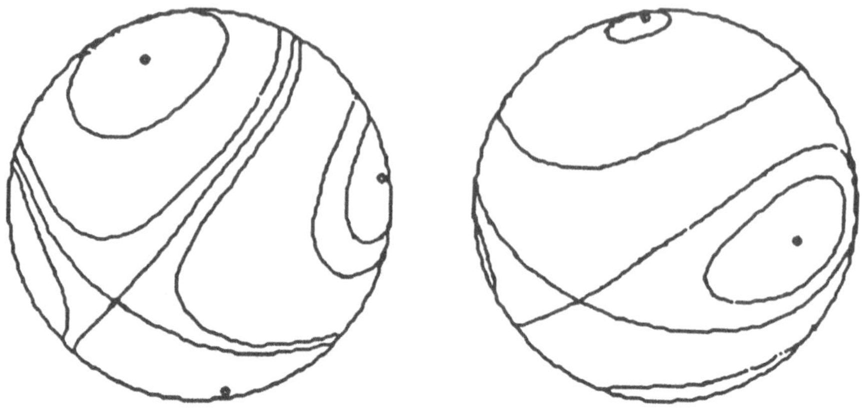

Fig. 2. Trajectories on the sphere when $\alpha = \frac{1}{9}, \beta = \frac{1}{10}, \eta = -1$.

5 Conclusions

Special purpose SAC systems in conjunction with elementary computer graphics put on the desk of an analyst most of the tools he needs to study in depth a wide class of dynamical systems in non-linear mechanics.

6 Acknowledgements

Dr. R. Dasenbrock of the Naval Research Laboratory has kindly provided us his processor for Poisson series. Dr. Deprit of the National Bureau of Standards suggested us the topic of this research. This work has been partially supported by CICYT Proy. No. PB87-0637 and by CAZAR.

References

[1] Deprit, A.: 1988, manuscript intended for publication in *Celes. Mec.*

[2] Dasenbrock, R.R.: 1982, *A FORTRAN-Based Program for Computerized Algebraic Manipulation* NRL Report 8611

COMPUTER ALGEBRA, LIE TRANSFORMS AND THE NONLINEAR STABILITY OF L_4

V.T. Coppola and R.H. Rand
Dept. of Theoretical and Applied Mechanics
Thurston Hall
Cornell University
Ithaca, NY 14853-1503

Although the circular restricted three-body problem and the stability of its triangular equilibrium point L_4 have been the subject of numerous investigations, the nonlinear stability question has only recently been completely answered [3]. The difficulty here is the tremendous quantity of algebra involved. Today, computer algebra systems can handle algebra-intensive calculations allowing investigations of problems previously intractable.

The hamiltonian H describing the system at L_4 is written as

$$(1) \qquad H = \frac{1}{2}\left[p_1^{\,2} + \omega_1^{\,2} q_1^{\,2}\right] - \frac{1}{2}\left[p_2^{\,2} + \omega_2^{2} q_2^{\,2}\right] + \text{h.o.t.}$$

which shows that the system to lowest order is comprised of two linear oscillators. The corresponding linearized equations have purely imaginary eigenvalues, and thus are inconclusive regarding nonlinear stability. Moreover, H is not positive-definite so Lyapunov's direct method cannot be applied using H. Nonlinear stability for hamiltonians of the form (1) can be proved, however, by appealing to a theorem of Arnold *when H is expressed in Birkhoff normal form.* This paper describes the transformation to Birkhoff normal form.

The Lie Transform method is a canonical near-identity perturbation method for hamiltonian systems. We use the method of Lie Transforms to transform hamiltonians of the form (1) into Birkhoff normal form. First, we make a canonical transformation from coordinates (q_m, p_m) to eigencoordinates (x_m, y_m) using the eigenvectors of the linearized equations. The hamiltonian H then becomes

$$(2) \qquad H = i\,\omega_1\, x_1\, y_1 - i\,\omega_2\, x_2\, y_2 + \Sigma\, H_{jklm}\, x_1^j\, y_1^k\, x_2^l\, y_2^m\,, \quad i = \sqrt{-1}$$

We take ω_1, ω_2, and H_{jklm} as symbolic coefficients (i.e., treated as parameters with no numerical values) in performing the transformation so that the transformation is done once for all hamiltonians with form (2).

Celestial Mechanics **45**: 103–104, 1989.

The Lie Transform near-identity transformation from (x_m, y_m) to (X_m, Y_m) transforms H into K (the Birkhoff normal form) given by

$$(3) \qquad K = \omega_1 J_1 - \omega_2 J_2 + K_{20} J_1^2 + K_{11} J_1 J_2 + K_{02} J_2^2 + K_{30} J_1^3 + K_{21} J_1^2 J_2 + K_{12} J_1 J_2^2 + K_{03} J_2^3$$

$$\text{and } J_1 = i X_1 Y_1, \quad J_2 = i X_2 Y_2, \quad i = \sqrt{-1}$$

where the K_{np} coefficients are now known functions of ω_1, ω_2, and the H_{jklm}. In this way, we derive formulas for K explicitly in terms of the original coefficients appearing in H. Moreover, nonlinear stability criteria formulated for K can now also be expressed solely in terms of the coefficients in H. We can then apply these results to any hamiltonian of the form (1) by simple substitution; specifically, we use these results to analyze the nonlinear stability of L_4.

We have implemented this Lie Transform algorithm in the computer algebra system MACSYMA. The program, however, fails to run (i.e., the Symbolics 3670 computer runs out of memory space) unless memory-saving programming techniques are employed. Even with these steps, many megabytes of memory are used in the calculations. Coppola and Rand [1] describe this procedure in greater detail.

As a check of our work, we determined the nonlinear stability of L_4. Deprit and Deprit-Bartholome [2] first computed the Birkhoff normal form (3) (through quadratic terms in J_1 and J_2) for L_4 as a function of the mass ratio parameter μ using D'Alembert series. They then applied the stability theorem to prove stability of L_4 for $0 < \mu < \mu_1$, $\mu \neq \{\mu_2, \mu_3, \mu_c\}$ where μ_1, μ_2, and μ_3 are mass ratios for which the linearized frequencies ω_1 and ω_2 are in 1:1, 1:2, and 1:3 resonance, respectively. The value $\mu_c = 0.0109136\ldots$ does not correspond to a resonance but rather to a mass ratio where higher order terms determine the stability. Meyer and Schmidt [3], using a computer algebra program written in PL/I, verified the work of Deprit and Deprit-Bartholome and computed the cubic terms in (J_1, J_2) for K. They then determined that μ_c is stable. Using our more general approach, we corroborate these results.

[1] Coppola, V. and Rand, R. (1989) "Computer Algebra Implementation of Lie Transforms: Application to L_4", to appear in ZAMM.

[2] Deprit, A. and Deprit-Bartholome, A. (1967) "Stability of the Triangular Lagrangian Points", *Astron. J.* 72(2), pp.173-179.

[3] Meyer, K. and Schmidt, D. (1986) "The Stability of the Lagrange Triangular Point and a Theorem of Arnold", *J. Diff. Eqns* 62(2), pp.222-236.

SOLUTION OF AN INFINITE NUMBER OF INEQUATIONS DEPENDING ON A CONTINUOUS PARAMETER AND APPLICATION TO THE SOLUTION OF EQUATIONS IN DYNAMICAL ASTRONOMY

J.-P. J. LAFON
Observatoire de Paris-Meudon
DASGAL - URA CNRS 663
92195 Meudon Principal Cedex
France

ABSTRACT. The investigation of a large class of problems in physics and astrophysics requires the dtermination of the ranges of some parameters z, E, ... for which inequations of the form $F(r;z,E,...) \geqslant 0$ are satisfied for all r in some interval)a,b(. The solution of this problem is given under the form of three general theorems and, resulting from them, a very simple numerical procedure. This can also be used to solve equations of the form $df/dt = 0$ where f is some function of variables and derivatives of these variables (functions of t) with respect to t, for instance Vlasov-type equations in dynamics of flat stellar disks.

1. THE PROBLEM

Consider a set of inequations of the form

$$F(r;x,y,...) \geqslant 0 \qquad \forall \quad r \in)p,R(\tag{1}$$

with $(x,y,...) \in R^n$

There are an <u>infinite number</u> of inequalities indexed by a <u>continuous parameter r</u> and the (x,y,...) belong to <u>continuous sets</u> (the variables x, y, ... belong to intervals of real numbers).

The solution lies in two theorems (Lafon , 1977) with which we can reduce the problem to a similar but much simpler problem i.e. with a set of countable values of r and, in most cases, with a finite number of values of r.

The resulting method of solution has many advantages over the usual empirical methods because it is universal and systematic, well fitted to numerical analysis and rigorous :

Such inequalities are usually solved with empirical methods depending on the particular form of the function, one for each case. The method proposed is independent of F for a large class of functions satisfying a few simple conditions (continuity and derivability) chosen so that they are satisfied in most cases. The method can easily be generalized even when some of these conditions are no longer satisfied.

The solutions are found progressively i.e; successively for all the intervals)p,a(or)a,R(with $p \leqslant a \leqslant R$. Thus, the number of nume-

Celestial Mechanics **45**: 105–110, 1989.

rical tests is greatly reduced since the interval)p,R(is traversed only once whereas other methods require many traverses. As a consequence, the numerical determination of the solutions is also more easy and rigorous.

For instance, inequations like (1) have to be solved when building models of finite, flat, axisymmetric, self-gravitating disks of stars (Lafon, 1976). The star motion is governed by Vlasov equations; in other words, the stars move in the gravitational potential that they generate collectively, along non collisional orbits characterized by two constant parameters, the angular momentum z and the energy E. A star belonging to the system travels along its trajectory between two radial distances lower than the radius of the system. A non collisional stellar orbit in the self gravitation field of the system is populated only if it does not pass beyond the external limit of the system; otherwise the 'potential' orbit is unpopulated.

Thus, the distribution functions of the stars which satisfy the Vlasov equations are functions f(r;z,E) equal to zero within the ranges of z and E for which inequalities like (1) are satisfied : the functions F(r;z,E) depend on the potential profile; each condition $F \geqslant 0$ expresses that a star moving along the orbit characterized by z and E could overcome the local potential barrier; at least R is greater than the radius of the system.

Details are given elsewhere (Lafon, 1976, 1986)

2. DEFINITIONS AND HYPOTHESES

For simplicity and to illustrate the theorems easily with figures, the problem will be concerned hereafter with two variables only x and y. However the method does not have to be always restricted to two variables (Lafon, 1977).

We shall assume that :

- F(r;x,y) is continuously derivable with respect to r,x,y
- $F'_x = \partial F/\partial x$ and $F'_y = \partial F/\partial y$ are never simultaneously zero
- $\forall$ r $\in$)p,R(the sets of points x,y such that $F \geqslant 0$ and $F \leqslant 0$ are connected.
- the solutions, if any, of the system F = 0 and $F'_r = 0$ are bounded and $F'_r = 0$ $\forall$ x and y such that F = 0 for at most a finite number of r $\in$)p,R(

These conditions are fairly weak; the generalisation to the case where they are not satisfied for countable sets of r,x,y is straightforward. The third assumption is used only to illustrate more easily the solutions by figures.

In terms of geometrical representation in the x,y plane these assumptions mean that, for each r, F = 0 is the equation of a continuous, differentiable, non singular curve P_r.

The following notations will be used : for F(s;x,y)

W^+_{ss} set of the points x,y such that $F > 0$

W^-_{ss} set of the points x,y such that $F < 0$

P_s set of the points x,y such that $F = 0$

W^+_{pr} set of the points x,y such that $F > 0$ $\forall$ $s \in)p,r($

W^-_{pr} set of the points x,y such that $F < 0$ $\forall$ $s \in)p,r($

N_{pr} set of the points x,y such that $F = 0$ for at least one s in)p,r(

$\overset{o}{N}_{pr}$ interior of N_{pr}

Q_{pr} $W^+_{pr} \cap W^-_{pr}$ boundary of W^+_{pr}

W^+_{rR} set of the points x,y such that $F > 0$ $\forall$ $s \in)r,R($

W^-_{rR} set of the points x,y such that $F < 0$ $\forall$ $s \in)r,R($

In terms of geometrical representation in the x,y plane the curve P_s divides the plane into two connected regions W^+_{ss} and W^-_{ss}. The regions W^+_{pr}, W^-_{pr}, W^+_{rR} and W^-_{rR} are the intersections of infinite sets of such regions.

Interms of 'shadow' ($F > 0$) and 'light' ($F < 0$), W^+_{pr} is the zone of full shadow, N_{pr} is the zone of penumbra and W^-_{pr} is the zone of full light.

NOTE The graphical representations are useful to visualize the results, in which there are only a few possible cases; this should not suggest that so it is with the demonstrations, for which many situations have to be envisaged before they can be reduced to these cases. Thus, we shall not go into details here : the demonstrations are detailed elsewhere (Lafon, 1977). We only display the results and the way to use them.

3. THEOREM 1

Lafon (1977) proved that, when the conditions stated in section 2 are satisfied :

$$N_{pr} = \left(\bigcup_{p \leq r \leq R} W^+_{rr} \right) \cap \complement \overset{o}{W}{}^+_{pR}$$

$$N_{pr} = \left(\bigcup_{p \leq r \leq R} W^-_{rr} \right) \cap \complement \overset{o}{W}{}^-_{pR}$$

$$\overset{o}{N}_{pr} \neq 0$$

4. THEOREM 2

Lafon (1977) also proved that, when the conditions stated in section 2 are satisfied, two exclusive cases are possible :

1/ There is some $L \in)p,R)$ and $l \in (p,R($ such that

$$W^+_{pr} = \phi \qquad \forall\, r \geq L$$

$$W^+_{rR} = \phi \qquad \forall\, r \leq l$$

2/ The interval $)p,R($ can be divided into a countable set of adjacent subintervals $)a_i,a_{i+1}($ such that $a_1 = p$, $a_{i+1} \leq R$ and such that one of the following four cases describes what occurs for $a_i < r < a_{i+1}$

**** a/ $F = 0$ and $F'_r \neq 0$ except for a finite number of values of r say $r_1, r_2, \ldots r_j$ for which $F'_r = 0 \; \forall$ x and y such that $F = 0$ (extremal curves P_r). Then,

$$W^+_{pr} = W^+_{pa_i} \cap W^+_{rr} \left(\bigcap_{k=1,2,\ldots j} W^+_{r_k r_k} \right) \qquad \text{Fig. 1}$$

**** b/ $F = 0$ and $F'_r = 0 \;\; \forall$ r for a finite number of fixed points x,y ('envelope points') . Then,

$$W^+_{pr} = W^+_{pa_i} \cap W^+_{rr} \qquad \text{Fig. 2}$$

**** c/ $F(s;x,y) = 0$ and $F'_r(s;x,y) = 0$ define some bounded continuous functions x(s) and y(s) (corresponding to arcs of envelopes of the curves P_s). Then,

$$W^+_{pr} = W_E \subset W^+_{pa_i} \cap W^+_{rr} \qquad \text{Figs 3, 4}$$

where W_E denotes that part of the region $W^+_{pa_i} \cap W^+_{rr}$ bounded by the set of the points x(s),y(s) for all s between a_i and r; in terms of geometry, W^+_{pr} is that part of the region $W^+_{pa_i} \cap W^+_{rr}$ reduced by the arcs of the envelope corresponding to s between a_i and r.

**** d/ Cases b and c appear simultaneously within the same interval a_i, a_{i+1}. Then W^+_{pr} is the intersection of the solutions obtained assuming pure b and c cases separately.

<u>NOTES</u> Other theorems :

There is also a countable set of adjacent subintervals h_{k+1}, h_k such that $h_1 = R$ and $h_{k+1} \geq p$ with similar properties for W^+_{rR}

Similar theorems can also be proven for W^-_{pr} and W^-_{rR}

In most of the practical cases, the countable sets of a_i and h_k are sets with a finite number of values.

CONCLUSION

oooooooooo

The determination of W^+_{pR} and W^-_{pR} and, consequently, of N_{pR} is finally reduced to intersections of <u>countable and usually finite</u> sets of regions in the x,y plane.

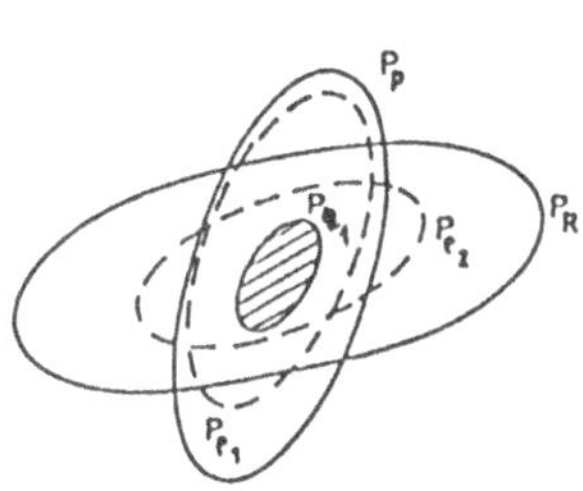

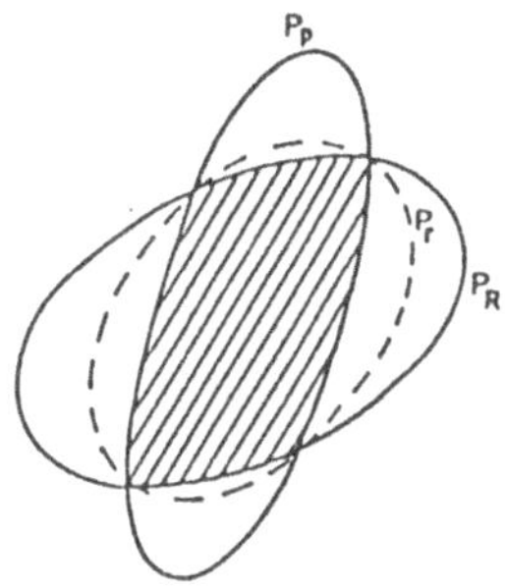

Fig 1 Interval of type a
Here P_a is an extremal curve

Fig 2 Interval of type b

Fig 3 Interval of type c

Fig 4 Interval of type c

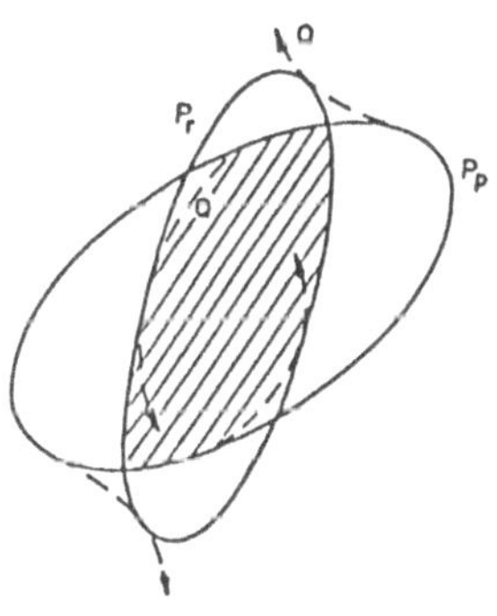

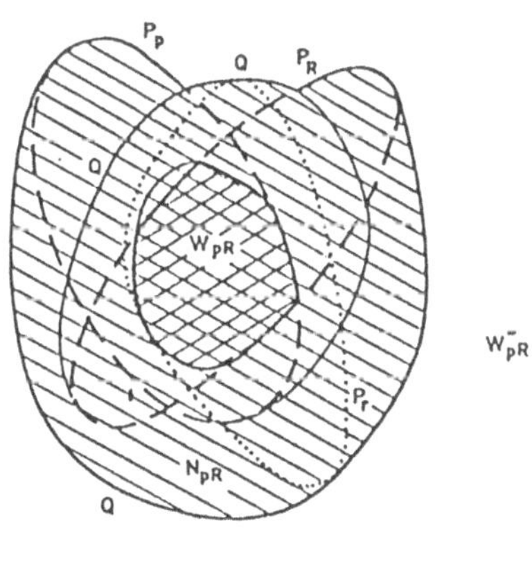

5. NUMERICAL TESTS FOR DETECTING THE LIMITS OF THE SUBINTERVALS

It may seem difficult to determine the a_i or h_k and so one could think that , in spite of its generality, the method that can be deduced from the above mentioned theorems requires complicated analyses for numerical computations. This is far from beeing true.

Indeed, in a numerical problem, it is never necessary to determine a_i or h_k with an accuracy better than the step characterizing the finite differences δr. Moreover, it is very easy to detect when one steps into a new subinterval.

In practice, it is only important to detect :

1/ cases of type a; one steps into such an interval when $Q_r \cap P_{r+\delta r} = \Phi$.

2/ other cases when some arcs of envelopes bound W_E and effectively reduce $W^+_{pa_i} \cap W^+_{rr}$. This occurs as long as $Q_r \cap P_r \in W^+_{r+\delta r, r+\delta r}$

6. REFERENCES

LAFON J.-P. J., Astron. Astrophys., 'On self gravitating stellar disks I A new model', 1976, 46, 461

LAFON J.-P. J., J. Math. Phys. , ' On a topological problem arising in physics', 1977, 18, 1178.

LAFON J.-P. J., Hevetica Physica Acta,'Missing mass and stellar dynamics in flat galaxies', 59, 1142

PLANET CROSSING ASTEROIDS AND PARALLEL COMPUTING: PROJECT SPACEGUARD

ANDREA MILANI
Department of Mathematics, University of Pisa
Via Buonarroti 2, I-56100 Pisa, Italy
and
Department of Astronomy, Cornell University
Ithaca, NY 14853, USA

ABSTRACT The orbits of the asteroids crossing the orbit of the Earth and other planets are chaotic and cannot be computed in a deterministic way for a time span long enough to study the probability of collisions. It is possible to study the statistical behaviour of a large number of such orbits over a long span of time, provided enough computing resources and intelligent post processing software are available. The former problem can be handled by exploiting the enormous power of parallel computing systems. The orbit of the asteroids can be studied as a restricted (N+M)-body problem which is suitable for the use of parallel processing, by using one processor to compute the orbits of the planets and the others to compute the orbits the asteroids. This scheme has been implemented on LCAP-2, an array of IBM and FPS processors with shared memory designed by E. Clementi (IBM). The parallelisation efficiency has been over 80%, and the overall speed over 90 MegaFLOPS; the orbits of all the asteroids with perihelia lower than the aphelion of Mars (410 objects) have been computed for 200,000 years (Project SPACEGUARD). The most difficult step of the project is the post processing of the very large amount of output data and to gather qualitative information on the behaviour of so many orbits without resorting to the traditional technique, i.e. human examination of output in graphical form. Within Project SPACEGUARD we have developed a qualitative classification of the orbits of the planet crossers. To develop an entirely automated classification of the qualitative orbital behaviour -including crossing behaviour, resonances (mean motion and secular), and protection mechanisms avoiding collisions- is a challenge to be met.

1. Chaotic dynamics and planet crossing asteroids

Non integrable dynamical systems contain chaotic orbits, which are unpredictable, that is, nearby orbits diverge as $exp(\chi t)$, where χ is the maximum Lyapounov characteristic exponent, and the smallest error in either initial conditions or physical model or orbit computation results in a completely different orbit after a time $>> 1/\chi$ (Lighthill, 1986). Actually, in a non-compact phase space —as is always the case in N-body problems— escape to infinity and/or collision is always possible and unpredictable, thus not even whether the orbit is bounded chaotic can be established.

Celestial Mechanics **45**: 111–118, 1989.

Once an orbit is found to be chaotic, is there anything more to be said about it? As far as a single orbit with a specific set of initial condition is concerned, no conclusions can be drawn on its long term behaviour. However statistical information on the behaviour of all the orbits belonging to the same chaotic region can be meaningful. In the case of a chaotic region which fulfill some special additional properties, an hyperbolic set, the following statements can be proven: 1) the orbit computed with a procedure introducing a small enough error at each step is close to some exact orbit, not necessarily with the same initial conditions *(shadowing lemma)*; 2) the average properties of a large enough number of orbits (e.g. the average value of some function and/or the proportion of time spent in some subregion) and the average over a single orbit for a long enough time are close *(ergodic lemma)*(Newhouse 1980).

The main problem in the investigation of chaotic orbits is the difficulty of evaluating a priori what *small enough, large enough,* and *long enough* in the previous paragraph actually mean in quantitative terms. The only possible approach appears to be an experimental one, that is, to compute larger and larger numbers of orbits, for longer and longer time spans with both numerical methods of increasing accuracy and increasingly realistic physical models, until the results appear to settle to a statistical equilibrium. It is clear that such an approach requires us to exploit, to its technological and economical limits, the computational power available.

It used to be believed that the orbits of the real objects in our solar system must be stable *because they are there*. While this might be true for the major planets for a timescale comparable to the age of the solar system (Nobili et al., 1988), such an argument should not be used for orbits whose dynamical age is unknown. Actually most of the phase space, with the exception of very low eccentricity and inclination orbits where the major planets are found, is formed by *planet crossing orbits*. This large chaotic sea is not empty; not only the comets cross it over their comparatively short lifetime, but over 400 planet crossing asteroids are known (Marsden, 1986). The dynamical lifetime of these asteroids, the probability of either physical collision with a major planet or hyperbolic escape, and the relationship with the comets are the main problems of interest for the planetary sciences –and for many other disciplines as well, given the available evidence of mass extinctions resulting from such a collision at the Cretaceous–Tertiary boundary (Snowbird, 1988).

Since all these orbits are chaotic, with Lyapounov exponents very large with respect to the timescales of planetary collisions ($\chi >> 1/100,000 yr^{-1}$), only a statistical investigation is possible. The special nature of such an investigation lies in that, although the purpose is to study the statistical properties of a population of real objects and their real collision rates with the Earth and the other planets, the computed orbits do not predict the specific fate of any single real asteroid. The initial conditions of real, observed objects can be used as a convenient way to simulate the probability distribution of the asteroids in the phase space; this procedure however does not fully remove the observational bias, e.g. more asteroids approaching the Earth are represented in the sample because they can be discovered at the close approach (Milani et al., 1988b).

2. Matching problems with technology: parallel computing of asteroid orbits

The simplest kind of parallel computer is formed by N processors connected in a network. The computational speed is increased by a factor γN, where γ is the *parallelisation efficiency,* the cost by a factor $<< N$, since the hardware costs are only a fraction of the cost of a computer system.

An outstanding example of this approach is the *Loosely Coupled Array of Processors (LCAP)* designed by E. Clementi (IBM Kingston, NY) and assembled by using only commercial hardware and existing operating systems, compilers and communication utilities. The operation of LCAP involves a set of virtual machines, or independent tasks, appearing to the operating system of the front end computer as sharing the same processor and the same memory; actually, to all but one a real attached processor is assigned; in the model LCAP–2 the attached processors are FPS–264, each a powerful pipelined 64–bit CPU. The code is prepared by a precompiler (FORTRAN in, FORTRAN out) which handles the calls to the communication utilities inside the front end, and also allows access to a shared memory from each of the attached processors; then the code is compiled by the usual IBM and FPS compilers. The theoretical peak speed of LCAP–2 is about 380 megaflops (Clementi et al., 1986).

LCAP is an experimental machine, but a similar approach to parallel computing inside multi–processor systems is becoming available on a number of top performance supercomputers; e.g. IBM has released as an official product a *Parallel FORTRAN* (IBM, 1988), which builds upon Clementi's experience. Therefore it is important to assess which computational problems in Celestial Mechanics are suitable for parallel computing. It turns out than N–body problems are not efficiently handled by most parallel computers; if each one of P processors computes the orbit of N/P bodies, the amount of computation to be performed by each processor grows like N^2/P but the amount of communication to be handled by the network grows like N^2P. Unless $N >> P$, which is not the case in planetary orbit computations, such a system requires a very high ratio between speed of communication and speed of computation to achieve a good parallelisation efficiency; in practice, unless all the CPU are on the same bus –as it is the case in the purpose–built *Digital Orrery* (Applegate et al., 1985)– the Bytes/Flops ratio is too low.

The case of the *restricted* $(N + M)$*–body problem*, with N massive bodies (Sun and major planets) and M massless test particles (asteroids, comets) can be handled in an efficient way if a *planet processor* handles the N massive bodies and each one of P *asteroid processors* handle M/P particles; then the planet processor needs to broadcast the positions of the planets to the asteroid processors, but the latter need not send any data (until real output is needed), and the communication only grows as NP. If the processors can communicate between themselves faster than with the front end –as is the case in shared memory networks such as LCAP– the parallelisation efficiency can be further improved by assigning the task of the *gatekeeper* to the planet processor, that is by dispatching all the output to the front end through the planet processor. In this way P processors are computing,

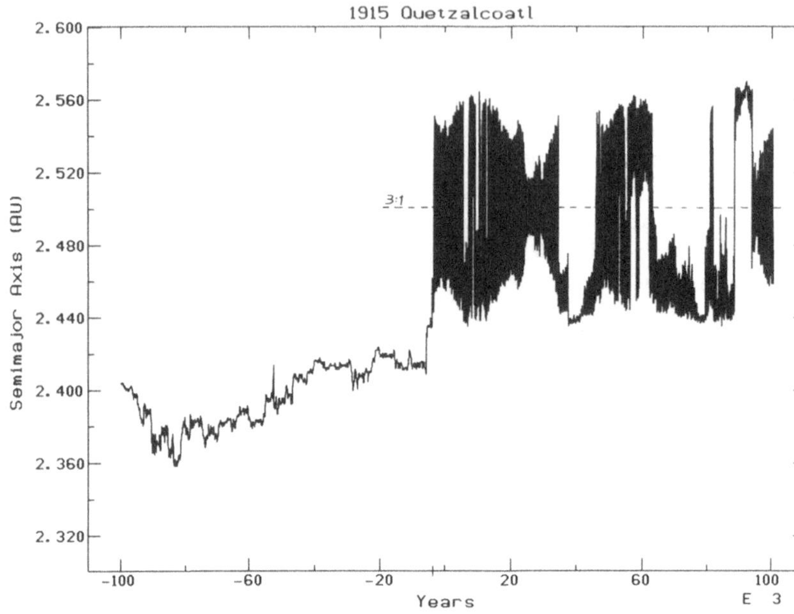

Figure 1: the evolution of the semimajor axis of 1915 Quetzalcoatl over the timespan of the SPACEGUARD integration. At time -4,000 years from present a qualitative change occurs, which is interpreted as a transition between the Geographos class of non resonant Earth-crossers and the Alinda class of Kirkwood gaps asteroids.

with minimal idle times, even though the program is I/O bound, i.e. the planet processor and the front end are handling output most of the time.

This procedure was devised and actually implemented during a series of visits to IBM Kingston in 1986–87; with LCAP–2, an $8+360$–body problem was integrated with a multistep predictor at a speed of about 1.5 milliseconds per step, corresponding to a sustained speed of about 92 megaflops and a parallelisation efficiency of about 0.81; the Bytes exchanged/Flops computed ratio was 0.96 for the planet processor, but only 0.013 for the asteroid processors (Milani et al, 1988a).

3. Project SPACEGUARD

This remarkable performance, and IBM sponsorship, allowed us to compute the orbits of all the asteroids that cross the orbits of either the Earth, Mars, or Venus (with the exception of the Mercury crossers, for stepsize reasons) discovered as of March 1986, for a total of 410 objects; the best available initial conditions were supplied by B. Marsden, Minor Planet Center. The orbits have been computed under the gravitational attraction of the Sun and the major planets Venus to Neptune for a time span of 200,000 years centered at the present time (Milani et al., 1988b).

It must be stressed again that the purpose of this large computation (amounting to 8.5 *Teraflops*) was not to establish the fate of each single object, since even the most stable of these orbits are chaotic, as it was checked within the same computation. To predict the occurrence of a close approach to a major planet in a deterministic sense is impossible, apart from short term predictions (e.g. no approach closer to Earth than 1 million kilometers in the next 120 years, as was checked within the first CPU minute of computations). The main purpose of our *Project SPACEGUARD* was to allow a statistical study of the dynamics of a supposedly representative sample of the planet crossing asteroid population.

The main challenge in such a project is to design the output of the main orbit computation. The *SPACEGUARD* output included time series of orbital elements for the planets and the asteroids, both sampled and digitally filtered to eliminate the less interesting short periodic features (see Nobili, 1988; Carpino et al., 1987). Moreover, each close approach of an asteroid to a major planet was recorded and the positions of both bodies were stored; the total stored output amounts to 2,270 *Megabytes*. Nevertheless, it may well turn out that we have not stored enough information because some of the close approaches are very dramatic events, which need to be monitored at a very high temporal resolution. Only a complete analysis of the output, which requires an enormous amount of human and computer time, can tell us how successful this project has been.

The main challenge lies in describing such a huge output in a meaningful way. Each asteroid shows a dynamical behaviour which could be studied by itself with the conventional methods of Celestial Mechanics, and indeed many have interesting features, such as resonances –both in mean motion and secular– with one or another of the planets, very significant secular changes in the orbital elements which bring them in and out of a crossing state with the Earth, and protection mechanisms preventing close approaches to some planet. However, to describe all these peculiarities of each orbit is not the point. What is needed is the definition of broad, qualitative classes of orbital behaviour based upon these features, in such a way that a statistical model of the entire population can be deduced from the output. We believe this has never been attempted before.

4. The SPAGEGUARD dynamical classes

The procedure to define classes of dynamical behaviour, as devised within the *SPACEGUARD* Project and as used so far for the 89 Earth approaching asteroids in the output, is as follows (Milani et al., 1988b).

First, the time series for the orbital elements were screened for resonances, the crossing behaviour (Shoemaker et al., 1979) with respect to each planet determined by means of an automated crossing–counter, and the number and minimum distance of close approaches computed. All this information was displayed also in graphical form (about 1,400 plots). Second, a rough classification was devised in a purely intuitive way. Having classified each orbit into these classes, analysis of the objective parameters of the members of each class allowed us to improve the definition of each class, and start a new iteration. Each improved classification was judged on the basis of two criteria: the sharpness of the boundaries, and the number of short–lived transitions; of course the classification is more useful, the more it describes in a non ambigous way long term behaviour. After a number of iterations, the orbits of the Earth approachers were classified in 6 + 1 classes as follows; each class is named after the best known –and/or most representative– example.

Geographos: Earth crossing and deep Earth approaching, with orbits piecewise regular between node crossing, semimajor axes changing because of close approaches to Earth (and possibly Venus).

Toro: Earth crossing but not Earth approaching, because of mean motion resonance with the Earth.

Kozai: perihelion below 1 AU, but not Earth crossing because of secular protection mechanisms.

Alinda: in deep mean motion resonance with Jupiter, eccentricity undergoing large changes in and out of Earth and Venus crossing values; no deep approach to Jupiter.

Eros: non Earth crossing nor deep Earth approaching; deep Mars crossing and deep Mars approaching.

Oljato: large scale chaotic orbits, with large changes in semimajor axes; very large eccentricity and Venus crossing, but not Jupiter crossing nor deep Jupiter approaching.

A supplementary class contains the objects which, although observed as asteroids, i.e. without obvious cometary features such as a tail, have orbits which are –or become within the integration timespan– typically cometary, i.e. Jupiter crossing and Jupiter approaching and/or deep Mercury crossing.

Transitions between these classes occur, both as a result of very close approaches and of the instability of the chaotic states corresponding to each class. Figure 1 shows an example of a Geographos to Alinda transition. The Alinda class, which includes all the asteroids resonant in mean motion with Jupiter (not only in the 3:1 resonance, but in the 4:1, 7:2, 5:2, 5:1, 7:3, 5:3, 11:4, 14:5 resonances as well), has an average residence time of $\simeq 100,000$ years. The Toro remain in their class for an average of only $\simeq 20,000$, while Eros and Kozai have average residence times far longer than the integration time span of 200,000 years.

The results of the *SPACEGUARD* classification can be summarised in Figure 2, where for each class a number of members (averaged over 200,000 years) is given beside the class name, and the arrows represent all the observed transitions. Such a figure already allows us to draw some interesting conclusions about the dynamical evolution of the Earth approaching population over timescales longer than the integration time span. As an example, we can say that the Alinda class (i.e. all the mean motion resonances with Jupiter, not only the 3:1) is an important source of Earth approaching objects (and of meteoritic material), but secular resonances and secular perturbations are of comparable relevance in producing direct Eros–Geographos transitions. Moreover, the Alinda as well as the Oljato are closely related to typically cometary orbits. Some of the features of Figure 2 indicate that a statistical steady state for the Earth approaching population cannot be computed without a longer integration; e.g. the Kozai class has no arrow pointing out.

5. Future work

It is not appropriate to draw too many conclusions out of a work which is far from complete; we should rather briefly list the major tasks which are either being tackled now, or will be in the future.

The *SPACEGUARD* database contains a record of $\simeq 600,000$ close approaches of asteroids, either to the inner planets within 0.1 *AU* or to Jupiter within 1 *AU*. The statistical analysis of these will soon provide a more reliable estimate of the collision probabilities for each dynamical class, and thus allow a better computation of the cratering flux. The correlation between orbital changes due to the close approaches and overall, long term orbital changes has to be investigated; e.g. the transition shown in Figure 1 does not result from a particularly close approach, while the changes in semimajor axis of *1915 Quetzalcoatl* while in a Geographos state do occur mostly because of the close approaches.

The investigation of the far larger Mars approaching population (321 orbits) is a much more challenging task, simply because the number of orbits is such that it would be impossible, and even unwise, to use the same human-controlled and intuition-driven procedure used so far. Automated pattern recognition in the detection of resonant behaviour and protection mechanisms needs to be used. The development of such software tools will take a long time, but we hope will lead to the availability of an entirely new field of investigation, namely the statistical analysis of qualitative orbital behaviour. Moreover, other groups appear to be aiming at the development of similar tools (Sussman, 1988). Only with the latter will the question of the origin and average fate of the planet crossing population, of the relationship between asteroids and comets, and of the shape of the asteroidal belt become possible to be solved.

References

Applegate, J.H., Douglas, M.R., Gürsel, Y., Hunter, P., Seitz, C.L. and Sussman, G.J. (1985) A Digital Orrery *IEEE Trans. on Computers* **C–34**, 822–831

Carpino, M., Milani, A. and Nobili, A.M. (1987). Long-term numerical integrations and synthetic theories for the motion of the outer planets. *Astron. Astrophys.* **181**, 182–194

Clementi,E., Dietrich, J., Chin, S., Corongiu, G., Folsom, D., Logan, D., Caltabiano, R., Carnevali, A., Helin, J., Russo, M., Gnudi, A. and Palamidese, P. (1986) Large scale computations on a scalar, vector and parallel Supercomputer, *IBM Technical report*, **KGN–77**

International Business Machines Corporation (1988) Parallel FORTRAN. Language and Library reference. Pubblication SC23-0431–0

Lighthill,J. (1986). The recently recognized failure of predictability in Newtonian dynamics. *Proc. Roy. Soc. London* **407**, 35–50

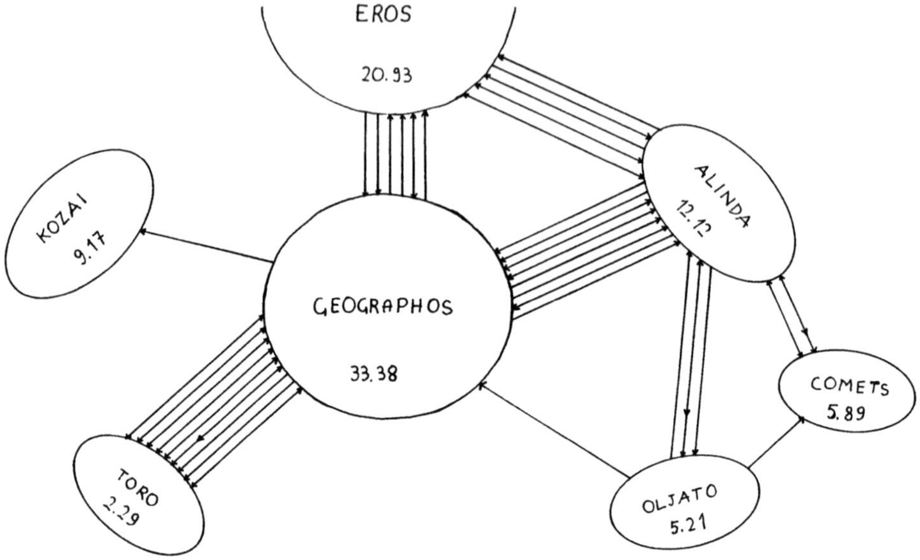

Figure 2: The SPACEGUARD classification of Earth-crossing and almost crossing orbits is represented schematically. For each class we give the number of objects -averaged over 200,000 years- and the transitions represented as lines with arrows (multiple arrows on the same line indicate multiple transitions by the same object).

Marsden, B.G., (1986) *Catalogue of orbits of unnumbered minor planets (second edition)*. Minor Planet Center, SAO, Cambridge MA.

Milani, A., Carpino, M. and Logan, D. (1988a). Project Spaceguard. Parallel Computation of Planet-Crossing Asteroid Orbits *Advances in Parallel Computing*, in press

Milani, A., Carpino, M., Hahn, G. and Nobili, A.M. (1988b) Project Spaceguard: Dynamics of planet-crossing asteroids. Classes of orbital behaviour *Icarus*, in press

Newhouse, S.E. (1980) in *Dynamical Systems*, Marchioro, C. ed., Liguori pub., Napoli

Nobili, A.M. (1989) *Celest. Mech.*, this issue

Nobili, A.M., Milani, A. and Carpino, M. (1988) Fundamental frequencies and small divisors in the orbits of the outer planets, *Astron. Astrophys.*, in press

Shoemaker, E.M., J.G. Williams, E.F. Helin, and R.F. Wolfe (1979). Earth-crossing asteroids: Orbital classes, collision rates with Earth and origin. In *Asteroids* (T.Gehrels, Ed.), pp. 253–282, Univ. of Arizona Press

Snowbird, (1988) *Global catastrophes in earth History*; Proceedings of the second Snowbird conference on Impacts, Volcanism, and Mass Mortality , Burke et al. eds., in press

Sussman, G.J. (1989) *Celest. Mech.*, this issue

IMPLEMENTATION OF AN N-BODY CODE IN A HP 1000 COMPUTER

D. D. Carpintero, J. C. Muzzio and M. M. Vergne
Facultad de Ciencias Astronómicas y Geofísicas de la UNLP
Paseo del Bosque
1900 La Plata
Argentina

ABSTRACT. We describe the implementation of Aarseth's NBODY2 code on a HP 1000 computer. We use the Extended Memory Array (EMA) feature with this code in order to investigate problems that include several hundreds of bodies, but the use of EMA requires some care in order to avoid large increases in computing time. The Vector Instruction Set (VIS) feature, a group of arithmetic subroutines that operates on arrays of floating point numbers and significantly reduces the computing time, turned out to be of little value for this application. We present the computing times demanded by two different problems for a variety of programs, including EMA and VIS. Finally, we present mass loss and mass accretion results for several simulations of galaxy-galaxy encounters performed with our implementation of the NBODY2 code.

I. INTRODUCTION

For the past few years we have been using the HP 1000 computer at La Plata Observatory to perform numerical experiments that simulate the evolution of galaxies in clusters (Muzzio, 1987). Since this machine is employed mainly for processes that demand little cpu usage, our programs run more efficiently in the HP 1000 than in the larger and faster machines available at the Computing Center of La Plata University, which are shared with other heavy cpu users from all over the University.

Since we are now interested in self-consistent problems, wich our former test-particle approach cannot handle, we obtained copies of S.J. Aarseth's codes that the author kindly allowed us to use.

In order to get acquainted with those codes, we began using the simplest, the NBODY1 program, with about 100 particles (Carpintero and Muzzio 1988). Now, in the first part of the present work, we describe the implementation of the NBODY2 code on our HP 1000 computer; this code uses the Ahmad-Cohen method (Aarseth 1985) and is fast enough to allow us to attack problems that involve several hundreds of particles in a reasonable (i.e., hours or a few days) time. The second part of the present work describes some of the results on galaxy-galaxy encounters that we are obtaining with the NBODY2 code and the HP 1000 computer.

Celestial Mechanics **45**: 119–124, 1989.

II. CODE IMPLEMENTATION

Aarseth advises us to define several double precision variables when using his program in 32-bit machines, in order to improve accuracy at little extra cost. In our 16-bit machine, however, the cost was high since, in addition to the larger storage space demanded, computing time was increased by about 20% to 25% when using double precision. Also, we are interested in the investigation of encounters of galaxies, where one uses large softening parameters and integrations are usually limited to intervals of less than 10 crossing times, so accuracy is not a major problem. Therefore, we decided from the start to use single precision variables only.

We preferred to follow the same approach we had followed in our previous work and to keep separate programs to create the initial conditions and to perform the numerical integration. Thus, we prepared a program that comprises Aarseth's subroutines MYDUMP, START and DATA, plus a small main program that just governs the flow through those subroutines; a few obvious changes were introduced in the subroutines too. With this program we create the initial conditions (DATA) and perform the initialization (START); the results are dumped in a file which can then be read by a second program that performs the integration. We will refer only to the second program, which is of course the most time consuming of the two, in what follows. Thus, we just close our description of the first program indicating that to create and initialize a 50-body problem takes between 30 and 40 seconds in our machine, while a 450-body problem demands between 11 and 14 minutes for the same.

We implemented the integration program again using Aarseth's subroutine MYDUMP, plus his subroutines MAINPT and INTGRT as well as his main program. All of them needed only small obvious changes. The result is a program that we call AARSE in what follows.

Our first problem was how to be able to include a number of bodies adequate for our purpose. We estimated that we would need about 400 to 500 bodies to investigate galaxy encounters, but less than 100 could be processed within the 64k directly addressable by our machine. The HP 1000 machine solves the problem of addressing large bodies of data with the Extended Memory Array (EMA) feature. The use of EMA is very simple and one just has to declare the large arrays in an EMA COMMON statement, and to be careful on how EMA variables are declared in subroutines. Nevertheless, the indirect addressing performed by EMA, although transparent to the user, is very time consuming. A reckless aplication of EMA, say, declaring all Aarseth's original COMMON statements in a single EMA COMMON statement can easily lead to a code that demands more than twice the computing time of the non-EMA version. For the present application we decided to exclude from the EMA declaration several large arrays that are frequently addresed (Aarseth's variables X, XDOT, TO, BODY and STEP) as well as, of course, all simple variables and small arrays. If one wishes to include more than 500 bodies, one will need to declare all the large arrays in the EMA COMMON, but that will increase the computing time by 15% to 20%.

An useful feature on the HP 1000 computer is the Vector Instruction Set (VIS), a group of arithmetic subroutines that operates on arrays of floating point numbers and significantly reduces the computing time. VIS is very easy to use, since one only has to change the DO loops of a program by the VIS subroutines, and we had successfully used it to reduce by 30% the computing time demanded by the NBODY1 code.

Unfortunately, the use of VIS requires breaking up the operations performed within the DO loops into separate simple vector operations (additions, multiplications and so on). To do this one may have to increment the number of addressings and to use auxiliary arrays. When, as in the present case, the auxiliary arrays have to be declared as EMA variables, the result is that the use of VIS may increase, rather than reduce, the computing time! Therefore, we had to limit the application of VIS to a few small DO loops and the reduction of computing time was very small indeed, less than 1%.

We used two different problems to test our implementation of Aarseth's codes: 1) We followed during one crossing time the evolution of 50 bodies that, at the start, were uniformly distributed within a sphere with random velocities; the softening parameter is just the one needed to prevent the formation of binaries. This is the kind of problem investigated by Aarseth (1985) and our results can be compared with those from his Table 1 (1.03 and 0.76 minutes, respectively, for NBODY1 and NBODY2). The number of integration steps we needed were very similar to the numbers quoted on that table so that the computing time differences just reflect the different speeds of our HP 1000 and of the VAX 11/780 he used. 2) The same problem, but with zero initial velocities, followed during 2.1 crossing times. The theoretical collapse time is 1.5 crossing times.

TABLE 1. CPU time demanded by NBODY1 and NBODY2

Code	Precision	EMA	VIS	Problem 1 minutes	Problem 2 minutes
NBODY1	Double	No	No	4.08	5.62
NBODY1	Single	No	No	3.36	4.57
NBODY1	Double	No	Yes	3.31	4.57
NBODY1	Single	No	Yes	2.34	3.22
NBODY2	Double	No	No	2.70	5.06
NBODY2	Single	No	No	2.19	4.17
NBODY2	Single	Yes	No	3.07	5.60
NBODY2	Single	Yes	Yes	3.06	5.55

From the data presented in our Table 1, we notice that the advantage of NBODY2 over NBODY1 is much greater for the first problem. Despite our care, the use of EMA increases the computing time by about 30% to 40%. In fact, although the EMA version of NBODY2 is still faster than the non EMA version of NBODY1 for the first problem, the latter is the fastest to solve the second problem. Besides, while the use of VIS sensibly improves the performance of NBODY1, it offers a negligibly small advantage for NBODY2. Nevertheless, since computing time increases as Nxx2.1 for NBODY1 and as Nxx1.6 for NBODY2 (Aarseth 1985) the latter has obvious advantages when the number of bodies is as large as 400 or 500 as we want to consider.

III. ENCOUNTERS OF GALAXIES

We are interested in investigating, with self-consistent simulations of galaxy encounters, how the galaxies lose and exchange matter due to their tidal interaction. We need a low relative velocity, so that one galaxy can capture the material lost by the other, but not too low, in order to avoid the merging of the two galaxies; to prevent merging the minimum separation of the galaxies cannot be too small either (Aarseth and Fall 1980). Muzzio (1987, 1988) showed that the brightest galaxy of a cluster captures matter (e.g., globular clusters) mainly from galaxies that are moderately fainter (i.e., less masive) than itself; therefore, we considered galaxies of moderately different masses.

We simulated the galaxies following the recipe of White (1978), and we let them evolve in isolation for five crossing times before using them for the encounter experiments. Masses are equal to the numbers of particles the galaxies have and radii are proportional to the square roots of the masses; the galactic half mass radius of a galaxy made up of 300 particles is about 9.8 units after it has evolved for five crossing times. The single exception is Case 10 (Table 2) where we adopted a law of radius proportional to mass and we reduced the radius of the smaller galaxy accordingly (half mass radius equal to 4.7 units). The second column (labeled "Iso.") of Table 3 of White (1978) gives the radial structure of this kind of model.

The encounter simulations start with the galaxies separated 100 units and the velocity is chosen in such a way that, if the galaxies were point masses, their relative velocity at infinity would have been one unit and their minimum distance 20, 30 or 40 units depending on the experiment. The evolution is followed, for those separations, through 6.5, 6.0 or 5.5 crossing times, respectively; the distances between the two galaxies after those intervals are about 100 units. Energy is conserved within 2 (-04). The sum of the two masses is always the same, 450, and three different mass ratios (1.5, 2.0 and 4.0) were considered.

Although our experiments start from mildly hyperbolic initial conditions, most of them end up with the galaxies in bound orbits after the encounter; thus, if the galaxies were isolated, they would end up merging in the long run. Nevertheless, apocentric distances were well beyond 100 units for almost all the experiments, and usually were much larger than that (i.e., much more than ten typical half mass radii). For galaxies in clusters those distances should be large enough to prevent merging, because the subsequent evolution would be altered by the cluster field and by encounters with other galaxies. The single exception is our case 4, whose apocentric distance turned out to be 80 units only, still fairly large compared to the galactic half mass radii.

To follow the evolution of isolated galaxies for five crossing times with program AARSE takes 1.2 and 3.9 hours, respectively, for galaxies made up of 150 and 300 bodies. To follow an encounter simulation (5.5 to 6.5 crossing times) takes between 12 and 15 hours; when the galaxies are distant, it takes about 1.1 hours per 0.5 crossing time, but the computing time almost doubles when the galaxies are very close.

Table 2 presents our results on mass loss and accretion. Several interesting properties are apparent from Table 2. First we notice that both galaxies lose and accrete matter. Losses are usually larger for the smaller galaxy (particularly if percentages are considered and, curiously enough, the mass accreted is similar for both galaxies, or even larger for the smaller galaxy if percentages are considered. The losses impose a heavy toll on small galaxies, however, and except for the marginal gain in case 3 and the rather special (small radius) case 10, the small galaxies suffer net losses that are very important in some cases (4 and 7). The balance is more favorable to the larger galaxies, wich suffer moderate net losses and even manage to get some very modest net gains in several cases (6, 7 and 8). We also notice that for these slow encounters a high proportion of escapes end up captured by the other galaxy and, again, the smaller galaxy is the one with the largest proportion of captures. A crude average, excluding case 10, suggests that 60% and 30% of the mass lost by the larger galaxy and by the smaller one, respectively, is captured by the other galaxy; the proportion is even larger if only the more distant encounters are considered.

TABLE 2. Mass lost and accreted during galaxy encounters

Case	M1	M2	Rmin	Lost1	Lost2	Gain1	Gain2
1	270	180	20	18	18	6	6
2	270	180	30	6	9	4	5
3	270	180	40	3	2	0	3
4	300	150	20	10	25	6	8
5	300	150	30	5	8	2	3
6	300	150	40	0	4	3	0
7	360	90	20	4	18	5	2
8	360	90	30	0	9	2	0
9	360	90	40	1	4	1	1
10	300	150	20	11	4	1	8

Contrary to the suggestion of Miller (1983), we think that the mass-radius law adopted can have important effects in the evolution of clusters of galaxies. The decrease of radius together with the mass undoubtedly prevents puny galaxies from suffering even heavier losses; the net gain experimented by the smaller galaxy in case 10 is the largest one of the whole Table 2 and also certainly a consequence of its small linear size.

Finally, although the material captured in every encounter shown in Table 2 amounts to only a small fraction of the galactic mass, we must recall that Muzzio (1987) estimated that the largest galaxy of a cluster suffers about six "relevant" encounters every 10 Gy. The cumulative effect can thus be significant and, besides, since the galaxies expand after each encounter, succesive encounters can result in losses (and captures) that are more and more important every time.

It seems advisable to perform additional experiments using galaxies with different radial distributions. The galaxies of our previous models (see Muzzio 1987 and references therein) had much more extended halos than those used here (White 1978) and should yield larger losses and gains. Internal kinematics are also important because, as noted by Dekel, Lecar and Shaham (1980), galaxies with predominantly radial orbits yield a smaller proportion of captures than galaxies with more circular orbits.

We are very grateful to S.J. Aarseth for allowing us to use his code. Special thanks go to G. Ginestet, S.D. Abal de Rocha and R.C. Leonardi for their technical assistance. J.C. Muzzio is a member of the Career of the Scientific Researcher of the Consejo Nacional de Investigaciones Científicas y Técnicas de la República Argentina (CONICET) and M.M. Vergne is on a fellowship from the same Council. The support from grants from the CONICET and the Comisión de Investigaciones Científicas de la Provincia de Buenos Aires is gratefully acknowledged.

REFERENCES

Aarseth, S.J. (1985) 'Direct Methods for N-Body Simulations', in J.W. Brackbill and B.J. Cohen (eds.), Multiple Time Scales, New York Academic, pp. 377-418.

Aarseth, S.J. and Fall, S.M. (1980) 'Cosmological N-Body Simulations of Galaxy Merging', Astrophysical Journal, 236, 43-57.

Carpintero, D.D. and Muzzio, J.C. (1988) 'Disintegration of a Satellite that Orbits around a Galaxy', II Regional Meeting on Extragalactic Astronomy (Córdoba, Argentina), in press.

Dekel, A., Lecar, M. and Shaham, J. (1980) 'N-body Simulations of Slow tidal Encounters and the Formation of Galactic Halos', Astrophysical Journal, 241, 946-964.

Miller, G.E. (1983) 'Effects of Galaxy Collisions on the Structure and Evolution of Galaxy Clusters. I. Mass and Luminosity Functions and Background Light', Astrophysical Journal, 268, 495-512.

Muzzio, J.C. (1987) 'Dynamical Evolution of Globular-Cluster Systems in Clusters of Galaxies', Publications of the Astronomical Society of the Pacific, 99, 245-264.

Muzzio, J.C. (1988) 'Cluster Swapping', in J.E. Grindlay and A.G. Davis Philip (eds.), The Harlow-Shapley Symposium on Globular Cluster Systems in Galaxies, pp. 297-309.

White, S.D.M. (1978) 'Simulations of Merging Galaxies', Monthly Notices of the Royal Astronomical Society, 184, 185-203.

COMPUTER APPLICATIONS TO LUNAR AND ARTIFICIAL SATELLITE LASER RANGING

R. L. RICKLEFS and P. J. SHELUS
McDonald Observatory and Department of Astronomy
The University of Texas at Austin
Austin, TX 78712-1083
USA

ABSTRACT. The University of Texas McDonald Observatory has long been a pioneer in acquiring laser ranging data, a data type which has substantially improved our knowledge of the dynamics of the earth-moon system as well as various aspects of geophysics and general relativity. (See Mulholland, 1980; Shelus, 1985; Shelus, 1987.) The McDonald Laser Ranging System (MLRS) is one of only 2 laser ranging stations world-wide having the capability of routine data acquisition on both lunar and artificial satellite targets (Shelus, *IEEE*, 1985). In this paper we discuss the current applications of modern computer technology to the problems of acquiring and reducing that ranging data. As technology continues to improve, the logical upgrade is the replacement of obsolescent station minicomputers with the resource-rich environment of micro-computers. The goal is to allow the automation of many station ranging functions as well as the enhancement of on-site data quality control, filtering, and analysis. Plans for such upgrades and their implications for dynamical astronomy are discussed.

1. Introduction

The Texas lunar data management software has evolved substantially over the 19 years since lunar ranges were first received from the Apollo 11 reflector in the summer of 1969. The first incarnation of that system put most of the non-acquisition prediction and data handling responsibility at locations which had the proper amount of computing power for the job. For example, topocentric predictions were created by researchers at the Jet Propulsion Laboratory and then mailed to McDonald Observatory on punched cards. Observations were accumulated over a lunation and then transmitted on magnetic tape to Austin for processing and distribution. This latter process might take as little as 2 days or as much as several weeks.

2. MLRS Today

When the dedicated lunar/artificial satellite ranging facility called MLRS was constructed in the early 1980's, the prediction task and a reasonable portion of the data reduction task could already be accomplished on site, albeit somewhat slowly. Electronic data communications had already improved considerably with the addition of modems and communications software; the use of mail and personal courier services could be partially eliminated. The dual Data General NOVA 4/X computer system and hard disk drives on

Celestial Mechanics **45**: 125–128, 1989.

site utilize an operating system capable of supporting FORTRAN with real-time extensions that is used for 90% of the programming, the other 10% being assembler code.

2.1 PREDICTIONS

As with many artificial satellite ranging stations, state vectors ('IRVs') for LAGEOS are received at the site from the University of Texas Center for Space Research (CSR) once every 12 - 18 months. There is one state vector per day, which is then integrated to provide tabular ephemerides for each of the several passes visible from the MLRS on that day. At this time, satellites lower than LAGEOS are not tracked. This should change within the next year.

Lunar predictions are handled in a different manner. The MLRS NOVA-based prediction software can presently accept either the JPL (Jet Propulsion Laboratory) Developmental Ephemeris series or the MIT (Massachusetts Institute of Technology) PEP lunar and planetary ephemerides as input. To this we augment the U. S. Naval Observatory's predictions for earth rotation and polar motion. Our software system then produces predictions which are limited only by the accuracy of the earth orientation parameters and ephemeris errors. (On-site MLRS predictions have been verified to agree with MIT predictions at the 30 psec level.) The output consists of predicted ranges and point-angle information to each of the 4 actively-observed reflectors, at 15 minute intervals. The point-angle information comprises station to center-of-moon angles as well as a libration matrix, which allow observers to offset between reflector sites and various lunar features, without having an individual ephemeris for each. This offset capability is used to aid in real-time tracking.

2.2 ACQUISITION

The data acquisition software has low level control over much of the station hardware. One main subsystem is the continuously-running 'monitor' program which interrogates the telescope encoders, the clocks, and the meteorological instrumentation to maintain a status display and to make this information available to other programs. The ranging software moves the telescope and sets the range data window using pre-calculated tabular ephemerides. It instigates the firing of the laser 10 times per second and then receives, stores, and displays the ranging data in real-time. The several real-time displays consist of a plot of (O-C) range residuals versus time and a histogram of calibration data. These graphs allow the operators to confirm that the system is functioning properly and that data is being received.

2.3 CALIBRATION PROCESSING

Calibrated ranges are always referred back to the first non-moving point in the telescope's optical train. This is accomplished by removing both electronic and geometric delays. Electronic calibration and actual ranging data are acquired simultaneously in real-time in a so-called 'internal calibration' mode, i.e., a feed-back calibration path is incorporated into the ranging equipment itself. This has the advantage of providing detailed information on any drifts in the system over the actual time that observations are being obtained, as compared with techniques in which external targets are used to calibrate the equipment before and after each data gathering pass. Due to software constraints, all data is taken, and later processed, in 3-minute 'bursts'. Bursts are separated by only a few seconds. Calibration data over a 3-minute burst is averaged, with standard deviation of the sample and of the mean being extracted. In the lunar case, the statistical first and second moments of the calibrations are saved to produce a calibration result for application to the normal-point.

Geometric corrections are provided by physical measurements of the actual path lengths inside the telescope. This is essentially a one-time measurement. Since the MLRS telescope is an X-Y (alt-alt) mounting in which the 2 axes of rotation do not intersect, there is a point-angle-dependent range correction term which is also a part of this geometric correction.

2.4 RANGE DATA PROCESSING

Artificial satellite data is pre-processed on site to produce a 'quick-look' data type that is transmitted to NASA laser ranging network data quality control at Goddard Space Flight Center within 24 hours of its reception. The quick-look data type consists of 50 returns, selected at random, which represent the density of returns as a function of time. This quick-look data is fully corrected for electronic and geometric effects as well as atmospheric refraction, and provide a useful data product, originally designed only for engineering support, but now used for many "quick-look" scientific studies, the foremost of these being the production of earth orientation parameters. Full-rate magnetic data tapes are mailed weekly to Goddard, to be fully processed and merged with satellite data from other stations. The full-rate data type can be described as nearly-raw data, provided in a standard format and with only low-level corrections to electronic calibration being present.

Lunar data is fully processed on site at the MLRS. Calibration data is reduced and the lunar data 'windowed' (to limit processing to a band of data 60 nanoseconds wide in time/range residual space) and formatted. This data is then filtered with a 'lunar data editor' which uses Poisson statistics and interactive graphics, allowing a user to identify the lunar data and handle abnormal data distributions. With this task completed, the filtered data is again compared with the nominal lunar orbit and compressed normal-points are formed. Sufficiently dense data is then used to compute earth orientation parameters (UT-0 and variation of latitude) (Ricklefs and Shelus, 1986). For this task, we require at least 2 strong normal-points (i.e., there are no doubts as to proper identification of returns) from the Apollo 15 reflector separated in time by at least 2.5 hours to produce an acceptable earth orientation result. Some attempts have been made to combine data over several days to produce UT-0 results, especially when there are times with only a single strong normal-point is available each day. Results have not been totally acceptable, and the practice has been temporarily shelved. Finally, selected normal-points and the computed earth orientation parameters are transmitted to other data users within hours of data collection. All normal-points are transmitted electronically via telephone lines to Austin twice per lunation for quality checking and transmittal to other data users. A more general data archiving of MLRS lunar laser ranging and ancillary data is accomplished via the NASA Crustal Dynamics Project Data Information Service on a substantially longer time base.

3. MLRS Tomorrow

While the current systems within the MLRS function well with the current data load, it is by no means optimal in light of present and envisioned computer technology. Processing is much too time consuming for the higher data rates we are expecting for the future. More sophisticated data validation and quality control techniques are not available due to the slow speed and memory limitations of the NOVA mini-computers. The requirement for real-time decision making for the optimal observing of 8 or more artificial satellites and 4 lunar reflectors is far beyond the capacity of this system. There is also a need to improve (or hide) file handling and data archiving to limit confusion and effort for processing personnel.

With all of this in mind, we have been anticipating the moving of the MLRS data processing and later real-time software to smaller but much more powerful and tool-rich micro-computers like the the Apple Macintosh II or the IBM-compatible 80386 computers. The NASA Crustal Dynamics Project, our primary funding agency, has just recently mandated the use of Hewlett Packard Series 9000/Model 360 workstations for use in all its laser ranging stations, to off-load some of their data processing tasks from their already overburdened data acquisition systems. This H-P system is Motorola 68030-based and employs the UNIX operating system. The Texas laser ranging group will be primarily responsible for porting and further developing lunar prediction and data reduction software for the affected network lunar ranging stations, as well as participating in certain other parts of artificial satellite related sub-projects. The software generated thereby should be available to the general user community in one to two years.

4. Conclusion

Laser ranging at McDonald Observatory has progressed from mini-computer data acquisition and main-frame oriented batch reduction modes to interactive on-site acquisition and reduction. Meanwhile, the accuracy of the ranging data itself has improved more than an order of magnitude from 30 cm to 2 cm rms. As we approach the sub-centimeter accuracy level required for the geo- and astro-dynamicist data user, the acquisition and reduction software must also increase in capability and speed.

5. Acknowledgements

This project is funded under National Aeronautics and Space Administration Contract NAS5-29404 and Grant NAG 5-754.

6. References

Mulholland, J. D. 1980 "Scientific Achievements from Ten Years of Lunar Laser Ranging", *Rev. of Geophys. and Sp. Phys.*, **18**, p. 549-564.

Ricklefs, R. L. and Shelus, P. J. 1986 "An Analysis of MLRS Near-Real-Time Earth Orientation Results", *Earth Rotation and Reference Frames for Geodesy and Geodynamics*, Coolfont.

Shelus, P. J. 1985 "MLRS: A Lunar/Artificial Satellite Laser Ranging Facility at McDonald Observatory", *IEEE Transactions on Geoscience and Remote Sensing*.

Shelus, P. J. 1985 "Fundamental Coordinate Ties Using Laser Ranging Data", *Cel. Mech.*, **37**, p. 243-246.

Shelus, P.J. 1987 "To the Moon ... and Back", *Discovery (Research and Scholarship)*, **10**, p. 33-37.

THE USE OF IMAGE PROCESSORS FOR ACQUISITION AND REDUCTION OF ASTROMETRIC DATA IN REAL TIME

J.-E. Arlot*, W. Thuillot*, F. Colas*, Ch. Allet+, D.T. Vu*

*: Bureau des Longitudes, 77 avenue Denfert-Rochereau, F-75014 PARIS

+: INSU-Mesures, 4 avenue Neptune, F-94100 SAINT MAUR.

ABSTRACT. In Dynamical Astronomy, the data used to fit theories are positional observations. If the data are reduced using computers, the measure itself is not, and informations are lost. The classical way is to use photographic plates measured by hand. The use of automatic machine permits to save time but not to increase the amount of data.

The arrival of silicium targets and CCD allows to acquire much more data directly stored on computer. We used this technique associated with array processors. So we succeeded to increase the accuracy of the data. This method was applied to the observation of the motion of some asteroids and of the Galilean satellites of Jupiter.

1. INTRODUCTION

The goal of this work is to improve the accuracy of astrometric observations by increasing the amount of data and to make the reduction in real time, using microcomputers. We also applied two-dimensional photometry during specific events for astrometric studies of the motion of fast moving objects of the solar system.

The material we use for these observations is intensified silicon target associated with digitizing boards and CCD receptors. We also use image processors and high-speed microcomputers for the reduction in real time.

2. THE METHOD

The image processing system uses an intensified camera associated with digitizing boards (from Data-translation) on a PC compatible Compaq 386 microcomputer. UTC is directly recorded with the images and a VCR 3/4 inches is used as an analogic memory of high capacity (figure n°1).

Celestial Mechanics **45**: 129–133, 1989.

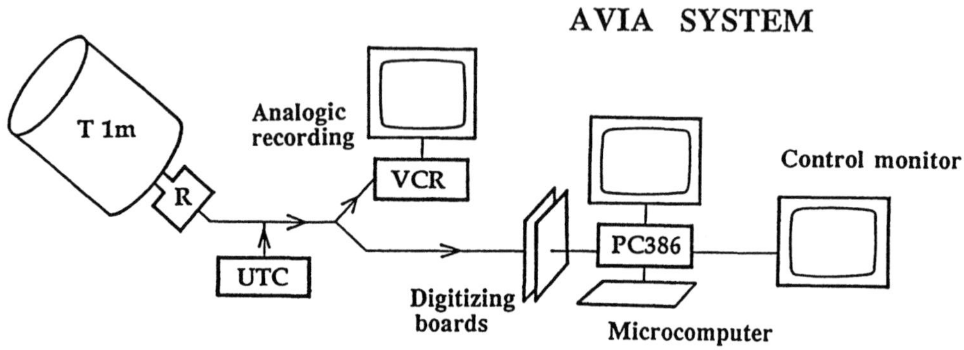

Figure n° 1
System using SIT-Vidicon camera
associated to digitizing boards

The used receptor permits observations of faint objects (magnitude 17 at the 1m telescope of the Pic du Midi Observatory) as well as mutual events of Galilean satellites. The images are simutaneously recorded on the VCR, and reduced in real time with the help of the microcomputer and the image processor. The two "Data Translation" boards permit to digitize and reduce images in real time. These boards are drived by the Compaq 386 microcomputer. The first board is devoted to the digitization of images at the video frequency (one image of 512 lines of 512 pixels on 8 bits in 1/25 second of time). The second board is an array processor devoted to special treatments in real time which is possible thanks to fast exchanges between the two boards. A technical description of the digitizing board (DT-2851) and the array processor (DT-2858) may be found in the Data-Translation handbook (1987). Our software allows to avoid that numerical images be stored. Relative flux or position measurements are directly issued from the treatment. Furthermore these computations are applied only on the small arrays defined by windows.

Images can be recorded also directly with CCD, and astrometric measurements can be made as Pascu et al. (1987) have shown. Such a system works at the Pic du Midi Observatory. The reduction in real time of the data is the same as in AVIA system.

3. OBSERVATIONS OF MUTUAL EVENTS OF THE GALILEAN SATELLITES

Mutual occultations and eclipses of the Galilean Satellites are rare and occur only every six years (Arlot, 1984). They permit observations with high astrometric accuracy very useful for dynamical studies. In 1985 we applied our AVIA system to these observations. We were able to reduce observations recorded on analog magnetic tapes from VCR. Such a method permits to reduce observations made even in bad sky conditions because of the simulaneous recording of a reference object and of the sky background around the objects. So, fast variations in the transparency of the sky are eliminated. Furthermore, this system avoids the use of a diaphragm but replaces it by software designed diaphragms.

The observation of relative motions of the Galilean Satellites during a close approach leads to astrometric data. The date of the minimum of distance between the satellites, the value of this distance and its orientation are to be compared with theoretical quantities. This type of purely astrometric measures can be made during the mutual events period, before and after a mutual occultation which is also photometrically observed. Out of the mutual events periods, we can also observe these close approaches, which may provide data of great precision comparable to the data issued from mutual events. The table n°1 gives numerical data concerning the comparison of the astrometric reduction with the photometric reduction of some mutual occultations observed in 1985. Previous campaign showed the great interest of mutual events for astrometric purpose because of the precision of this type of observation. The goal of the observation of close approaches of the Galilean Satellites is to reach a similar precision out of the mutual event period.

Event	Prediction UTC			Observations photometry photoelectric				video astrometry			video photometry		
	h	m	s	h	m	s		h	m	s	h	m	s
85-09-07													
1o3	20	33	54	20	34	01	(1)	20	34	25	20	34	25
				20	34	00	(2)						
85-06-17													
3o4	1	44	55	1	44	40	(3)	1	44	49	1	44	51

Table n°1
Comparison of astrometric and photometric reduction
(1): OHP (France); (2): Catania (Italy); (3): Granada (Spain).

4. OBSERVATIONS OF OCCULTATIONS OF STARS BY ASTEROIDS

The two-dimensional photometric observation is well adapted to the observation of occultations of stars by asteroids. During such events, the astrometric interpretation of the observation leads to a better understanding of what has been observed. During such events unpredictible phenomena may occur: the secondary events. These short occultations are generally explained by the presence of material near the asteroid (satellite, dust, ...) and the image processing systems can be very useful to reduce such an occultation.

The main advantages of the two dimensional method is to avoid the use of diaphragms, and to permit simultaneously the astrometric and the photometric reduction. Such an observation put into light a secondary event that we have observed at Meudon Observatory (near Paris) in 1982 during a close approach of the asteroid 146 Lucina with a star. A 0.6s event has been observed. Our astrometric reduction was confirmed by a spanish observation of the main occultation. The complete reduction of the Meudon observation leads to assume that the secondary event observed was due to a possible satellite of Lucina (Arlot et al., 1985).

The occultations of stars by asteroids lead to accurate measures of the shape, of astrometric data related to the motion of the asteroid, and can put into light the presence of material in their vicinity. The observation of the occultation of stars by asteroids are made by a survey network. We participate to such a network in Europe in relation with I.O.T.A. and G.E.O.S. associations. The use of two-dimensional receptors for this type of observations permits to analyse other observations made by observers of the network.

The figure n°2 shows the results that we have obtained on 1988 March 24 during the occultation of a star by the asteroid 14 Irene. At Meudon Observatory we observed a close approach of the asteroid and the star. We deduced the position labelled "M" on the figure n°2 from the astrometric reduction. A hungarian observer timed the occultation from which we deduced the position of the asteroid labelled "H". Furthermore last minute observations made at Bordeaux Observatory lead to predicted positions labelled "P1 and P2" (Rocher, 1988). This result shows a good agreement of the observation of an occultation with the last minute predictions and with our two-dimensional observation. This put in light the importance of two-dimensional receptors in such

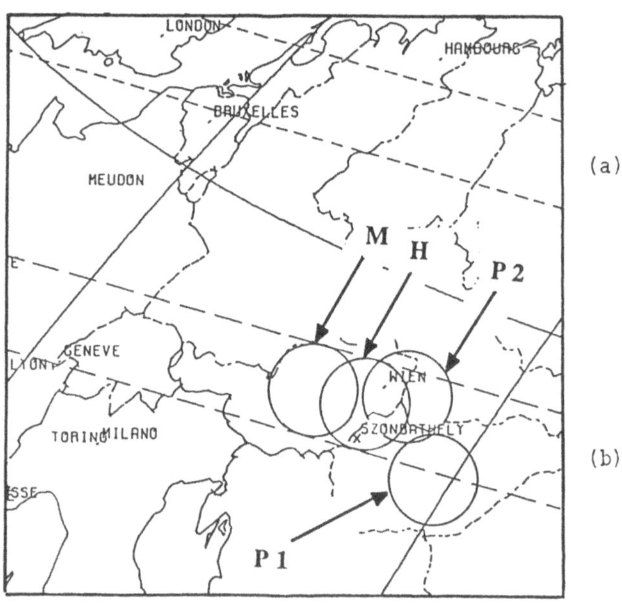

Figure n° 2
Observation of 14 Irene, on March 24, 1988
(a): predicted path of the umbra of (14) Irene
(b): path deduced from the hungarian observation of the occultation and from Meudon observation

observations. Note that the reduction of such a close approach is the same as for the galilean satellites (cf paragraph 3).

5. CONCLUSION

So, our successful results concern either application of photometry to astrometry (mutual events of the Galilean Satellites which may be observed even with bad sky conditions, occultations of stars by asteroids which have to be catched very fast), or direct astrometric measurements (close approaches between natural satellites or between asteroids and stars; determination of relative positions of natural satellites). Depending on the speed of analysis we want, on the accuracy on photometry or astrometry we wish, we may use either SIT Vidicon camera or CCD target.

6. REFERENCES

Arlot J.-E. : 1984, Astron. Astroph., 138, 113-120.

Arlot J.-E, Lecacheux J., Richardson Ch., Thuillot W. : 1982, IAUC 3692 et 3699. Occultation by (146) Lucina.

Arlot J.-E, Lecacheux J., Thuillot W. : 1983, IAUC 3776. Occultation of AGK3+11°201 by (29) Fortuna.

Arlot J.-E, Lecacheux J., Richardson Ch., Thuillot W, Baruch J.O.: 1983, in Asteroids, Comets and Meteors, University of Upsala, Lagerkvist, Rickman eds., 11.

Arlot J.-E, Thuillot W. : 1984, Vidicon photometry and astrometry applied to objects of the Solar System, in Astrometric Techniques, IAU Symposium 109, Gainesville, Eichhorn ed.

Arlot J.-E, Lecacheux J., Richardson Ch., Thuillot W. : 1985, A possible satellite of (146) Lucina, Icarus 61, 224

Data translation: 1987, Applications Handbook.

Pascu D., Seidelmann P.K., Schmidt R.E., Santoro E.J., Hershey J.L.: Astrometric CCD observations of Miranda: 1981-1985. Astron. J. 93 (4), 963.

Rocher P.: 1988, Private communication.

Thuillot W., Laques P.: 1987, Supp. Annales de Physique, 12, 195.

Thuillot W., Arlot J.-E, Fettig S., Colas.: 1988, Video technique applied to astrometry, Cel. Mechanics, (in press).

Modeling of Asymmetric Light Curves of Eclipsing Binaries on the Cyber 205 Supercomputer

E. F. MILONE, G. GROISMAN, D. J. I. FRY and S. J. SCHILLER*
Physics Department, University of Calgary,
Calgary, Alberta, T2N 1N4. CANADA.
*present address: Physics Dept., South Dakota State University, CEH 2219, RR 4, Brookings, SD 57006. U.S.A.

ABSTRACT. A major program at the University of Calgary's Rothney Astrophysical Observatory (RAO) is the study of eclipsing close-binary stars, which, because of circularization effects, are all cases of the restricted three-body problem. The physical properties of such systems are derived from modeling of the light and radial velocity curves. Uniqueness questions notwithstanding, careful use of the technique is found to yield reliable elements of the orbit and parameters of the component stars, even in the face of light curve perturbations. This work requires a greatly enlarged data base over previous work, and more complex modeling procedures, necessitating the use of a supercomputer. A version of the generalized synthetic light curve program of Wilson (1979) with star spot simulations has been adapted to and optimized for the University of Calgary's Cyber 205 supercomputer and further improvements are underway. With the use of personal computer graphics software, results have been transformed into three-dimensional, rotating models which help visualize the overcontact and perturbation conditions.

1. Purposes

The development of machine codes has greatly facilitated light curve synthesis and element determination of the orbits and components of eclipsing close-binary systems. Most current techniques explore parameter space either through differential corrections or through the simplex method (Kallrath 1987) so that the analysis requires testing the completeness of the investigation of that space. Even one high quality light curve can yield such orbital elements as the eccentricity or the inclination. However, each additional light curve helps to determine the spatial and spectral flux distributions of the light of the component stars and to establish the uniqueness of the adopted model. Infrared (>1 μm) or far ultraviolet passband data may be essential for modeling the properties of a secondary component which is of greatly different temperature than the primary. In addition, radial velocity curves are valuable in helping to fix the mass ratio, which is needed to determine the potentials which describe the shapes of the stars. In the case of contact systems, we can then obtain the filling factor - a

Celestial Mechanics **45**: 135–139, 1989.

measure of the overfill of the inner Lagrangian surface. It has been argued that in certain marginal overcontact cases, the narrow neck region may be choked off, resulting in thermal discontinutity. Coeval multicolor light curves are even more important when the light curves are perturbed. The literature reveals that asymmetries, especially at maximum light (the O'Connell Effect: Davidge and Milone 1984), are common phenomena in close binary light curves, and that they have multiple origins. The particular mechanism in any given case - whether thermal or nonthermal - can be determined by examining the spectral distribution of the residuals between the synthetic light curves and the observed ones. Careful modeling can also help to place the physical locations of the perturbations - on the stars or elsewhere - in the co-rotating frame of the system. Finally, 'third light' may be investigated for each passband, and, with cross-correlation spectroscopic, astrometric, and/or other data, elucidate the dynamics of three-body systems.

The addition of more and fuller light (and radial velocity) curves, perturbation-modeling artifices, and finer grids involve nontrivial amounts of computer time and necessitate the use of large computers.

2. Methods

We have adapted the eccentric-orbit version of the Wilson-Devinney (Wilson 1979) synthetic light curve program to the Cyber 205 and are using it to obtain fundamental parameters even when the light curves themselves are perturbed. The code correctly treats the shapes of the stars as potential surfaces and incorporates a spot-handling routine that permits superposition of higher and lower temperature regions on surface elements of the stars. This artifice permits simulation of sunspots, plages, and other features. The spot-modeling elements for each star are: temperature factor, spot radius, and longitude and latitude. The existing atmospheres flux ratios in the Wilson code are currently being replaced with newer, more accurate values (Kurucz 1979, 1988). Monitoring of the fitting process has been made much easier by a program (written by GG) which takes the output file from the Cyber 205 and produces hard-copy graphs of the light-curves and radial-velocity plots.

The Wilson-Devinney program has two main subroutines: LC which produces the model light- and radial velocity curves, and DC which performs the least squares differential corrections to the input model elements. Using as test input a 5-color light curve and 2-radial velocity curve set, a moderate-sized (30 element) grid for each component star, but no perturbation-modeling parameters, the LC program requires about 15 minutes, and DC about 90 minutes CPU time on the Honeywell-Multics computer of the University of Calgary. The CDC Cyber 205 supercomputer requires less than 3 minutes to run DC with the same data, and under the same constraints. Improving the precision of the fit by using twice the number of full light-curves, three spots, and twice the number of area elements would increase the running time for DC to 4 or 5 hours on the Cyber 205, but to an impractical duration on the Honeywell. The Cyber 205 has two vector pipes, is capable of 400 megaflops for fully vectorized code, and has 4 million 64-bit words of memory with bandwidth 51.2 gigabits/second.

3. Results

We have determined the intervals of time spent in the various subroutines, and optimized the existing code. Table I summarizes the System Time Units savings for several optimization trials. The optimization was accomplished through a variety of procedures: the reduction of the number of called subroutines, the elimination of unnecessary redefinitions of variables and the optimization and vectorization of loops. In a run involving no star spots, no radial velocities and only two light curves, a total savings of nearly 10% in STU was achieved, and 15% was achieved in cases involving three light curves, two spots, and two radial velocity curves. Further improvements in running time are expected through the rewriting of the most heavily used loop structures to permit the full benefit of pipeline processing.

Fig. 1 illustrates a synthetic light and radial velocity curve fit to the data of D. Bradstreet for the contact system binary TY Bootis. For each spot, an optimum temperature factor has been determined from the Σwr^2 parabola which is generated by a succession of DC runs of otherwise identical elements. The same technique has been applied to optimize the size and location of the spot regions. Two spots were used in the modeling of this system. The light curve asymmetry varies from season to season, but the elements of the system are recoverable if the perturbations are properly treated. Fig. 2 depicts the contours of the Roche potential for TY Boo. Fig. 3 demonstrates the appearance of such a system to an observer close enough to spatially resolve surface features. A movie displaying the rotation of the contact system was produced by one of us (GG) with the DANCAD graphing package. It may be run on any IBM compatible personal computer which has a hard disk.

This work has been supported by grants from the Research Grants Committee of the University of Calgary and from the Natural Sciences and Engineering Research Council of Canada to EFM, which are gratefully acknowledged. It is a pleasure for EFM to thank D. Bradstreet for the use of his data, and R. E. Wilson for much help and encouragement during a sabbatical leave at the University of Florida in 1986, and for the use of his program. This is RAO Publication No. 54.

4. References

Davidge, T. J., and Milone, E. F. 1984, Astrophys. J. Suppl., **55**, 571.

Kallrath, J. 1987, "Analyse v. Lichtkurven enger Doppelstern-systeme", Diplomarbeit. (Bonn: Sternw. d. Univ. Bonn).

Kurucz, R. L. 1979, Astrophys. J. Suppl., **40**, 1.

____________ 1988, private communication.

Milone, E. F., and Hrivnak, B. J. 1981, Astron. J., **86**, 1546.

Milone, E. F., Wilson, R. E., and Hrivnak, B. J. 1987, Astrophys. J., **319**, 325.

Wilson, R. E. 1979, Astrophys. J., **234**, 1054.

Table 1. Optimization trials for the Cyber 205 version of the Wilson-Devinney code

Trial/Options[1]	Time (STUs)[2]	Δt %	Σ Δt %
Original	312.00	....	....
DRS	303.79	2.63	2.63
V	303.55	0.08	2.71
variables eliminated	303.45	0.03	2.74
subroutines embedded	282.34	6.77	9.51

[1]Options: D = optimization of DO loops;
R = removal of redundant code;
S = scheduling instructions;
V = vectorization of DO loops where possible.

[2]System Time Units

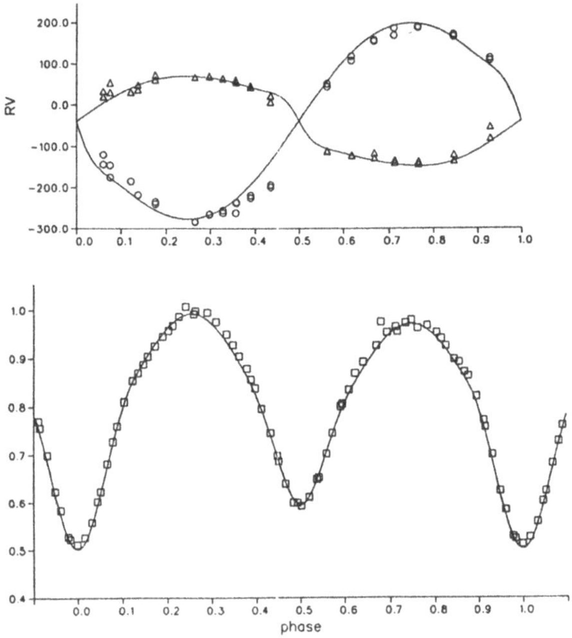

Figure 1. Graphics output showing modeled light and radial velocity curves of the W UMa binary TY Bootis. This model has two spots. Details of this and other models will be published elsewhere.

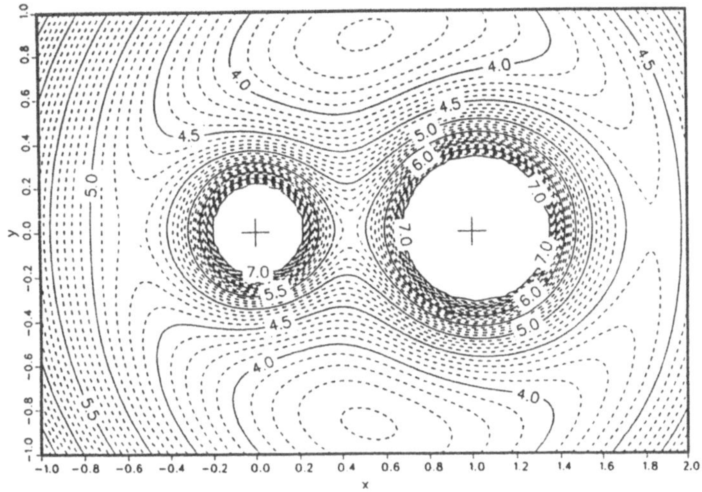

Figure 2. Contoured Roche potentials of the TY Bootis system in the orbital plane.

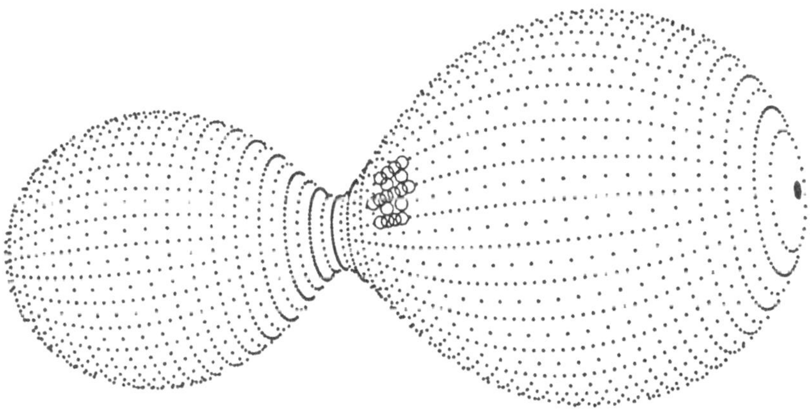

Figure 3. The model of the contact system TY Bootis. One of the spots which was used to model a light curve perturbation is visible on the larger component. As is the case for other W-type W UMa systems, the larger component is both the cooler and the more massive star in the system.

GALAXIES IN THE CONNECTION MACHINE

JUNICHIRO MAKINO
Department of Earth Science and Astronomy
College of Arts and Sciences, University of Tokyo
3-8-1 Komaba, Meguro-ku, Tokyo 153, Japan

PIET HUT
Institute for Advanced Study
Princeton, NJ 08540, U.S.A.

ABSTRACT. Many problems in galactic dynamics require computer speeds orders of magnitude larger than what can be achieved on current single-processor machines. In the near future such speeds are likely to become available through computer architectures based on large-scale, fine-grained parallelism. An example of a highly parallel computer is the Connection Machine, with up to 65,636 processors. We have benchmarked gravitational N-body algorithms on the Connection Machine, and compared those with similar benchmarks which we have obtained on more traditional vector supercomputers. Our conclusions are: (1) The direct summation algorithm, with of order N^2 interactions for N particles, can be made to run with high efficiency on either type of computer. As a result, the Connection Machine clearly wins in speed over all supercomputers tested, with the exception of an 8-processor ETA, which shows a comparable performance. (2) A more efficient tree algorithm reduces the growth of the number of interactions from N^2 to $N \log N$. However, the greater complexity of this algorithm causes a considerable degradation of efficiency, by a factor which is larger on the Connection Machine than on vector supercomputers. As a result, our tree code runs at comparable speeds on both types of machines, with the notable exception of the 8-processor ETA, which has an extrapolated speed for running our tree code which is higher than any of the other machines we have tested.

1. Introduction

The Connection Machine (Hillis 1985) is a highly parallel computer with 65,536 processors and a peak speed of 10 Gflops, higher than that of most vector supercomputers. Both types of machine share a SIMD (single instruction multiple data) architecture, which in practice implies that top speeds can only be achieved by applying the same operation on a large, ideally homogeneous, data set. A simple example of such an operation is the following DO loop:

```
do i=1,n
   a(i)=b(i)+c(i)
enddo
```

On vector supercomputers, this DO loop is written in Fortran, and is compiled to a sequence of vector instructions. On the Connection Machine we have a different way of expressing such constructs, in a parallel extension of C or of Lisp (*cf.* Hertz &

Celestial Mechanics **45**: 141–147, 1989.

McMillan 1988, Makino & Hut 1988 and references therein), with a parallel version of Fortran being under development.

Figures 1 and 2 show the speed we have measured for the above simple `DO` loop, as a function of the loop length n, on different computers. Both the Connection Machine and vector machines show a higher speed for larger n, but the Connection Machine requires much larger n values to reach optimal performance. For applications in galactic dynamics, this difference is not important because the number of particles we use is typically large enough ($> 10^4$). Moreover, we can run several calculations simultaneously to keep all processors busy. Thus we can expect either type of machine to be efficient in running large N-body simulations. Let us look at two representative types of algorithm.

2. Direct Summation Algorithms

The implementation of the direct summation algorithm is quite straightforward, both on supercomputers and on the Connection Machine. On supercomputers we had to make some modifications to the basic algorithm to make sure that the source program written in Fortran is properly vectorized by the compiler. On the connection Machine, we had to develop an algorithm somewhat different from that which is used on conventional machines. In the algorithm we tested, we used N processors to execute a N-body simulation (Depending on the number of particles preferred, other mappings are possible as well; *cf.* Makino & Hut 1988).

Figure 3 summarizes the speed of the direct summation algorithm as measured by us on various supercomputers and on the Connection Machine. We decided not to present the data in terms of the usual Mflops, but instead in a more practical unit of Mfocs (million force calculations per second). The reason is that a single square root calculation dominates the computational cost of a single interparticle force calculation. Since different computers differ vastly in efficiency of square root calculations, a presentation in terms of Mflops would penalize machines with efficient square root evaluators.

In figure 3, solid bars essentially indicate actual measurements. The only exception is that for the Connection Machine and for the ETA-10, at the time of our benchmarking, the clock speed was not yet up to the speed advertised for the later, commercially available product. For these machines, therefore, we simply multiplied our test results by the expected ratio of the future clock speed over clock speed at the time of measurements. In contrast, dashed lines indicate extrapolations of our measurements to configurations which differ from those tested by us. For multiprocessor systems, we simply multiplied the speed obtained on a single processor by the number of processors available. As long as the number of scientists using a given supercomputer vastly exceeds that of the number of processors, this seems to be a reasonable way of estimating effective speed.

Note how the Connection Machine and the ETA-10 clearly stand out from among all machines tested. With a typical computational cost of one Mfocs being equivalent to a few tens of Mflops, we find the net useful performance of these two machines to be of order a few Gflops, an impressive speed.

3. Tree Algorithms

The most efficient algorithm for simulating large systems of self-gravitating particles, such as star clusters or galaxies, has a hierarchical force calculation pattern based on a tree structure (*cf.* Barnes & Hut 1986). The tree structure is highly irregular and rapidly changing in time. The tree algorithm therefore presents an extreme challenge for hardware as well as software of fast computers.

The implementation of our tree algorithm turned out to be significantly more complicated than implementing the direct interaction algorithm. The simplest description of the force calculation algorithm is in recursive form, which is difficult to implement efficiently, both on vector processors and on the Connection Machine. A more efficient algorithm results from a rewriting of the recursive form to a purely iterative form (Makino 1988). By implementing this iterative algorithm on the Connection Machine we enabled all forces in a N particle system to be computed in parallel using N processors. Also for vector processors we have developed a similar algorithm.

Figure 4 summarizes the measured speed of the tree algorithm, on various computers. Note that for this more complicated algorithm, the performance of the Connection Machine is comparable to that of most vector computers. Some new vector supercomputers, such as the Hitachi Hitac S-820 and especially the ETA-10G, outperform the Connection Machine. However, the extremely low efficiency of the tree code on the Connection machine is partially due to limitations in the software available at the time of our benchmarking. We estimate a possible improvement of the performance of our tree code on the Connection Machine of about a factor three, as a consequence of future software and firmware improvements, before the current hardware will act as a limiting factor.

4. Conclusions

From a hardware point of view, vector pipeline supercomputers and fine-grained parallel computers such as the Connection Machine are very different. Nevertheless, the structure of optimal algorithms and the efficiency of their performance is remarkably similar. The similarity in the structure of efficient algorithms is a reflection of the fact that both types of computer have SIMD architectures. The rather low efficiency is the result of the fact that these machines are designed for optimal performance on relatively simple algorithms, *e.g.* finite difference methods using a regular grid. More detailed results of our study can be found in Makino & Hut (1988).

Switching from a direct summation code to a tree code becomes advantageous for a large number of particles, of order 10^3 - 10^4. In comparison, on scalar machines such as mainframes and workstations, the cross-over particle number lies in the range $N \sim 10^2 - 10^3$. This difference is a direct reflection of the loss in efficiency, of about an order of magnitude, when running tree codes on current versions of SIMD machines.

Acknowledgements

We thank Joshua Barnes for useful discussions. We also thank Danny Hillis, Brewster Kahle, Jacob Katzenelson, Gerald Sussman, and Feng Zhao for stimulating discussions concerning the Connection Machine.

We acknowledge the following institutions for providing us with computer time: the Computer Center of the University of Tokyo (Hitachi S-820); the Institute for Supercomputing Research (Cray X-MP; Facom VP-400); the John von Neumann National Supercomputer Center (Cyber 205; ETA-10); Thinking Machines Corporation (Connection Machine). We appreciate the hospitality of the above organizations, as well as the Institute for Advanced Study, the Massachusetts Institute for Technology, and the University of Tokyo, where parts of this research have been carried out.

This work was supported in part by the National Science Foundation, Grant Number INT 86-13539, under the US-Japan Cooperative Science Program.

References

Barnes, J. & Hut, P., 1986, *Nature* **324**, 446.

Hertz, P. & McMillan, S.L.W, 1989, in this volume.

Hillis, W. D., 1985, *The Connection Machine*, (Cambridge: MIT Press).

Makino, J., 1988, submitted to *J. Comp. Phys.*

Makino, J. & Hut, P., 1988, submitted to *Comp. Phys. Repts.*

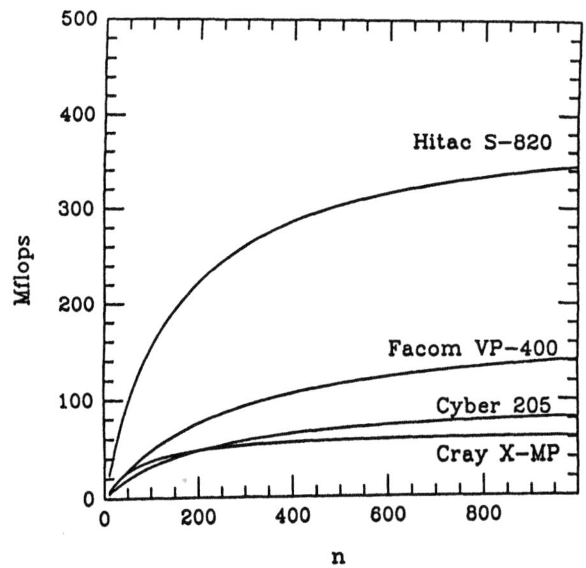

Fig. 1 Performance Characteristics of several vector processors. The speed for calculating a loop over `a(i)=b(i)+c(i)` is plotted as a function of the loop length n, in unit of Mflops.

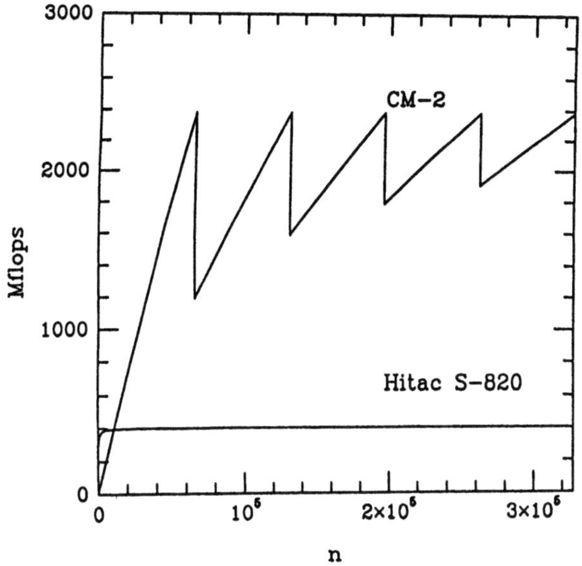

Fig. 2 Performance Characteristics for the Connection Machine. The curve for Hitac S-820 is reproduced from the previous figure for comparison. Note the much larger n scale in this figure, which reflects the much larger n values needed to reach top speed on the Connection Machine, compared with vector machines.

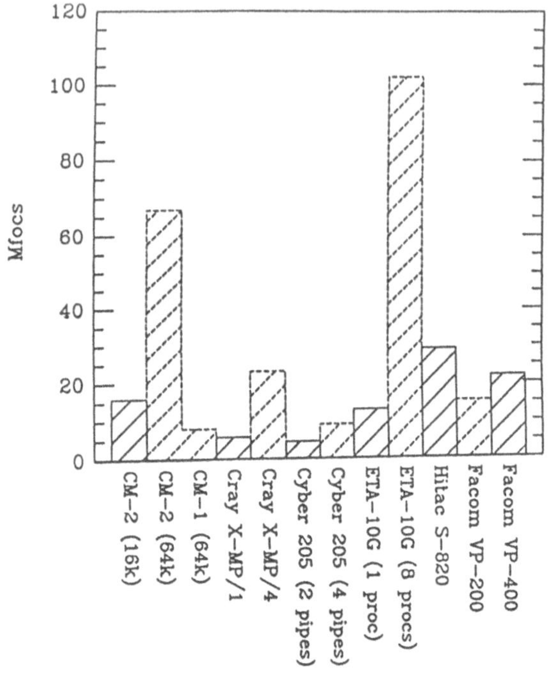

Fig. 3 Performance of the direct summation algorithm on various computers, represented here in units of Mfocs (million force calculations per second). Solid lines indicate our results by direct measurement, corrected only for anticipated changes in clock speed for the Connection Machine and ETA, since both machines were just coming on the market at the time of our testing. Dashed lines indicate extrapolations of our measurements to larger configurations. For multi-processor systems, we simply multiplied by the number of processors, on the grounds that in the foreseeable future the number of scientists per supercomputer will remain larger than the number of processors, which will make independent use of processors the preferred mode in most cases. Note that 1 Mfocs $\simeq$ 30 Mflops, but with a range of about a factor two, dependent on the efficiency of the square root operation which is at the heart of a force calculation.

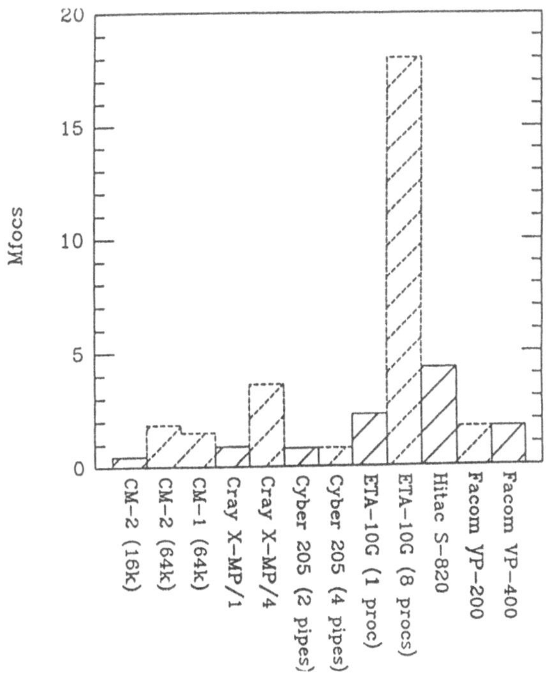

Fig. 4 Performance of the tree algorithm on various computers. Notation as in the previous figure. Note the change in scale of Mfocs of nearly an order of magnitude.

SPECIALIZED CELESTIAL MECHANICS SYSTEMS FOR SYMBOLIC MANIPULATION

V.A.Brumberg, S.V.Tarasevich and N.N.Vasiliev
Institute of Applied Astronomy of the USSR Academy of Sciences Leningrad, USSR

ABSTRACT. Construction and application of the current high accuracy analytical theories of motion of celestial bodies necessitates the development of specialized software for the implementation of analytical algorithms of celestial mechanics. This paper describes a typical software package of this kind. This package includes a universal Poisson processor for the rational functions of many variables, a tensorial processor for purposes of relativistic celestial mechanics, a Keplerian processor valid for the solutions of the two body problem in the form of a Poisson series, Taylor expansions in powers of time and closed expressions, and an analytical generator of celestial mechanics functions, facilitating the immediate implementation of the present analytical methods of celestial mechanics. The package is completed with a numerical-analytical interface designed, in particular, for the fast evaluation of the long Poisson series.

1. INTRODUCTION

Thirty years ago the specialists in celestial mechanics were among the first who expressed interest in using computers for performing analytical (symbolic) manipulation (Davis, 1958). A year later there appeared the first program to deal with Poisson series analytical operations by computer (Herget and Musen, 1959). During the ensuing years the specific requirements of analytical transformations in celestial mechanics involving Poisson series with many thousands of terms stimulated the development of specialized computer algebra systems (CAS). These specialized CAS allowed astronomers efficient use of computers available at that time (Jefferys,1970,1971a; Barton and Fitch,1972).

Today rich experience is gained in elaborating and applying CAS in different branches of science. The total number of CAS runs into the hundreds (Grosheva et al., 1983; Grosheva, 1985). Along with specialized CAS a demand arose for large universal CAS such as MACSYMA, REDUCE, etc. These systems, designed for the most powerful computers, enable one to manipulate symbolic quantities of different types. The explosive advancement of mini- and microcomputer made it possible to utilize modified versions of super-CAS such as MACSYMA,REDUCE, VAXYMA, etc. on micromainframes (Fateman,1984). The wide availability of personal computers facilitates the use of super-CAS in astronomy, for example, for inverting matrices of moderate size and manipulating with not too cumbersome objects.

However, the attempts to solve the large problems of celestial mechanics with the aid of universal CAS even by super-computers resulted in unjustifiably high expenses of computer

Celestial Mechanics **45**: 149–162, 1989.

time. The increasing requirements for accuracy in celestial mechanics theories lead to the outrunning growth of the number of terms in the expansions in powers of small parameters. Therefore, the use of specialized CAS by all existing types of computers seems to be best suited for the problems of celestial mechanics.

The best use of the specialized CAS can be made when 1) one has to deal with cumbersome uniform objects by means of the repeated operations of the same kind and 2) there exist well elaborated algorithms of symbolic manipulation leading to the solutions of problems in question. Both these conditions are typical for celestial mechanics. The main objects of relevant transformations are Poisson series, (preserving their form under the demanded algebraic operations), where the analytical algorithms of celestial mechanics are well elaborated (see, for example, Brumberg, 1980).

This paper deals mainly with the specialized CAS elaborated by the group of co-workers from the Institute of Theoretical Astronomy (Leningrad) including, in addition to the authors of the present paper, Drs. I.O.Babaev, T.V.Ivanova,V.I.Skripnichenko (presently this whole group is working at the Institute of Applied Astronomy, Leningrad) and A.V.Kuzmin. Section 2 is devoted to three CAS: UPP,SASM and GRATOS and to the development of the Keplerian processor in different forms and the analytical generator of functions and methods of celestial mechanics. These tools cover all basic kinds of symbolic manipulation in celestial mechanics. Section 3 concerns the numerical -analytical interface designed,in particular, for the fast evaluation of the long Poisson series. Some considerations of future development and application of CAS are given in the conclusion.

2. SPECIALIZED CAS

2.1 UPP (Universal Poisson Processor)

This system, developed in 1979 (Tarasevich, 1979; Babaev et al., 1980) is available as a FORTRAN based subroutine library. Also, there is a special command language PS-FORTRAN (Babaev, 1978) extending FORTRAN by introducing new data types such as power and trigonometric variables, and series and rational numbers.The preprocessor converts arithmetic expressions of the user program written in PS-FORTRAN into a chain of calls to the appropriate subroutines of UPP.The preprocessor yields the output file valid for the usual FORTRAN compiler. The use of the PS-FORTRAN language facilitates the application of UPP and makes the user program easily readable.

UPP represents a specialized CAS designed for manipulating long Poisson series of the form

$$S=\sum C_j^i x^i \begin{matrix}\sin\\ \cos\end{matrix}(jy) \quad (1)$$

x $(x_1,\ldots,x_m)$, y $(y_1,\ldots,y_n)$ being power and trigonometric variables respectively. Summation in (1) is performed over all integer (positive and negative) values of indices $i=(i_1,...,i_m)$ and $j=(j_1,...,j_n)$. Coefficients C_j^i may be specified by the user as rational numbers or floating-point numbers, as well as polynomials of a single variable with such coefficients. The system is universal in the sense that unlike many other Poisson processors the numbers of power and trigonometric variables and the ranges of the associated indices are arbitrary and may be prescribed by the user for any specific problem.Each term of the series is characterized by the analytical order of smallness which is defined as the sum of the power indices multiplied by the weight functions of the corresponding variables. This characteristic permits manipulation with different expansions of the given order with respect to the small parameters. The Poisson series occurs most often in the problems of celestial mechanics but may be used as well in a variety of problems in theoretical mechanics and physics.

In addition to the arithmetic operations the list of basic operations of UPP includes differentiating and integrating, raising to an integer power, raising to a negative fractional power (-1/N) or (-N/2), binomial and Taylor expanding up to a prescribed order, substituting other series in place of power variables or trigonometric functions of any trigonometric variable, linear changing of variables, fast evaluating of the series in fixing values of some variables, etc. A provision is made to convert series, that is to find reciprocal relation $x=x'+q(x')$ proceeding from the transformation of variables of the form $x'=x+p(x)$ for vectors x,x' p,q with components of p and q being Poisson series. Such a series conversion reduces to iterative solving of the equation $q(x) = -p(x+q(x))$ provided that p and q are small.

There are also service routines such as different sortings, renaming of variables and trigonometric functions, retrieval of some terms of the series, declaration of single term series and constant series, etc.

To perform any undocumented operation the user may merely write a FORTRAN subroutine executing the required action on each term of the series and generating an indicator for including this term into the resulting series. The name of this subroutine is used then as the argument in calling the system scanning procedure. UPP provides the output and input of the series with any backing storage in format or unformatted mode. The series may be typed in the natural mathematical form. The list representation of the series adopted in UPP for the dynamic memory allocation makes it unnecessary to use the garbage collector in execution time. The size of the working storage area and other system characteristics are prescribed in the user program.

The hierarchical architecture of UPP enables an easy modification of the system adapting it, for example, for manipulating the exponents instead of the trigonometric functions. To do this it is sufficient to replace several subroutines manipulating just these functions. UPP is described in the form of a library of compiled codes or text files. The basic subroutines of UPP are supplied with formal specification on the magnetic tape.

2.2 SASM (System for Algebraic and Symbolic Manipulation)

SASM (Kuzmin, 1982) is designed for a BESM-6 computer to manipulate multi-power and multi-trigonometric series whose coefficients are rational functions of many variables. SASM is divided into several subsystems each of which represents a set of subroutines for manipulating mathematical objects of one of the following types: integers and rational numbers of arbitrary accuracy, polynomials and rational functions of several variables, and multi-power and multi-trigonometric series. SASM is based on the system LRFOR (Kuzmin, 1980) manipulating integer and rational numbers of arbitrary accuracy and using some algorithms of SAC-1 system (Collins, 1971, 1973) . In producing analytical transformations by computer the use of arbitrary accuracy arithmetic has often the advantage over floating-point calculation.

The basic objects of SASM are represented by two types of series

$$S = \sum C_j \; R_j(X) \; {\cos \atop \sin}(jy) \; , \quad S = \sum C_j \; R_j(x) \; y^j \quad (2)$$

j $(j_1, \ldots, j_m)$ and $y(y_1, \ldots, y_n)$ being the vectors of indices and variables respectively. The summation is taken over the integer values of indices j. Each coefficient Cj is a rational number. Rj(X) represents the rational function of n variables $x(x_1, \ldots, x_n)$, i.e. the ratio of the polynomials of the form

$$P = \sum A_i \; X^i \quad (3)$$

A_I being an integer of the arbitrary length and the summation is produced over non-negative integer values of indices I ($i_1, \ldots, i_n$). The numbers m and n as well as the ranges i and j are arbitrary.

All these objects are stored in SASM using the linked list representation. For example, every object occurring repeatedly in (2) and (3) is stored once with the counter of links from other structures. Combinations of the linked list and balanced tree structures (Fateman, 1974) for the internal representation of polynomials and series result in efficient utilization of the main memory.

Finding the greatest common divisor of polynomials of several variables is performed in SASM with the aid of the modular algorithm by Brown (1971) and Collins (1973). The relevant subroutines of modular arithmetic, together with the subsystem of manipulation of polynomials and rational functions (Kuzmin, 1981), may be used separately in numerous applications, for example, in developing polynomial matrix algebra.

Along with arithmetic operations SASM allows differentiating of polynomials, rational functions and series, integrating series, different substitutions, and input-output operations. There are also more complicated series operations such as, binomial and Taylor expansions, conversions of series, etc. SASM is supplied by the special command language (Babaev and Novicov, 1979) which, similar to PS-FORTRAN, represents an extension of FORTRAN by introducing new data and operation types appropriate to SASM.

SASM is the computer-oriented system. Although the SASM subroutines are called from the user program in FORTRAN, most of these subroutines have been written in BESM-6 assembly language.

2.3 GRATOS (General Relativity Analytical Tensor Operation System)

The tensorial processor GRATOS (Tarasevich, 1980; Tarasevich et al., 1987) represents an interpreter of the specialized language GRATOS designed for the description of analytical expressions involving tensorial and scalar quantities, standard and arbitrary functions of several variables, and rational numbers. The system is primarily oriented for solving the problems of general relativity and relativistic celestial mechanics requiring cumbersome tensor manipulation. Unlike some other specialized systems for general relativity such as CLAM, which can calculate a fixed set of standard general relativity quantities, GRATOS is able to perform calculations by arbitrary formulas specified by the user. This brings GRATOS closer to the universal CAS and opens up possibilities for using it in different branches of physics.

A program written in the GRATOS language consists of a sequence of sentences of two types. The sentences of the first type are declarations, i.e. the usual mathematical formulas written according to some simple rules. The declaration represents an equality containing the subject to be declared (variable, constant or function) in the left-hand member and some analytical expression in the right-hand member. The sentences of the second type are directives. GRATOS permits writing only linear programs.

GRATOS involves analytical operations as follows: substitution of expressions, polynomial algebra, reduction of similar terms, summation over repeated indices, usual and covariant differentiation, Taylor expanding in powers of small parameters, and so on. Provisions are made for operations with standard elementary functions (exp, ln, sin, cos, tg, ctg, sh, ch, th, cth) as well as with arbitrary functions defined by the user. The output information may be printed in the common mathematical form using three levels for placing the upper and bottom indices.

The basic language of GRATOS is FORTRAN. Computer-dependent subroutines come to less than 10% of the total system. The system works either in batch or interactive mode. It is possible to combine these modes when solving a specific problem. The user may write his own subroutines in FORTRAN for some non-standard transformations and call them from the basic program written in GRATOS. Such subroutines are stored in magnetic data medium and form a user library.

Examples of applying GRATOS may be found in the paper by Brumberg and Tarasevich, (1983), which deals with determining perturbations of the spherically symmetrical metric due to the small masses.

2.4 Keplerian expansion processor

The Keplerian processor is the system of subroutines for manipulating functions depending on the coordinates of the unperturbed two body problem. As a rule, one considers only the case of the elliptic motion of the most importance for applications. The Keplerian processor is constructed on the basis of a Poisson processor or any universal CAS, and may be regarded as an illustration of practical uses of these systems.

It is well known that the solution of the two body problem may represented in three basic forms such as 1) trigonometric series in the mean anomaly, 2) series in powers of time and 3) closed expressions using true or eccentric anomalies. The second form is valid over some finite time interval. The choice of these forms can be expressed as the Keplerian processor in Poisson series form, Taylor series form, or closed form, respectively.

The Keplerian processor in the Poisson series form derives by computer the classical expansions of the elliptic motion. Most of these expansions are available now in the form of tables (Jarnagin, 1965), but for solving the problems of celestial mechanics by computer the subroutines to obtain the typical expansions are needed. A version of the Keplerian processor has been proposed in particular by Broucke (1970, 1974) . Our version (Ivanova, 1979; Brumberg, 1980) is based on three auxiliary subroutines yielding the expansions for the equation of the center, the solution of the Kepler equation and, the functions $(r/a)^n \exp(\sqrt{-1}mw)$, w being the true or eccentric anomaly (involving Hansen coefficients or Bessel functions as the coefficients of the expansions). All these expansions are Poisson series with one polynomial variable, eccentricity e, and one trigonometric variable, mean anomaly M. These subroutines permit writing a more general subroutine for producing the classical expansions

$$(x+\sqrt{-1}y)\, r^n/a^{n+1} = \sum X_k^{n+1,1}(e)\,(\cos^2(i/2)\exp\sqrt{-1}(\Omega+\omega+kM) \quad (4)$$
$$+ \sin^2(i/2)\exp\sqrt{-1}(\Omega-\omega-kM))\ ,$$
$$zr^n/a^{n+1} = \sin(i)\sum X_k^{n+1,1}(e)\,\sin(\omega+kM) \quad (5)$$

with three trigonometric variables, M, longitude of the node Ω, argument of the pericenter ω, and three polynomial variables, e, sin(i/2), cos(i/2). If inclination i is small it is suitable to replace sin(i/2), cos(i/2) by one polynomial variable sin(i) to be used as the parameter of expansion. Our version includes also the subroutine converting (4) and (5) into power series in terms of conjugate variables K, $\bar{K}$, L, $\bar{L}$ for the functions $(x+\sqrt{-1}y)\, r^n/a^{n+1} \exp\sqrt{-1}\lambda)$, zr^n/a^{n+1} with

$$K = e\exp\sqrt{-1}M,\quad L = -(\sqrt{-1}/2)\sin(i)\exp\sqrt{-1}(M+\omega),\quad \lambda = \Omega+\omega+M\ . \quad (6)$$

The coefficients of these power series are Newcomb operators. Also, there is the subroutine transforming these series into

Poisson series with one trigonometric variable λ and four polynomial variables

$$k=e\cos(\Omega+\omega),\quad l=e\sin(\Omega+\omega),\quad p=\sin(i)\cos\Omega,\quad q=\sin(i)\sin\Omega \tag{7}$$

The Keplerian processor in the Taylor series form is used less than the two other forms. Nevertheless, Taylor expansions in powers of time are employed in practical problems dealing with the representation of motion over short time intervals as well as in theoretical investigations. One of the first applications of CAS in classical celestial mechanics was with the Taylor series Keplerian processor (Sconzo et al., 1965). This processor has dealt with the universal solution (excepting the case of rectlinear motion) in the form

$$\underline{r}(t)=\underline{r}_0F(t)+\dot{\underline{r}}_0G(t)\ ,\quad F(t)=\sum_{n=0}^{\infty}f_nt^n/n!\ ,\quad G(t)=\sum_{n=0}^{\infty}g_nt^n/n! \tag{8}$$

with functions f_n and g_n being polynomials with integer coefficients of the local invariants $\mu=r^{-3}$, $\sigma=\underline{r}^{-2}\,\underline{r}\dot{\underline{r}}$, $\varepsilon=r^{-2}(\dot{\underline{r}}^2-r^{-1})$ (under the conventional choice of units) calculated for the initial moment of time. In our version of the Keplerian processor only the elliptical case is considered. The orbital coordinates are expanded in powers of the mean anomaly

$$X=\cos E-e=\sum_{k=0}^{\infty}a_kM^{2k},\quad Y=(1-e^2)^{1/2}\sin E=\sum_{k=0}^{\infty}b_kM^{2k+1} \tag{9}$$

with a_k and b_k of the form (Brumberg, 1980)

$$a_k=(-1)^k(1-e)^{-3k+1}\tilde{a}_k/(2k)!,\quad b_k=(-1)^k(1-e^2)^{1/2}(1-e)^{-3k-1}\tilde{b}_k/(2k+1)! \tag{10}$$

$\tilde{a}_k$ and $\tilde{b}_k$ (k=1,2,...) being polynomials in e of power k-1 with positive integer coefficients. These coefficients have been computed by means of the system LRFOR (Kuzmin, 1980).

2.5 Analytical generator of special functions and methods of celestial mechanics

Poisson and Keplerian processors permit the construction of an application package for performing typical analytical expansions in celestial mechanics. The subroutines of this package may be classified under three levels. The subroutines of first level expand the special functions of celestial mechanics (and their derivatives) such as Hansen coefficients, Kaula inclination functions and their generalizations, Laplace coefficients, etc. The subroutines of the second level are employed mainly to compose the right-hand members of the equations of perturbed motion, primarily to expand the perturbed function in its three basic forms as the satellite perturbing function due to the third body attraction, the satellite perturbing function due to the non-sphericity of the central body, and the planetary perturbing function. The main characteristic property of the package is the derivation of all these expansions, starting with the most elementary formulas. For example, with the aid of the Keplerian processor, one may immediately derive an expansion for the cosine of the central angle between two moving bodies

$$\cos H = (xx' + yy' + zz')/(rr') \qquad (11)$$

This expansion is the Poisson series with 6 trigonometric variables, $\Omega, \Omega', \omega, \omega'$, M, M' and 4, or 5 or 6 polynomial variables e,e' sin(i), or sin(i/2), cos(i/2) and similarly for i' . Then, using the recursion relation

$$C_n^{k/2}(x) = \frac{2n+k-2}{n} f(x) C_{n-1}^{k/2}(x) - \frac{n-k-2}{n} C_{n-1}^{k/2}(x) \qquad (12)$$

one can construct the expansions for Gegenbauer polynomials ($f(x)=x$, $C_0^{k/2}(x)=1$, $C_1^{k/2}(x)=kx$), Legendre polynomials (the preceeding case for k=1) and cos(nx) ($f(x)=\cos(x)$, k=2, $C_0^1(x)=1$, $C_1^1(x)=\cos(x)$).

This results in the expansion of the satellite perturbing function due to the third body attraction. It is to be noted that the coordinates x', y', z', of the disturbing body may be presented not only by the Keplerian expansions but also by any analytical theory of motion. In this case the perturbing function is expanded in series in terms of the satellite Keplerian elements and the parameters of the analytical theory of the disturbing body (for example, the use of the lunar analytical theory in the problem of artificial lunar satellites or distant Earth satellites). To construct the expansions of the satellite perturbing function due to the non-sphericity of the primary it is suitable to apply the algorithm of (Cunningham, 1970; Giacaglia, 1975). The same algorithm based on the recursion determination of the spherical functions may be employed for the perturbing function due to the third body attraction. As an initial expansion for the planetary disturbing function we use power series in terms of the variables p, p, p', p', w, w' being deviations from the complanar circular motions of the planets

$$x + \sqrt{-1}y = a(1-p)\exp\sqrt{-1}\lambda, \quad z = aw . \qquad (13)$$

The coefficients of these series depend on the ratio of the semi-major axes and the difference of the mean longitudes. From this it is easy, with the aid of Poisson and Keplerian processors, to derive any currently employed expansions of the disturbing function.

The subroutines of the third level of the analytical generator are designed for algorithmization of the basic methods of the celestial mechanics theory of perturbations, including the techniques by Zeipel, Krylov-Bogoljubov, Hori-Deprit, Kolmogorov-Arnold, etc. At present, all these techniques have been elaborated with such accuracy (in particular, by Musen, Broucke, Deprit, Holland and many others) that their computer realization involves no difficulties. Our version of the specialized software includes the complete package of subroutines of the first two levels

(Ivanova, 1979) and some subroutines of the third level. All algorithms are described in detail in (Brumberg, 1980).

2.6 Closed form Keplerian processor

At present, non-standard methods of expanding functions in celestial mechanicsattract particular interest. This involves the attempts to use elliptic functions in planetary and satellite problems (Williams et al., 1987), and the simultaneous application of true and eccentric anomalies as arguments of expansions (Deprit and Ferrer,1987), etc. Elaboration of the specialized CAS to deal with the elliptic functions remains to be done in the near future. As for using different anomalies of the elliptic motion, the corresponding algorithmic and programming software is already available. Simultaneous use of different anomalies permits the development of a closed form Keplerian processor, avoiding expansion in infinite series.

In using only one of the different anomalies (true or eccentric) the necessity for expansion in series arises when integrating the appropriate functions with respect to time. The first algorithm to integrate functions of elliptic motion in the closed form was developed by Jefferys (1971b). In our version of the closed form Keplerian processor based on UPP another, faster algorithm is applied (Vasiliev, 1982). In this algorithm any integrand $F(r,\dot{r},v,E)$ of polynomial form in r, $\dot{r}$ and trigonometric form in true anomaly v and eccentric anomaly E is presented first as $F=F_1(r)+\dot{r}F_2(r)$ where $F_k(r)$ (k=1,2) are polynomials of r (containing negative powers as well). Integrating the second part presents no difficulties in reducing to the quadrature of $F_2(r)dr$. As for the first part the relevant integral is presented in the form

$$\int F_1(r)dr = bv+cE+\dot{r}Q(r) \qquad (14)$$

Representing

$$F_1(r)=\sum a_i r^i \qquad\qquad Q(r)=\sum d_i r^i \qquad (15)$$

one can easily obtain the recursion relations to determine b, c, d_i with the known values of a_i. If the right-hand member of (14) contains no secular terms, i.e. c=b=0, then converting to the trigonometric form in terms of v and E may be performed by means of the standard formulas . Actually there is no secular term in the case $c = -b \neq 0$ as well. However it is impossible herewith to retain the closed form using classical expansions. Therefore, it is suitable to reject these expansions in favour of the uniformly best approximation for the quantity $(v-E)/\dot{r}$ as functions of r in the interval [1-e , 1+e] . In fixing maximal and minimal powers of the approximation polynomial, the construction of the approximation reduces to the problem of linear programming. Having found such a polynomial, one can construct the trigonometric finite sum for v-E using standard formulas expressing r, $\dot{r}$ in terms of v and E. This technique may be applied to any function of r.

3. NUMERICAL- ANALYTICAL INTERFACE

With the increasing development of the symbolic manipulation in celestial mechanics, the problem of constructing an efficient numerical-analytycal interface is of particular important. Such a problem has two aspects:
1) communication of CAS with the front -end software environment involving the use of the internal package of numerical algorithms in combination with symbolic manipulation, and 2) the use of the results obtained by CAS for the numerical applications. It is assumed that these results are not subjected to symbolic transformations and serve only for evaluation of corresponding analytical expressions in substituting numerical values of occuring variables.

The first aspect is solved quite satisfactorily by CAS, which contains the necessary facilities for their extension by the user. For example, GRATOS contains a special directive for introducing non-standard symbolic operations. Such operations may call any external subroutines. As for UPP, the user program may call any FORTRAN subroutine, since the UPP internal language represents an extension of FORTRAN.

The second aspect is more complicated. Its solution involves highly efficient algorithms of substituting numerical values into symbolic expressions, as well as special programming facilities for evaluating expressions outside CAS.

All more or less advanced CAS include the internal facilities to substitute numerical values into symbolic expressions. But such facilities usually are not too effective since the objective of CAS internal representation is to perform symbolic (not-numerical!) manipulation. For this reason the internal facilities should be applied only for occasional evaluation of expressions. The evaluation problem in celestial mechanics is of particular importance due to the very large number of terms of the relevant expressions. The symbolic expressions derived in CAS are often the right-hand members of the ordinary differential equations to be integrated numerically. As it is well known that the evaluation of the right-hand members is the most time consuming operation in the numerical integration programs. Another example is furnished by discussion of observations when the large data arrays are to be substituted into some analytical theory.

In recent years several methods were suggested for the fast evaluation of Poisson series. For example, the method developed in Coffey and Deprit, (1980), reduces the evaluation of Poisson series to numerous uses of the trigonometric addition theorems, provided that some tables of sines and cosines of specific linear combinations of trigonometric variables are computed in advance. Chapront (1982, 1984) succeeded in evaluating Poisson series by Fourier-Chebyshev approximations preserving the fast variables under the sign of trigonometric functions and using Chebyshev polynomial approximations for the slowly changing variables. This technique results in significant extension of the approximation interval. Constructing this approximation may be facilitated by using the specialized CAS including manipulation with Chebyshev polynomial expansions. Such a system is available in our package (Vasiljev,1980) .

From our point of view, the best tool to set an efficient interface between symbolic manipulation and numerical calculation is provided by generation of the evaluations programs. Such generation tools are available, for example, in the universal CAS such as MAXYMA, REDUCE, etc. But these tools of universal CAS are not so efficient when dealing with the extended symbolic expressions. The best efficiency may be achieved in the specialized evaluation systems manipulating a specific class of expressions like polynomials or Poisson series. We describe here the basic principles underlying the specialized system SPRINT designed to generate effective programs of evaluation of the polynomial-trigonometric series (Vasiljev, 1983). Algorithms realized in programs synthesized by SPRINT are very efficient and possess the non-improvable complexity estimate (a number of employed arithmetic operations) in the class of algorithms without preliminary adaptation of coefficients.

Let us remark first that any Poisson series of the general form may be reduced to the power series representation using complex coefficients and variables. Therefore, we may confine ourselves by the algorithm of evaluation of a polynomial of many variables. Such polynomials often turn out to be very sparse so that usual techniques like Horner method are not of much use here. Our algorithm of evaluating the sparse polynomial involves the binary tree representation. SPRINT transforms this representation into the program of evaluation.

Polynomial P is represented by a recursive procedure such as a binary tree with marked nodes. All end nodes correspond to the coefficients of P. All non-end nodes are related with some multi-indices. These multi-indices are not, in general the multi-indices of terms of P. The recursive procedure begins by finding the common subindex of all terms of P. The root of the constructing tree is marked with this multi-index i. This subindex is cancelled from all terms of P, i.e.all multi-indices should be diminished by i. P is split in two polynomials P_1 and P_2, P_1 being the initial maximal segment of terms of P having the non-trivial common subindex, and P_2 is the remainder of P. P_1 and P_2 are taken as left and right subtrees respectively. The same procedure is applied now to P_1 and P_2, and so on. If any polynomial reduces to a single free term, then it corresponds to the end node.

In this manner evaluating P consists of 1) simultaneous calculation of all monomials X_i of the non-end nodes of the tree and 2) tree traversal with multiplication at every node of the relevant monomial X_i by the sum of the values of polynomials represented by the appropriate subtrees.

The performance of this algorithm depends on the technique of simultaneously computing the set of monomials as well as the preliminary sorting of P. The algorithmic complexity of the first problem is determined in Pippenger, (1980). The calculation of the set of monomials is performed in SPRINT in an optional manner (Vasiljev, 1983). As for sorting, we have used the lexicographic sorting in the first version of SPRINT. In the second version a more complicated and more efficient sorting is applied (Vasiljev, 1985).

SPRINT enables the user to synthesize a program for simultaneous evaluation of any set of Poisson series. Moreover, the simultaneous evaluation of expressions results in substantially faster processing, for example, in numerical integration of the differential equations with the right-hand members obtained by some CAS. It is to be noted that SPRINT is not related with any specific CAS and may be used to synthesize evaluation programs in combination with a great variety of CAS. In calling SPRINT it remains only to give information about the representation of the Poisson series used. This system has been used to compute the lunar ephemerides given by the theory ELP-2000, the ephemerides of Mercury and Venus based on the theory VSOP-82, etc. Application of the synthesized programs reduced computer time by 10-20 times as compared with the standard algorithms.

4. CONCLUSION

As was already mentioned in the introduction, the trend today is toward applications of CAS by a whole variety of computers from micro- to supercomputers. This looks promising. Its efficiency depends largely on the CAS interface organization with the user and between different CAS. This approach demands the elaboration of the unique basis of different specialized systems including data type definition, command language, data base architecture (algorithms, formulas, values, etc.) and access modes, CAS generation methods for new objects and operations, numerical-analytical and graphic interface, etc. (Tarasevich, 1985) . One of the promising ways for technical realization of such an approach is producing the distributed local area networking environment (Heikkila, 1985) and elaborating specialized analytical microprocessor sets. The future CAS should include new interactive modes with the aid of a menu, a mouse and other systems. Rapid development of computer graphics should also enlarge the possibilities of using CAS and interacting with the user. The combination of the CAS facilities with a windowing system and a stack architecture in the FORTH language seems to be very productive.

As for the celestial mechanics aspects of CAS, most actual problems are related to elaborating an elliptic function processor and extending a closed form Keplerian processor, since the current tendencies of constructing analytical theories of motion require these tools of symbolic manipulation.

REFERENCES

Babaev I.O. 1978. *Algorithms of Celestial Mechanics,* No. 20, ITA, Leningrad

Babaev I.O. and Novikov F.A. 1979. *Algorithms of Celestial Mechanics,* No. 22, ITA, Leningrad

Babaev I.O., Brumberg V.A., Vasiliev N.N., Ivanova T.V.,

Skripnichenko V.I. and Tarasevich S.V. 1980. *Internat. Conf. on Systems and Techniques of Analytical Computing and Their Applications in Theoretical Physics*, p. 80, Dubna

Barton D. and Fitch J.P. 1972. *Rept Prog Phys.*, **35**, 235

Broucke R. 1970. *Celest. Mech.*, **2**, 9

Broucke R. 1974. *Lect. Notes Math.*, 362, 237

Brown W. S. 1971. *J. of ACM*, **18**, 478

Brumberg V.A. 1980. *Analytical Algorithms of Celestial Mechanics*, Nauka, Moscow

Brumberg V.A. and Tarasevich S.V. 1983. *Conf. on Systems and Techniques of Analytical Computing and Their Applications in Theoretical Physics*, p. 149, Dubna

Chapront J. 1982. *Celes. Mech.*, **28**, 415

Chapront J. 1984. *Celes. Mech.*, **34**, 165

Coffey S. and Deprit A. 1980. *Astron. Astroph.*, **81**, 310

Collins G.E. 1971. *Univ. Wisconsin Comp. Sci. Dept. Techn. Rep.* No. 129

Collins G.E. 1973. *Univ. Wisconsin Comp. Sci. Dept. Techn. Rep.* No. 156

Cunningham L.E. 1970. *Celes. Mech.*, **2**, 207

Davis M.S. 1958. *Astron. J.*, **63**, 462

Deprit A. and Ferrer S. 1987. *Celes. Mech.*, **40**, 335

Fateman R.J. 1974. *Celes. Mech.*, **10**, 243

Fateman R.J. 1984. *SIGSAM Bull.*, **18**, 10

Giacaglia G.E.O. 1975. *Univ. Texas at Austin Appl. Mech. Res. Lab. Paper* No.1073

Grosheva M.V. et al. 1983. *Inst . of Appl. Math. Informator* No. 1, Moscow

Grosheva M.V. 1985. *Internat. Conf. on Computer Algebra and Its Applications in Theoretical Physics*, p. 7, Dubna

Heikkila V.S. 1985. *Internat. Conf. on Computer Algebra and Its Applications in Theoretical Physics*, p. 17, Dubna

Herget P. and Musen P. 1959. *Astron. J.*, **11**

Ivanova T.V. 1979. *Algorithms of Celestial Mechanics*, No. 20, ITA, Leningrad

Jarnagin M.P. 1965. *Astron. Papers Am. Ephemeris*, **18**, 1

Jefferys W.H. 1970. *Celes. Mech.*, **2**, 474

Jefferys W.H. 1971a. *Comm. ACM.*, **14**, 538

Jefferys W.H. 1971b. *Celes. Mech.*, **3**, 390

Kuzmin A.V. 1980. *Algorithms of Celestial Mechanics*, No. 33, ITA, Leningrad

Kuzmin A.V. 1981. *Algorithms of Celestial Mechanics*, No. 34, ITA, Leningrad

Kuzmin A.V. 1982. *Algorithms of Celestial Mechanics*, No. 40, ITA, Leningrad

Pippenger N. 1980. *SIAM. J. Comput.*, **9**, No 2

Sconzo P., LeSchack A.R. and Tobey R. 1965. *Astron. J.*, **70**, 269

Tarasevich S.V. 1979. *Algorithms of Celestial Mechanics*, No.27, ITA, Leningrad

Tarasevich S.V. 1980. *Internat. Conf. on Systems and Techniques of Analytical Computing and Their Applications in Theoretical Physics*, p. 116, Dubna

Tarasevich S.V., Krivov A.V. and Titov V.B. 1987. *Algorithms of Celestial Mechanics*, Nos. 53, 54, ITA, Leningrad

Tarasevich S.V. 1985. *Internat. Conf. on Computer Algebra and Its Applications in Theoretical Physics*, p. 33, Dubna

Vasiliev N.N. 1980. *Internat. Conf. on Systems and Techniques of Analytical Computing and Their Applications in Theoretical Physics*, p. 186, Dubna

Vasiljev N.N. 1982. *Bull. ITA*, **15**, 142

Vasiljev N.N. 1983. *Conf. on Systems and Techniques of Analytical Computing and Their Applications in Theoretical Physics*, p. 120, Dubna

Vasiljev N.N. 1985. *Internat. Conf. on Computer Algebra and Its Applications in Theoretical Physics*, p. 154, Dubna

Williams C.A., Van Flandern T. and Wright E.A. 1987. *Celes. Mech.*, **40**, 367

ON A RESTRICTED (2n + 3)-BODY PROBLEM

A. OLLONGREN
Dept. of Computer Science
Leiden University
P.O. Box 9512
2300 RA Leiden
The Netherlands

ABSTRACT. We consider the case of (2n + 1) bodies (n ≥ 0) each of mass m, which are placed on a circle with radius r, such that they form a regular polygon: an equilateral (2n + 1)-angle. In the centre of the circle a body of mass b times m is placed, where b is chosen large enough to ensure stability of the system; only gravitational interaction is considered. Each of the bodies rotates uniformly around the centre with angular velocity ω. In addition to the (2n + 2) bodies, considered to be point masses, we have another point mass with negligible mass compared to the former ones; we are then interested in the motion of the small body in the gravitational field of force generated by the large ones, moving themselves in an equilibrium configuration reacting to each other's fields of force but not to the (2n + 3)-d body.

1. Generalized Third Law of Kepler

We begin the treatment by determining the angular velocity ω of the system of (2n + 2) bodies described above; this velocity must be such that the (2n + 1) bodies placed at the vertices of the regular polygon rotate uniformly about the central mass under the influence of the gravitational attractions. So we have the criterion that the gravitational attracting force on each body in the circle must be balanced exactly by the centrifugal force connected with its motion. The time variable is denoted by t'. Figure 1 shows the situation at time t' = 0 for some arbitrary n. Note that we place one body on the positive X-axis, and that there are consequently n bodies above it, and n bodies in mirror positions below it.

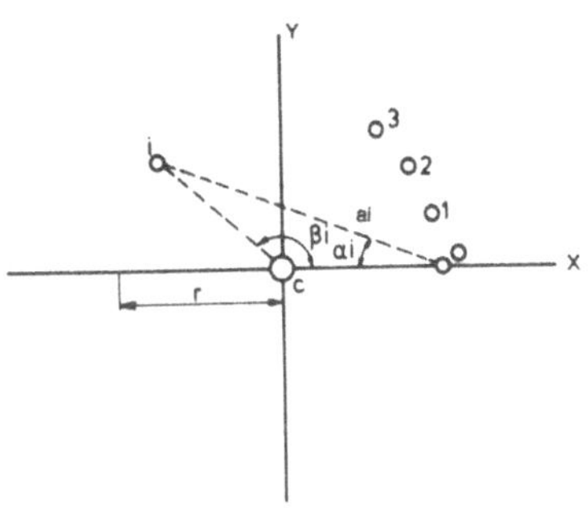

The geometry of the problem
in the fixed orbital plane

Figure 1

Celestial Mechanics **45**: 163–168, 1989.

The component along the X-axis of the gravitational attraction between body i (> 0) above the X-axis and body 0 is

$$G\,(m^2 / a_i^2)\,\cos \alpha_i \qquad i = 1,2,\ldots,n$$

where G is the gravitational constant, m is the (point) mass mentioned before and the other variables are as shown in figure 1. The gravitational attraction between the central body (with mass b.m) and body 0 is

$$b.G\,(m^2 / r^2)$$

where r is the radius of the circle; the vectors associated with these expressions are directed along the X-axis from the position of body 0 to the centre.

The centrifugal force acting on body 0 is $m.r.\omega^2$.
So, by summing over all bodies in the circle, using the balancing criterion and using for the distances a_i and the angles α_i the relations

$$a_i = 2\,.\,r \cos \alpha_i \qquad \alpha_i = \pi(1/2 - i/(2n+1)) \qquad i = 0,1,\ldots,2n$$

we obtain immediately a generalized third law of Kepler:

$$G.m\,[\,b + (1/2)\sum_{i=n}^{n} (\sec(\pi i / (2n+1)))\,] / r^3 = \omega^2 \tag{1}$$

For n = 0 this reduces to the third law of Kepler - but then for circular motion because of the assumptions. For n = 1 the law for the restricted five-body problem (A. Ollongren 1987, [2]) is retrieved.

2. The Potential in The Rotating Frame

We introduce now the coordinate system (u,v) with the same origin as that of the (fixed) (X,Y)-system, but rotating uniformly around the origin with angular velocity ω; the latter velocity is chosen in such a way that Kepler's third law (previous section) is satisfied - which means that the bodies in the circle are at rest in the (u,v) frame. Let Rc denote the distance of the "small" body to the centre and Ri (i = 0,1,...,2n) the distances from it to the bodies on the circle. In the rotating frame the small body is subjected to gravitational forces which are derivable from a potential function U_{grav} of the form

$$U_{grav} = f\,.\,(b/Rc + \sum_{i=0}^{2n} (1/Ri)) \tag{2}$$

This observation follows from general considerations which we shall briefly present below. In addition we derive the form of the constant factor f. It is evident that it must depend on the parameters of the problem: the mass-ratio b and the units of length and time.

For n = 1 (the bodies in an equilateral triangle) we have shown previously using the length of a side on the trangle as a unit (A. Ollongren 1987, [2])

$$f = 1 / (3 + 3.\, b \sqrt{3}), \text{ or } \ f = 1 / (\sqrt{3} / 3 + b) \qquad (3)$$

had we chosen the radius of the circumscribed circle of the triangle as the unit of length as we do in the present article.

In order to derive the required equations of motion we use the standard derivation, which runs along the following lines (see for example H.C. Plummer 1918, [3]). In the fixed system of coordinates (X,Y) the gravitational potential generated by a body of mass b.m (b > 1) situated at (X1,Y1) is

$$K = G.b.m\,(1/R1)$$

where for the distance we have

$$R1^2 = (X - X1)^2 + (Y - Y1)^2$$

A body of negligible mass (a test particle) moves in the field of force of the other, subject to the following differential equations:

$$d^2 X/dt'^2 = \delta K/\delta X \qquad d^2Y/dt'^2 = \delta K/\delta Y$$

where t' stands for the time parameter. The transformation to the system (u,v) is written

$$X = u \cos(\omega.t') - v \sin(\omega.t') \qquad Y = u \sin(\omega.t') + v \cos(\omega.t')$$

For the positions X1 and Y1 we have

$$X1 = r \cos(\omega.t' + \varphi) \qquad Y1 = r \sin(\omega.t' + \varphi)$$

where r is the distance of the point (X1,Y1) from the origin and φ is an angle measured in the positive direction in the (X,Y) plane (for a point on the X-axis $\varphi = 0$).
Further (u,v) is transformed to dimensionless coordinates (x,y) and the time variable t replaces the time variable t' by:

$$x = u/r \quad y = v/r \quad t = \omega.t'$$

It is a straightforward matter (possibly using computer algebra) to show that the equations of motion in x and y become

$$d^2x/dt^2 - 2\, dy/dt = (G.b.m/(\omega^2.r^3))(-x + r \cos\varphi)/R'^3 + x$$

$$d^2y/dt^2 + 2\, dx/dt = (G.b.m/(\omega^2.r^3))(-y + r \sin\varphi)/R'^3 + y$$

Here R' is the distance R1 measured in units r (the radius of the circle previously introduced in the (X,Y)-plane).

On the other hand it is directly verified that for

$$U = (x^2 + y^2)/2 + (G.b.m/(\omega^2 . r^3))/R' \tag{4}$$

the following holds:

$$\delta U/\delta x = (G.b.m/(\omega^2 . r^3))\ (-x + r \cos\varphi) / R'^3 + x$$

$$\delta U/\delta y = (G.b.m/(\omega^2 . r^3))\ (-y + r \sin\varphi) / R'^3 + y$$

which means that

$$d^2x/dt^2 - 2\,dy/dt = \delta U/\delta x \quad d^2y/dt^2 + 2dx/dt = \delta U/\delta y \tag{5}$$

In case there are two or more bodies in the (X,Y)-plane, one uses a superposition (and strictly speaking also an induction)-principle to obtain from (4) a form for U in which G.b.m/R' is replaced by a sum of terms of the form $G.b_i.m/R'_i$.

We have obtained a position now from which it is an easy matter to derive the equations of motion of the (2n + 3)-body problem: they are equations (5) with for U (from (4))

$$U = (x^2 + y^2)/2 + (G.m/(\omega^2 . r^3))\ .\ (b/Rc + \sum_{i=0}^{2n} 1/Ri) \tag{6}$$

denoting by Rc as before the distance from the centre.
Until now the 2n + 1 bodies are not arranged in any specific way in the plane. Now, however, we place them in the vertices of a regular polygon; then we can use the generalized form of the third Kepler equation (1) and replace the factor G.m/(ω**2 . r**3) by

$$f = 1 / [b + (1/2) \sum_{i=1}^{n} (\sec(\pi i/(2n + 1)))] \tag{7}$$

3. Power Series Expansions

Formulas (6) and (7) determine completely the potential function in the (uniformly rotating) (x,y)-plane. For n = 2 (the seven-body problem) we show in figure 2 schematically some equipotential curves. In the five-body problem there exist three Lagrangian points of equilibrium, which are globally stable (i.e. they are surrounded by closed oval-shaped Hill curves: curves of zero velocity which cannot be crossed by 'low energy'-particles moving in the vicinity of them). So we expect here the existence of five stable Lagrangian points of equilibrium. We are interested in their locations.

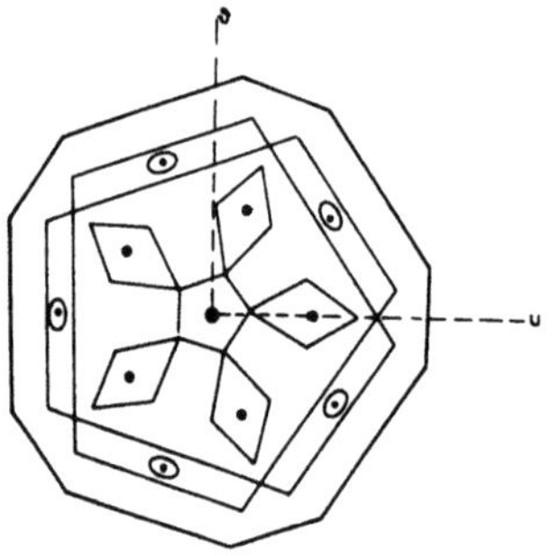

Equipotential curves in the (x,y)-plane for the seven-body problem (schematic)

Figure 2

Because of the symmetries we need only find the (negative) value of x on the axis y= 0 for which the partial derivative of U with respect to x vanishes. The determination of these values for various n is the subject of this paragraph.
We begin by writing U as defined by (6) and (7) in a different form. U contains Ri for i = 0,1,...,2n and since we are interested in U as a function of x and y we begin by writing

$$Ri = ((x - \cos \beta_i)^2 + (y - \sin \beta_i)^2)^{1/2}$$

where for the angle β_i we have (see also figure 1)

$$\beta_i = (2 \pi . i) / (2n + 1) \qquad i = 0,1,...,2n \tag{8}$$

These relations follow immediately from the geometry of the problem, see also figure 1. Using this we rewrite U into the following form

$$U = (x^2 + y^2)/2 + b . f . (x^2 + y^2)^{(-1/2)} + f . \sum_{i=0}^{2n} ((x - \cos \beta_i)^2 + (y - \sin \beta_i)^2)^{(-1/2)} \tag{9}$$

remembering the form (7) of f and the values (8) of the angles β_i.
For y = 0 we obtain the following partial derivative

$$\delta U / \delta x = x - b . f . x . (x^2)^{-3/2} - f. \sum_{i=0}^{2n} (x - \cos \beta_i)/(x^2 - 2 x \cos \beta_i + 1)^{(3/2)} \tag{10}$$

At first sight it seems a rather horrendous task to solve for x as a function of b for a given n such that this non-linear form vanishes. The observation, however, that x is analytic in b helps out. For large b we expect x to tend to -1. So we express the analyticity by assuming that x can be writen as a power series in 1/b

$$x = \sum_{i=0}^{\infty} x_i . b^{-i}, \quad x0 = -1 \text{ (expected)} \tag{11}$$

Most of this paragraph is devoted to the determination of the coefficients x0,x1,...
To begin with: substituting 1/b = 0 in (10), it is easily seen (using f→0) that
x0 - x0/((x0**2)**(3/2)) = 0, which corroborates the expected value -1 of x0.
Next we write the form (10) as the sum of three terms: T1 + T2 + T3

T1 = x is expanded by (11).
In T2 and T3 we find factors which are functions of x. Denote such a factor by G(x), and write G(x0 + h) where

$$h = x1/b + x2/b^2 + x3/b^3 + ...$$

Since G(x) is analytic in x it can be expanded in a Taylor series:

$$G(x0 + h) = G(x0) + \sum_{i=0}^{\infty} (h^i/i!).(d^i G(x)/dx^i)_{x0}$$

So all we need to do is compute the literal forms of the derivatives of G(x) with respuct to x, multiply them with the correct powers of h, and collect the coefficients of (1/b)**i for i = 0,1,.... This is evidently best done using computer algebra for such factors as occurring in T3. For the factor x / ((x**2)**(3/2)) in T2 we find immediately (x0 = -1)

$$(x0+h)/(((x0+h)**2)**(3/2)) = -1 -2.h/1! - 6.h^2/2! - 24.h^3/3! - ...$$

In the report [1], we give a REDUCE script which can be used to find x1, x2 and x3 for the five-body problem.
The script delivers expressions in res(1), res(2) and res(3) (the coefficients of the powers -1, -2 and -3 of b); they are linear in x1, x2 and x3. By taking them equal to zero, x1, x2 and x3 are immediately solved as real numbers. Similar scripts for the seven- and nine-body problems yield x1, x2 and x3 for those cases. In order to display the results we collect in the following table values of xi' defined by

$$xi' = xi / (2n + 1)^i \qquad i = 0,1,2, ...$$

In this way the mass ratio b' defined by (central mass / total mass in vertices of polygon) becomes the parameter of the problem.

n # of bodies	1 5	2 7	3 9
x0'	-1	-1	-1
x1'	-.01918	-.0406	-.0500
x2'	-.00123	-.0137	-.0420
x3'	+.00010	-.0033	-.0316

Table 1

4. Conclusions

The stable Lagrangian equilibrium points for the (2n+3)-body problem lie outside but close to the unit circle: the more massive the central body, the closer. When more bodies are placed on the unit circle (so increasing n), the number of "L4 points" increases of course (linearly with n), but at the same time they recede from the unit circle for a given mass ratio b or b/(2n+1).

ACKNOWLEDGEMENT. The author thanks Dr. P.J. Message for valuable suggestions for improvements to the original manuscript of this paper.

References

[1] A. Ollongren 1988, On a restricted (2n+3)-body problem, Dept. of Computer Science, Leiden University, Report No. 88-21, 1988.
[2] and [3]: see [1].

THE SIMPLEX METHOD FOR NONLINEAR MASS DETERMINATIONS

RICHARD L. BRANHAM, JR.
Centro Regional de Investigaciones Científicas y Tecnológicas
Casilla de Correo 131
5500 Mendoza
Argentina

ABSTRACT. The Gauss-Newton method, and calculating a mass that minimizes the variation of residuals are standard techniques for determining planetary masses, but both may fail under certain circumstances. The Gauss-Newton method, in particular, may diverge, and when it converges may converge to a local, rather than global, minimum of the nonlinear regression problem. The simplex method of nonlinear optimization needs no initial estimate for the solution and can be made to converge to a global minimum. It may also be used with non-least squares criteria, such as the L_1 criterion, for greater robustness. But the simplex method achieves these advantages at a high computational price. To test the method as a tool for dynamical astronomy, over 12,000 observations of Neptune were used to calculate Pluto's mass. From an initial estimate of 1/1,812,000 the Gauss-Newton method diverged. The simplex method converged to a more satisfactory 1/22,000,000 with a range of 1/47,000,000 to 1/14,000,000 as indicated by the mean error. Because the simplex method is considerably slower than competing methods, it should be reserved for refractory problems that do not yield facil solutions when tackled by other methods.

1. INTRODUCTION

When a satellite cannot be used, determinations of a planet's mass almost invariably rely on one of two techniques: a differential correction, based on linearized equations of condition, to an assumed mass; or calculation of residuals for a perturbed planet using a series of masses for a perturbing planet whose mass is sought and adoption of the mass that minimizes the sum of the squares of the residuals (if we are using the least squares criterion). As an example of the former see Janiczek's (1971) determination of Jupiter's mass from observations of the minor planet 33 and of the latter O'Handley's (1969) determination of Jupiter's mass from observations of planet 65.

Both Janiczek and O'Handley obtained good results, but the techniques they used sometimes fail. My study of the orbits of five minor planets (Branham, 1979) resulted in good corrections to a number of

Celestial Mechanics **45**: 169–174, 1989.

constants, such as the FK4 equinox, but differential corrections to Jupiter's mass always gave values close to 1/1046.9 rather than the accepted 1/1047.355. Seidelmann et al. (1971) calculated Pluto's mass from a series of assumed mass values and perturbations on Neptune. The best-fit mass that minimized the residuals in latitude and longitude for Neptune, 1/3,000,000, is far from the 1/130,000,000 known since the discovery of Pluto's satellite Charon as more realistic.

Why do these two techniques, differential corrections and minimizing residuals, sometimes work and sometimes fail? A differential correction, when applied more than once, is known in statistics as the Gauss-Newton method for nonlinear regression, one of a class of methods that uses the gradient to iteratively calculate the solution. Like all gradient methods the Gauss-Newton method may diverge if the first approximation differs greatly from the true solution or our assumed mathematical model is a poor representation of the actual physical situation. Even when the first approximation is good, noise in the data, coupled with the granularity of the floating-point number system, may prevent convergence. And when the method does converge to a minimum of the nonlinear regression problem, the minimum may be local rather than global.

When one uses a series of masses and adopts the one that minimizes the variation of the residuals, one assumes that a unique minimum exists. Because of the nonlinearity of the problem such an assumption may be unjustified. Although O'Handley (1969) obtained a good result for Jupiter's mass, his procedure, fitting a parabola to residuals in $\Delta\alpha \cos \delta$ given by four values of Jupiter's mass and taking the parabola's minimum as the best-fit mass, is unreliable unless one knows beforehand that the mass values straddle the global minimum of the nonlinear function.

2. THE SIMPLEX METHOD

Some techniques do not require linearizing equations of condition or assuming a series of values for the desired mass. Of these non-gradient techniques for nonlinear optimization, Nelder and Mead's simplex method may be applied to the nonlinear regression problem. The method, like Dantzig's simplex algorithm for linear programming, adopts a concept from topology called a simplex. Without going into details, which can be found in Press et al. (1986) along with a FORTRAN subroutine, we can say that a simplex in a space of n-dimensions is a closed geometrical figure with n + 1 vertices, a non-degenerate polytope in mathematical terminology: in one-dimension a simplex is a line segment, in two-dimensions a triangle, in three-dimensions a tetrahedron, and so forth.

The simplex method tests the response function, the sum of the squares of the residuals for nonlinear regression problems, at each of the vertices. The worst vertex, with highest sum square of the residuals, is rejected and a new one, with better response, is substituted for it. Press et al. (1986) and Nelder and Mead's original article (1965) discuss the mechanisms, technically known as non-degeneracy conserving expansions, contractions, and reflections, for performing the substitution. Nearing the minimum the simplex shrinks in size until it becomes smaller than a tolerance specified by the user or comparable to the machine epsilon of the computer's floating-point arithmetic. To check for a global rather than a local minimum one restarts the algorithm at the minimum, but with, once again, a large initial simplex. If the minimum is global the method converges there again; otherwise, the simplex wanders off in a new direction and converges to a different minimum, where the check can be applied again.

3. AN EXAMPLE

As a test of the simplex method and to compare it with one of the standard techniques, the Gauss-Newton method, I decided to calculate Pluto's mass using the over 12,000 observations of Neptune, made between 1847 and 1970, that Jackson (1974) reduced to the system of the FK3. Seidelmann et al. (1971) employed these same observations in their determination of Pluto's mass.

The differential corrections for Pluto's mass were based on numerically integrated partial derivatives obtainable from the non-relativistic equations of motion for a planet i,

$$\ddot{\underset{\sim}{r}}_i = -k^2 (1+m) \frac{\underset{\sim}{r}_i}{r_i^3} + \sum_{j=1,\, j\neq i}^{n} k^2 m_j \left(\frac{\underset{\sim}{\rho}_i}{\rho_i^3} - \frac{\underset{\sim}{r}_j}{r_j^3} \right) . \qquad (1)$$

In Eq. (1) k^2 is the Gaussian gravitational constant, $\underset{\sim}{r}_i$ the heliocentric distance of planet i of mass m, and $\underset{\sim}{\rho}_i$ the distance between perturbed planet i and perturbing planet j. Because we are interested in inferring Pluto's mass from its perturbations on Neptune, i=8, j=9, and the required partial derivatives follow upon our differentiating Eq.(1) with respect to m_9, Pluto's mass. Partials for correcting the Earth's and Neptune's osculating, rectangular coordinates and velocities at epoch JD 2420000.5, close to the mid-point of the observational time span, are also derivable form Eq. (1). See Janiczek (1971) for details.

My n-body Cowell program to generate the coordinates of the Earth and Neptune needed to calculate the (O - C)'s uses Eq. (1) augmented by a relativistic correction term based on a one-body Schwarzschild isotropic metric. A subroutine calculates the necessary partials but aplies no relativistic correction. The program uses 13-th order predictor -corrector formulas followd by a 13-th order deferred correction for the final pass. The coordinates were checked by comparing them with coordinates interpolated from DE200 and precessed to 1950.0; the partials were checked by the usual difference quotient method. To generate genuine Earth coordinates it was necessary to carry the Moon as a separate body. The step-size w was, therefore, small, $0\overset{d}{.}25$. The program needed about seventeen hours to calculate the coordinates, velocities, and partials for the period 1847-1970.

I formed equations of condition for the thirteen unknowns, corrections to the Earth's and Neptune's osculating rectangular coordinates and velocities and the correction to Pluto's mass, and calculated the solution by the QR transformation, accumulating the equations of condition in blocks of one thousand and reducing them by sequential Householder triangularizations. At first I included additional unknowns, such as corrections to the FK3 equinox and its centennial variation, but because their inclusion resulted in a poorly conditioned matrix of the equation of condition --6.5 · 10^7 even with column scaling-- they were suppressed. With their exclusion the condition number decreased to a more satisfactory 1.3 · 10^5. Pluto's initial mass to be corrected was m_9=1/1,812,000, high compared with 1/130,000,000 but a more or less standard value before Charon's discovery and one that would have been used for a differential correction.

The first differential correction gave $m_9 = 0.93646 \cdot 10^{-6} \pm 0.93186 \cdot 10^{-7}$, with mean error of unit weight $\sigma(1) = 1''.06$, indicating that Pluto's mass should be increased to 1/671,890. A second differential correction increases the mass even more, to the absurdly high m_9 = 1/242,787. The Gauss-Newton iterates for the mass correction are apparently divergent. Whether they are diverging to infinity or wandering is unimportant. What is important is to not take a single differential correction, along with its formal mean error, as the result, as one often finds in the literature. One must demonstrate that the Gauss-Newton iterates are, in fact, converging. Otherwise, one runs the risk of taking the first term of a divergent series as a meaningful result.

To adopt the simplex method to the problem of determining Pluto's mass we need some way for m_9 to enter as a parameter to be determined. From Eq. (1) if instead of multiplying the second term on the right-hand-side by m_9 to integrate Neptune's coordinate $\underset{\sim}{r}_8$, we leave out the term in the sum corresponding to m_9 and integrate it separately, we obtain the coordinate not as a single value $\underset{\sim}{r}_8$ but as the sum of two values $\underset{\sim}{r}_{8,0} + m_9 \underset{\sim}{r}_{8,1}$. Pluto's mass now enters the problem as a free parameter. To integrate separately the term

$$k^2 \left(\frac{\underset{\sim}{\rho}_8}{\rho_8^3} - \frac{\underset{\sim}{r}_9}{r_9^3} \right)$$

we use a previously determined $\underset{\sim}{r}_8$ to calculate $\underset{\sim}{\rho}_8$.

To apply this technique we need initial values for $\underset{\sim}{r}_{8,0}$, $\dot{\underset{\sim}{r}}_{8,0}$, $\underset{\sim}{r}_{8,1}$, $\dot{\underset{\sim}{r}}_{8,1}$. For the initial $\underset{\sim}{r}_{8,0}$ and $\dot{\underset{\sim}{r}}_{8,0}$ I set Pluto's mass to zero, integrated Neptune's coordinates, and performed the differential corrections of the coordinates to give the best fit with the observations. The corrected coordinates at epoch were taken as $\underset{\sim}{r}_{8,0}$ and $\dot{\underset{\sim}{r}}_{8,0}$. For $\underset{\sim}{r}_{8,1}$ and $\dot{\underset{\sim}{r}}_{8,1}$ the process was repeated, this time using an assumed value for m_9 to obtain $\underset{\sim}{r}_8$ and $\dot{\underset{\sim}{r}}_8$. Then

$$\underset{\sim}{r}_{8,1} = (\underset{\sim}{r}_8 - \underset{\sim}{r}_{8,0})/m_9 \, , \tag{2}$$

with a similar expression for the velocities. As a check numerically integrated $\underset{\sim}{r}_8$ and $\dot{\underset{\sim}{r}}_8$ were compared with numerically integrated $\underset{\sim}{r}_{8,0} + m_9 \underset{\sim}{r}_{8,1}$ and $\dot{\underset{\sim}{r}}_{8,0} + m_9 \dot{\underset{\sim}{r}}_{8,1}$ for the entire time period. The coordinates agreed with one another within the limits of statistically accumulated round-off error.

With coordinates available, calculated right ascensions (α) and declinations (δ) are found from

$$\alpha_c = \tan^{-1}\left[(y_{8,0} + m_9\, y_{8,1} - y_3)/(x_{8,0} + m_9\, x_{8,1} - x_3)\right] \quad (3)$$

and

$$\delta_c = \tan^{-1}\left[(z_{8,0} + m_9\, z_{8,1} - z_3)/\{(x_{8,0} + m_9\, x_{8,1} - x_3)^2 + (y_{8,0} + m_9\, y_{8,1} - y_3)^2\}^{1/2}\right], \quad (4)$$

where the subscript 3 refers to the Earth. The nonlinear expression to be minimized to determine m_9 is

$$\sum_{i=1}^{m'} \cos^2\delta\, (\alpha_0 - \alpha_c)^2 + \sum_{j=1}^{m''} (\delta_0 - \delta_c)^2 = \text{min.}, \quad (5)$$

where m' is the number of equations of condition in α, m'' the number in δ, and α_0 and δ_0 are observed. (Because α_0 and δ_0 are apparent they must be corrected for aberration, precession, and nutation.)

The simplex method, applied to Eq. (5) to calculate m_9, required 4.8 hours of CPU time to find m_9=1/21,995,096, still too high but moving in the direction of lower mass. The method was restarted with this value of m_9, but did not move towards another minimum, indicating that the minimum is indeed global. The mean error of the result follows from the $\sigma(1)$ of 1"24 and equation

$$\sigma^2(\underset{\sim}{X}) = \sigma^2(1)\left[J(\underset{\sim}{X})^T \cdot J(\underset{\sim}{X})\right]^{-1}, \quad (6)$$

where $\sigma^2(\underset{\sim}{X})$ is the vector of the dispersions of the solution and $J(\underset{\sim}{X})$ is the Jacobian of the response function, Eq. (5), evaluated at the minimum. With Eq. (6) we calculate $m_9 = 0.45465 \cdot 10^{-7} \pm 0.24375 \cdot 10^{-7}$, giving a range of m_9=1/47,000,000 to m_9=1/14,000,000.

As a further test of the simplex method I included four additional unknowns, corrections and centennial variations to the FK3 equinox and equator. After thirty hours of CPU time the method still had not converged to the specified tolerance, but was nevertheless close to a solution. The additional unknowns were being poorly determined and their effect on m_9 was parasitic. With neither linearized equations of condition nor a nonlinear method can we obtain reliable equinox and equator determinations from 123 years of observations of Neptune, which do not even cover one period of the planet.

4. CONCLUSIONS

For large problems the simplex method is extremely slow, but produces a solution when the Gauss-Newton method fails. Furthermore, the method can be made to converge to a global minimum. Even when the Gauss -Newton method converges, it may converge to a local minimum. Eq. (5), derived from the usual expression that relates spherical and rectangular coordinates that one differentiates to obtain linearized equations of condition, is clearly bimodal. Furthermore, because m_9 enters Eq. (5) via the complicated trigonometric expressions of Eqns. (3) and (4), Eq. (5) likely possesses many local minima with respect to m_9. Unless the first approximation is close to the true solution, the Gauss-Newton method, when it converges, could easily converge to an extraneous minimum. For refractory problems it may be worth-while to pay the computational price and use the simplex method.

This research was supported by Grant E 1501 of the Consejo Nacional de Investigaciones Científicas y Técnicas (CONICET).

REFERENCES

Branham, R.L. Jr. (1979) 'The Orbits of Five Minor Planets and corrections to the FK4 Equator and Equinox', Astron. Papers Amer. Ephem. 21, Part 3.

Jackson, E.S. (1974) 'A Discussion of the Observations of Neptune', Astron. Papers Amer. Ephem. 22, Part 2.

Janiczek, P.M. (1971) 'The Orbit of Polyhymnia and the Mass of Jupiter', Astron. Papers Amer. Ephem. 21, Part 1.

Nelder, J.A., and Mead, R. (1965) 'A Simplex Method for Function Minimization', The Computer Journal 7, 308-314.

O'Handley, D.A. (1969) 'Determination of the Mass of Jupiter from the Motion of 65 Cybele', Astron. Papers Amer. Ephem. 20, Part 3.

Press, W.H., Flannery, B.P., Teukolsky, S.A., and Vetterling, W.L. (1986) Numerical Recipes: the Art of Scientific Computing, Cambridge U. Press, Cambridge, Sec. 10.4.

Seidelmann, P.K., Klepczynski, W.J., Duncombe, R.L., and Jackson, E.S. (1971) 'Determination of the Mass of Pluto', Astron. J. 76, 488-492.

AN OPTIMUM METHOD FOR CALCULATING RESTRICTED THREE-BODY ORBITS

MARC A. MURISON

Dept. of Astronomy
Univ. Wisconsin-Madison
475 N. Charter St.
Madison, WI 53706
USA

ABSTRACT. The restricted three-body problem (RTBP) has in the past played an essential role in many different areas of dynamical astronomy, and indications are that this will continue. As the state of the art in computing becomes more advanced, larger numbers of integrations and longer durations are attempted. Thus, computational efficiency and accuracy are becoming more important. Also, the use of the RTBP in many different areas leads to the desire for a general integration method.

In order to maximize the efficiency of orbit calculations, comparisons are made of different methods of integration. The results can be summarized as follows: 1. The Bulirsch-Stoer extrapolation method is extremely fast and accurate, and is the method of choice. 2. Regularization of the equations of motion is essential. 3. When applicable, a manifold correction algorithm, originally due to Nacozy (1971), reduces numerical errors to the limits of machine accuracy, and at a cost of only 1 to 3 percent in cpu time.

1. Qualities of B-S Integration, Regularization, and Manifold Correction

There are three ways to improve speed and accuracy in integrating the restricted three-body equations. First, use a fast and efficient algorithm. The Bulirsch-Stoer (B-S) extrapolation method (see Press *et al.*, 1986) is a very fast and efficient algorithm for solving ordinary differential equations. Second, regularize the equations of motion. In the following, two-body regularization was incorporated (see, *e.g.*, Stiefel and Scheifele 1970, Bettis and Szebehely 1971, Murison 1988). Third, take advantage of integrals of the motion for error checking and correction.

Nacozy (1971) found a clever way to use constants of the motion to monitor and correct errors in position and velocity. When an integral exists, the motion is confined to a surface in the phase space. When a numerical error occurs, the motion jumps to a different surface. Nacozy's correction scheme uses Lagrange multipliers to find the least-squares shortest path back to the original manifold. Murison (1988) has applied this manifold correction to both the unregularized and regularized restricted three-body equations.

To illustrate the usefulness of these techniques, the orbit shown in figure 1 was integrated several different ways. This orbit is difficult,

Celestial Mechanics **45**: 175–180, 1989.

in the sense that it has many close approaches to the primary m2. For this illustration, the integration continued for 10 orbital periods of the primaries. During the integration, the error in the Jacobi constant C was monitored.

Figure 2 shows the error in C during an integration using an efficient, fourth-order Runga-Kutta method with variable step size (the curve labeled "RK"). Notice the steady increase of error with time. The CPU time needed by a VAX 11/780 was 617 seconds. This figure also shows the error when the B-S method was used. Discrete jumps, caused by close approaches, are larger. But the CPU time was a factor of 10 less (61.6 sec).

Figure 3 shows the effect of manifold correction. The B-S method was used again. Notice the drastic collapse of the errors. Yet the increased cost in CPU time was only 2.8 percent (to 63.3 sec)!

To illustrate the effect of regularization, figure 4 shows the result of using the B-S method with regularized equations of motion (no manifold correction). This is the curve labeled "BS+R". Note that *the vertical scale has been magnified by a factor of 1000.* The errors are incredibly small, compared to the previous figures. This is a result of trying to keep the amount of machine effort about the same (CPU time was 57.5 sec).

Finally, figure 4 also shows the result of combining the B-S method, regularization, and manifold correction (the curve labeled "BS+MC+R"). The CPU time was 59.0 sec. The errors have been reduced to effectively nothing (!), with a net *savings* of CPU time, compared to the B-S method alone (figure 2).

Thus, we conclude that the B-S method of integration is very fast, and, when combined with regularization and especially manifold correction, extremely accurate. We also note the both importance of regularization and the small cost of manifold correction.

2. Backwards Integration Check

As a check on the B-S integrator, an orbit very similar to the one of figure 1, but with even closer approaches to m2, was integrated forward and then backward in time. The errors in the final values of x, y, x′, and y′ were noted as a function of length and accuracy of the integration. Both normal and regularized equations were used. Figure 5 shows the distance from m2 as a function of time for this orbit (escape occurs at about T = 22.8 primary periods). One can see that close approaches are frequent. When regularization was turned on, the integration switched to regularized equations when $r2 \le 0.11$. This limit is shown in figure 5. Thus, most of the orbit was regularized when regularization was turned on.

The difference between the initial and final x coordinate is a good indicator of how well the integrator is able to retrace the orbit backwards. The log of Δx is plotted as a function of forward integration time in figure 6. The B-S accuracy parameter (see Press *et al.*, 1986) is denoted by ε. One sees that, for this particular orbit, (1) regularization improves the accuracy by a factor of from 30 to over

1000, (2) for moderate integration times $T \simeq 50$ the orbit is believable *in detail* (not just in a statistical sense, as is commonly the case in these kinds of calculations), and (3) values of ε smaller than $\sim 10^{-12}$ are required.

3. Application to Satellite Capture

The ability to integrate the restricted three-body equations quickly and accurately enables us to explore numerically the restricted three-body problem to a much further and more general extent than has previously been possible. Murison (1988) contains examples of such extensive calculations, in the framework of satellite capture.

4. Conclusions

A. The B-S extrapolation method by itself is very fast, and fairly accurate.
B. Regularization of the equations of motion is essential.
C. Manifold correction, when applicable, reduces the errors to effectively zero and is very inexpensive.
D. The combination of the above three allows one to make extensive numerical explorations which were previously not feasible.

These methods should find application to systems other than the restricted three-body equations as well.

References

Bettis D.G. and Szebehely V. (1971). *Astrophys. Space Sci.* **14**, 133.
Hénon M. (1969). *Quart. Appl. Math.* **27**, 291.
Huang T.-Y. and Innanen K.A. (1983). *Astron. J.* **88**, 1537.
Murison M.A. (1988). Ph.D. Thesis, Univ. of Wisconsin-Madison.
Nacozy P.E. (1971). *Astrophys. Space Sci.* **14**, 40.
Press W.H., Flannery B.P., Teukolsky S.A. and Vetterling W.T. (1986). *Numerical Recipes: The Art of Scientific Computing.* Cambridge Univ. Press.
Stiefel E. and Scheifele G. (1970). *Linear and Regular Celestial Mechanics.* Springer-Verlag.

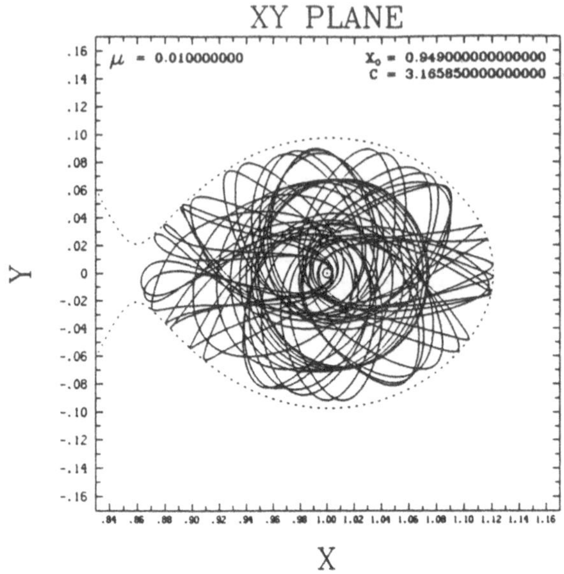

Figure 1. Orbit used for comparison of integration methods. The smaller primary m2 is located at (x,y) = (1.0,0.0). The dashed curve is the zero-velocity curve for this orbit. Notice several close approaches to m2.

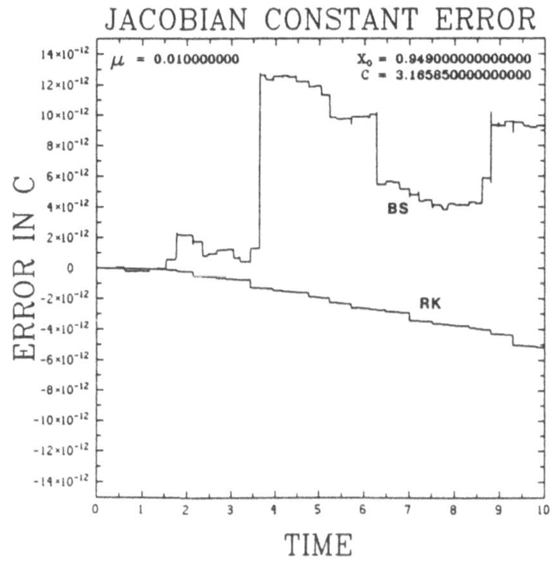

Figure 2. Error in Jacobi constant C vs. time in units of primary orbital periods. BS = Bulirsch-Stoer, RK = fourth-order Runga-Kutta. The RK calculation took ten times more cpu time than the BS calculation.

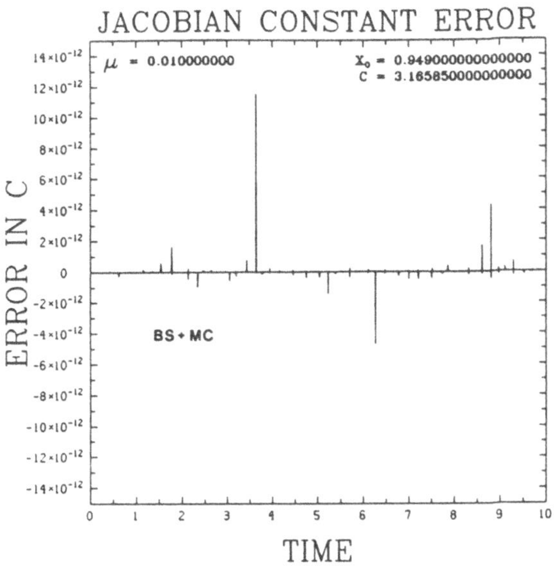

Figure 3. Error in Jacobi constant C vs. time. Bulirsch-Stoer, with manifold correction. Note drastic reduction of errors (compare with BS curve of figure 2).

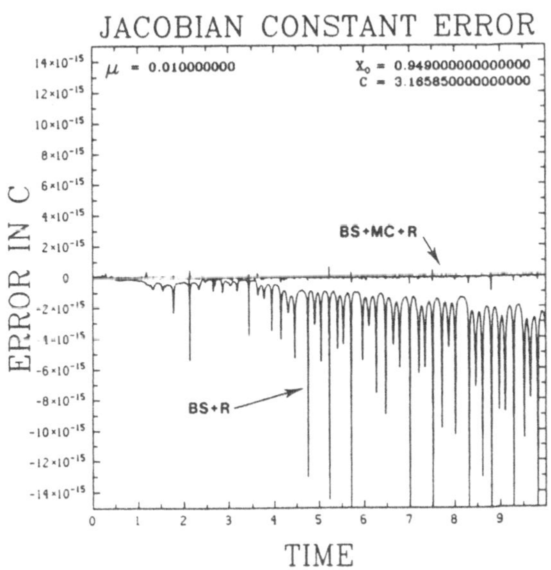

Figure 4. Error in Jacobi constant C vs. time. Curve labeled BS+R is Bulirsch-Stoer and regularized equations. BS+MC+R is combination of Bulirsch-Stoer, manifold correction, and regularization. Notice factor of 1000 change in scale.

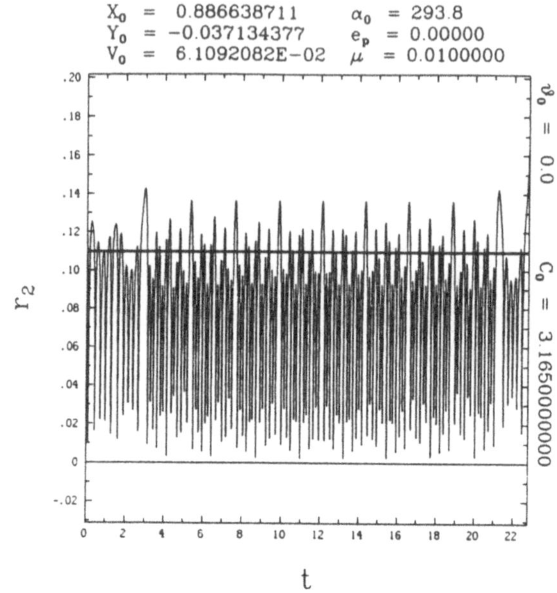

Figure 5. Distance from m2 vs. time for orbit used in backward integration check. Solid line at r2 = 0.11 shows limit below which regularization was in effect during the "regularized" calculation (see text). Note the several very close approaches to m2.

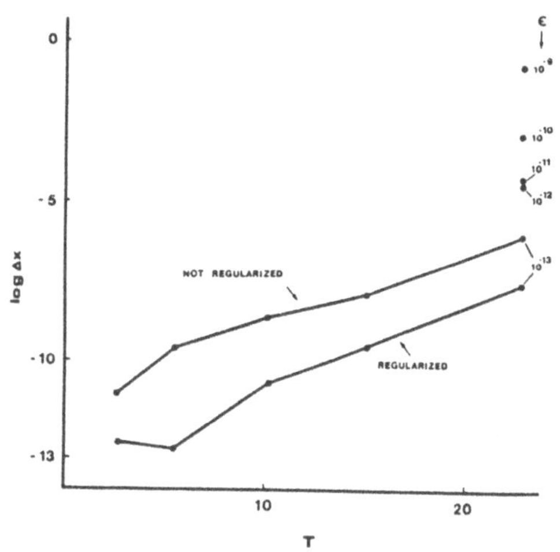

Figure 6. Log of difference in x position (Log Δx) between start and end of backwards integration check, as a function of forward integration time. Curves are shown for regularized and unregularized calculations with $\varepsilon = 10^{-13}$. Other points for different ε, at T = 22.8, represent unregularized calculations.

THE CONSTRAINED NORMAL FORM ALGORITHM

Richard Cushman
Mathematics Institute, Rijskuniversiteit Utrecht
3508TA Utrecht, the Netherlands

and

Jan A. Sanders
Department of Mathematics and Computer Science, Vrije Universiteit
1067MC Amsterdam, the Netherlands

Abstract

This paper describes an algorithm which brings a regularizable polynomial perturbation of a three degree of freedom Kepler problem into a normal form which Poisson commutes with the Kepler Hamiltonian. We illustrate the alogrithm with an example: the quadratic Zeeman effect. For other applications of this algorithm see [1],[4], and [5]. The authors have written a program in MAPLE which implements the constrained normal form.

1. The Quadratic Zeeman Effect

The Zeeman effect describes the splitting of the spectral lines of the hydrogen atom under the influence of an axial magnetic field of moderate strength. For very high strength fields an extra quadratic nonlinear term is needed. The classical analog of the quadratic Zeeman effect (see [2], [3]) is described by the Hamiltonian

$$F(x,y) = \frac{1}{2}|y|^2 - \frac{\mu}{|x|} - \epsilon\langle e_3, x \times y\rangle + \frac{1}{2}\epsilon^2 |e_3 \times x|^2 \tag{1}$$

where

$$(x,y) \in (\mathbf{R}^3 - \{0\}) \times \mathbf{R}^3 = T_0\mathbf{R}^3$$

has the standard symplectic form $\omega = \sum_{i=1}^{3} dx_1 \wedge dy_i$ and ϵ is small. Here $\langle , \rangle$ is the Euclidian inner product on $\mathbf{R}^3$ with norm $|\,|$ and $\times$ is the vector product. In rotating co-ordinates $(x,y) = (R_{\epsilon t}\xi, R_{\epsilon t}\eta)$ where R_s is rotation about the e_3 axis through an angle s, F becomes

$$K(\xi,\eta) = \frac{1}{2}|\eta|^2 - \frac{\mu}{|\xi|} + \frac{1}{2}\epsilon^2({\xi_1}^2 + {\xi_2}^2). \tag{2}$$

Our goal is to compute the constrained normal form of K through terms of order ϵ^2. We follow [4]. First we write $(K, T_0\mathbf{R}^3, \omega)$ as a perturbation of the Hamiltonian $G_0(q,p) = |p|$ for the geodesic flow on $T^+S^3 = \{(q,p) \in$

Celestial Mechanics **45**: 181–187, 1989.

$T\mathbf{R}^4|\ |q|^2 = 1, \langle q,p\rangle = 0, p \neq 0\}$ (which is the tangent bundle of the 3-sphere minus the zero section). We begin by restricting K to the level set $K^{-1}(-\frac{1}{2}k^2)$ and changing time scale by

$$\frac{ds}{dt} = \frac{k}{|x|}.$$

Then we apply Moser's mapping $\mathcal{M} : T^+(S^3 - np) \longrightarrow T\mathbf{R} : (q,p) \longrightarrow (\xi,\eta)$ where $np = (0,0,0,1)$ and

$$\xi_i = -\frac{1}{k}(p_i + S_{i4})$$
$$\eta_i = \frac{kq_i}{1-q_4} \qquad \text{for } i = 1,\ldots,4.$$

Here $S_{ij} = q_i p_j - q_j p_i$ for $1 \leq i,j \leq 4$. In fact $\mathcal{M}$ is the tangent of stereographic projection from the north pole $(0,0,0,1)$ of S^3 followed by interchanging positions and the negative of momenta. Thus we obtain the Hamiltonian

$$\begin{aligned} G(q,p) &= |p| + \frac{1}{2k^4}\epsilon^2|p|(1-q_4)\left(p_1^2 + p_2^2 + 2(p_1S_{14} + p_2S_{24}) + S_{14}^2 + S_{24}^2\right) \\ &= G_0(q,p) + \frac{1}{2}\epsilon^2 G_1(q,p). \end{aligned}$$

Since G is defined on <u>all</u> of T^+S^3 and not just on $T^+(S^3 - \{(0,0,0,1)\})$, which is the domain of Moser's mapping, we have regularized the collisions in (2). Next we extend G to a smooth function

$$\begin{aligned} H(q,p) &= \left(|q|^2|p|^2 - \langle q,p\rangle^2\right)^{\frac{1}{2}} + \frac{1}{2}\epsilon^2 G_1(q,p) \\ &= H_0(q,p) + \frac{1}{2}\epsilon^2 H_1(q,p) \end{aligned} \tag{3}$$

on $T\mathbf{R}^4 - \{H_0 = 0\}$. Note that $H|T^+S^3 = G$ and T^+S^3 is an invariant manifold of X_{H_0}. Since all the orbits of the flow φ_t of X_{H_0} on T^+S^3 are periodic of period π, we may apply the constrained normal form algorithm to H (see [4] for more details). Up through terms of order ϵ^2, the normal of H <u>on T^+S^3</u> is

$$\mathcal{H} = G_0(q,p) + \frac{1}{2}\epsilon^2\overline{H}_1(q,p)$$

where $\overline{H}_1$ is the average of H_1 over the orbits of X_{H_0}. Finally to compute $\overline{H}_1$ we use the following facts:

(a). The functions $|q|^2, |p|^2, \langle q,p\rangle, S_{ij}$ for $1 \leq i,j \leq 4$ are integrals of X_{H_0} and hence are equal to their average over the orbits of X_{H_0};

(b). All odd degree monomials in q and p have zero average.
We obtain

$$\overline{H}_1 = \frac{1}{k^4}|p|\left(\overline{p_1^2} + \overline{p_2^2} + S_{14}^2 + S_{24}^2 - 2\overline{q_4 p_1} S_{14} - 2\overline{q_4 p_2} S_{24}\right) \tag{4}$$

Because

$$2\overline{q_i p_j} \simeq S_{ij} \quad \text{and} \quad 2\overline{p_i^2} \simeq \sum_{\ell=1}^{4} S_{\ell i}^2$$

where two smooth functions f, g are $\simeq$ if their difference when restricted to T^+S^3 is zero, we get

$$\overline{H}_1 \simeq \frac{1}{2k^4}|p|\left(2S_{12}^2 + (S_{13}^2 + S_{23}^2) + 5(S_{14}^2 + S_{24}^2)\right)$$

which completes the calculation of the constrained normal form of $\mathcal{H}$. $\mathcal{H}$ is Liouville integrable with Poisson commuting integrals $\mathcal{H}$,$|p|$ and S_{12}.

Using the reduction process to remove the S^1 symmetries generated by X_{G_0} and $X_{S_{12}}$ (see [1] for more details) we obtain for $G_0 = \ell(=\frac{\mu}{k})$ and $S_{12} = c$ the second reduced phase space $P_{\ell,c}$:

$$\pi_2{}^2 + \pi_3{}^2 = (\ell^2 - \pi_1{}^2)(\ell^2 - (2c - \pi_1)^2), \; \pi_1 \in \begin{cases} [-\ell + 2c, \ell], & \text{if } 0 < c < \ell; \\ [-\ell, \ell + 2c], & \text{if } -\ell < c < 0; \\ [-\ell, \ell], & \text{if c} = 0. \end{cases}$$

When $|c| < \ell$, $P_{\ell,c}$ is diffeomorphic to a 2-sphere;while when $c = 0$, it is only homeomorphic to a 2-sphere (having two cone-like singularities at $(\pi_1, \pi_2, \pi_3) = (\pm\ell, 0, 0)$.The second reduced Hamiltonian on $P_{\ell,c}$ is

$$H_{\ell c} = 3(\pi_1 - c)^2 + 2\pi_2.$$

Following the analysis in [5] we obtain

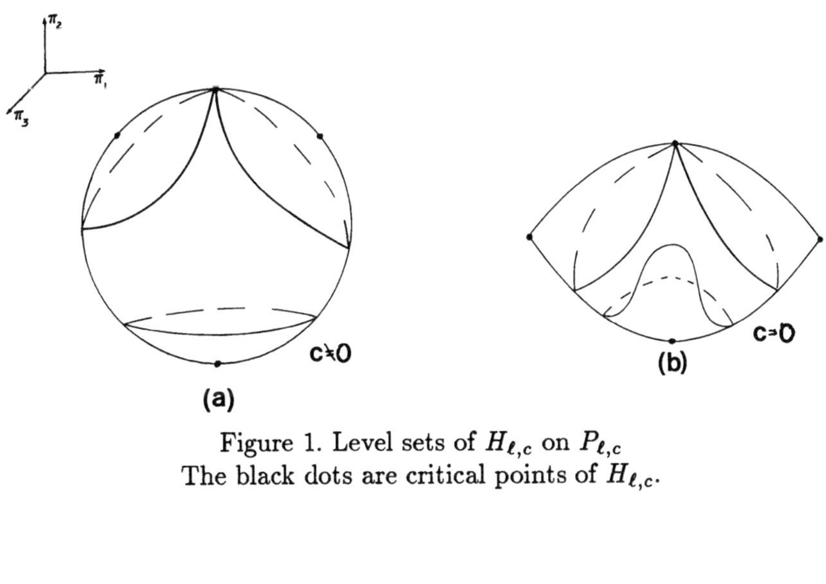

Figure 1. Level sets of $H_{\ell,c}$ on $P_{\ell,c}$
The black dots are critical points of $H_{\ell,c}$.

The critical points of $H_{\ell,0}$, which are singular points of $P_{\ell,0}$, correspond to (elliptic) periodic orbits Γ of $X_{\mathcal{H}}$ on T^+S^3,which are limits of two dimensional tori $T^2_{\ell,c}$ (as $c \to 0$) corresponding to the symmetric pair of critical points of $H_{\ell,c}$. These tori are invariant under the flows of the vectorfields X_{G_0} and $X_{S_{12}}$ (and hence $X_{\mathcal{H}}$). Thus they are filled up with periodic orbits. Also Γ are limits $\big($as $h = H_{\ell,0} \to 3(\ell - c)^2\big)$ of three dimensional tori which are invariant under $X_{\mathcal{H}}$. (This can be read off from figure 1(b)). These orbits Γ cannot be found in [3].

2.Constrained normal form:Abstract theory

We now turn to discussing the basic mathematical structure behind the constrained normalization algorithm: namely, the Poisson algebra of formal series. On $\mathbf{R}^{12}$ we have the co-ordinates $x = (x_1, x_2, ..., x_{12}) = (\tilde{x}, x_{11}, x_{12})$ where in Table 1 we give the correspondence with the variables of the example in §1.

x_1	x_2	x_3	x_4	x_5	x_6	x_7	x_8	x_9	x_{10}	x_{11}	x_{12}
q_1	q_2	q_3	q_4	p_1	p_2	p_3	p_4	$\lvert q\rvert^2 - 1$	$\langle q, p\rangle$	H_0	$\lvert p\rvert$

Table 1. Correspondence of variables.

Let $\mathcal{F}$ be the set of formal series $f = \sum_{\ell \geq 0} \epsilon^\ell f_\ell$ where f_ℓ is a finite sum of monomials $\tilde{x}^i x_{11}^{j_1} x_{12}^{j_2} k^{j_3}$ and the last three components of the exponent (i, j) may be negative integers. Defining multiplication of monomials by addition of exponents and extending this by linearity gives a multiplication $\bullet$ on $\mathcal{F}$. Thus $(\mathcal{F}, \bullet)$ is an algebra. Define a Poisson bracket $\{\ \}$ on $\mathcal{F}$ by

$$\{f, g\} = \sum_{1 \leq i < j \leq 12} \frac{\partial f}{\partial x_i} \frac{\partial g}{\partial x_j} W_{ij}$$

where $(W_{ij}) = (\{x_i, x_j\})$ is an antisymmetric matrix. Here $\{x_i, x_j\}$ $1 \leq i, j \leq 8$ is given by the standard Poisson bracket on $T\mathbf{R}^4$ with co-ordinates $(x_1, \ldots, x_8)$ and the other entries in W are given by computing the usual Poisson bracket on $T\mathbf{R}^4$ with the functions

$$\begin{aligned} x_9 &= x_1^2 + x_2^2 + x_3^2 + x_4^2, \\ x_{10} &= x_1 x_5 + x_2 x_6 + x_3 x_7 + x_4 x_8, \\ x_{11}^2 &= x_{12}^2(1 + x_9) - x_{10}^2, \\ x_{12}^2 &= x_5^2 + x_6^2 + x_7^2 + x_8^2. \end{aligned}$$

Since $(\mathcal{F}, \{\ \})$ is a Lie algebra and

$$\{f, g \bullet h\} = \{f, g\} \bullet h + g \bullet \{f, h\} \quad f, g, h \in \mathcal{F},$$

$(\mathcal{F}, \bullet, \{\ \})$ is a Poisson algebra.

Let $\mathcal{G}$ be the subset of $\mathcal{F}$ with $f_0 = x_{11}$. The flow of the Hamiltonian vectorfield $X_{x_{11}} = ad_{x_{11}} = \{x_{11}, \cdot\}$ associated to x_{11} is $(exp\ tad_{x_{11}})x = \varphi_t(x) = \xi$. From the definition of W, we find that

$$\xi_i = -x_i x_{10} x_{11}^{-1} \sin 2t + x_i \cos 2t + x_{4+i}(x_9 + 1)x_{11}^{-1} \sin 2t$$
$$\xi_{4+i} = -x_i x_{12}^2 x_{11}^{-1} \sin 2t + x_{4+i} x_{10} x_{11}^{-1} \sin 2t + x_{4+i} \cos 2t$$

for $1 \leq i \leq 4$, and

$$\xi_9 = x_9, \quad \xi_{10} = x_{10}, \quad \xi_{11} = x_{11}, \quad \xi_{12} = x_{12}.$$

Clearly φ_t leaves the constraint

$$\mathcal{M} : x_9 = x_{10} = 0 \text{ and } x_{11} = x_{12}$$

invariant. $f \in \mathcal{G}$ is in <u>constrained normal form</u> if

$$\{x_{11}, f\}|\mathcal{M} = 0.$$

3.Constrained normal form:Algorithm

The constrained normal form algorithm is a process for finding a formal series $R \in \mathcal{F}$ such that the automorphism $\varphi_R = exp\ ad_R$ of the Poisson algebra $(\mathcal{F}, \bullet, \{\})$ leaves $\mathcal{M}$ invariant and transforms a given formal series into constrained normal form.

The algorithm is described as follows:

(0) <u>Input data</u>: A finite formal series $f \in \mathcal{G}$; a fixed integer M (the order of normalization).

(1) <u>Initialization</u>: $\ell := 1$, $NF := 0$, $R := 0$.

(2) <u>Find Coefficient</u>: f_ℓ of ϵ^ℓ in f.

(3) <u>Restrict f_ℓ to $\mathcal{M}$</u>: Set $x_9 = 0, x_{10} = 0, x_{11} = x_{12}$ in f_ℓ.

(4) <u>Split f_ℓ</u>: Compute

$$\bar{f}_\ell := \frac{1}{\pi} \int_0^\pi f \circ \varphi_t \, dt$$

and

$$R := \frac{1}{\pi} \int_0^\pi t(f_\ell - \bar{f}_\ell) \circ \varphi_t \, dt$$

(5) <u>Modify R</u>: $R^* := R - \frac{1}{2}\{R, x_{10}\}x_9 + \{R, x_9\}x_{10}$.

(6) <u>Update</u> NF and R: $NF := NF + \epsilon^\ell f_\ell$, $R := R + R^* \epsilon^\ell$.

(7) <u>Increment</u> ℓ: $\ell := \ell + 1$.

(8) Test : If $\ell > M$, then output NF (the constrained normal form to order M),and R (the generating function of the normal form to order M);otherwise, compute

$$f := \sum_{0 \leq j \leq \left[\frac{M}{\ell}\right]} \frac{1}{j!} ad^j_{\epsilon^\ell R^*} f.$$

(9) Return to step 2.

We now discuss the above algorithm. Only steps 4,5 and 8 need further explanation, the others being entirely straightforward. In step 4 the integrals give no trouble since the integrand of $\overline{f_\ell}$ is a finite series in $\sin 2t$ and $\cos 2t$, while the integrand of R is a finite series in $t \sin 2t$ and $t \cos 2t$. In step 5 the formal series R is modified to R^* so that $\mathcal{M}$ is an invariant manifold of X_{R^*}, that is, $\{R^*, x_9\}|\mathcal{M} = \{R^*, x_{10}\}|\mathcal{M} = 0$. Thus $exp\, \epsilon^\ell ad_{R^*}$ maps $\mathcal{M}$ into itself. In step 8 we have replaced f by the terms up through order M in $(exp\, \epsilon^\ell ad_{R^*})f$.

Actually step 3 can be refined by first straightening f_ℓ with respect to the Gröbner bases

$$\{x_1^2 + x_2^2 + x_3^2 + x_4^2 - 1 - x_9, x_1x_5 + x_2x_6 + x_3x_7 + x_4x_8 - x_{10}\}$$

of the ideal defining T^+S^3. Here we use the lexicographic ordering on monomials induced by $1 < x_{12} < x_{11} < ... < x_1$ (see [6] for details on the notion of straightening).

The only point remaining to be discussed is: how do we write NF as a formal series in $k, x_{12}(= |p|)$ and $x_i x_{4+j} - x_j x_{4+i}(= S_{ij})$ for $1 \leq i,j \leq 4$? We answer this by noting that $\varphi_t|\mathcal{M}$ is the flow of the harmonic oscillator on $T\mathbf{R}^4$ which in complex conjugate co-ordinates reads:

$$\begin{aligned}\zeta_j &= \xi_{11}\xi_j + i\xi_{4+j} \\ &= e^{-2it}(x_{11}x_j + ix_{4+j}) = e^{-2it}z_j\end{aligned}$$

$j = 1,2,3,4$. Since NF is invariant under $\varphi_t|\mathcal{M}$, NF is a formal power series in the monomials $z^j \bar{z}^\ell$ with $|j| = |\ell|$. Because every such monomial may be written as a product of quadratic monomials of the form $z_r \bar{z}_s$ for some $1 \leq r,s \leq 4$ (see [7]) and

$$\begin{aligned}z_r\bar{z}_s &= (x_{11}^2 x_r x_s + x_{4+r}x_{4+s}) + ix_{11}(x_{4+r}x_s - x_r x_{4+s}) \\ &\simeq \sum_{\ell=1}^{4} S_{\ell r}S_{\ell s} + i|p|S_{sr},\end{aligned}$$

we can write NF as a power series in S_{ij} with coefficients which are formal series in $|p|$ and k.

References

[1] Cushman,R. *A survey of normalization techniques applied to perturbed Keplerian systems,* preprint # 496, Rijksuniversiteit Utrecht,1988.

[2] Reinhardt,W. and Farelly,D.,*The quadratic Zeeman effect in hydrogen,* J.de Physique, colloquium # 2, Supplement to vol.43,no.11(1982),29-43.

[3] Deprit,A.,*The quadratic Zeeman effect,*(to appear in Ann. N.Y. Acad. Sci.).

[4] J.-C. van der Meer and Cushman,R.,*Constrained normalization Hamiltonian systems and perturbed Keplerian motion,* J.Appl. Math. and Phys. (ZAMP) 37 (1986),402-424.

[5] J.-C. van der Meer and Cushman,R.,*Orbiting dust under radiation pressure,* in Proc XV conf. on differential geometric methods in theoretical physics,eds. H. Doebner and J.D. Henning, World Scientific,Singapore, 1987,403-414.

[6] Sturmfels,B. and White,N.,*Combinatorial decompositions of rings,*preprint, Institute of Mathematics and its applications,1988.

[7] Billera,L.,Cushman,R.,and Sanders,J.,*The Stanley decomposition of the harmonic oscillator,*preprint, Vrije Universiteit, Amsterdam,1987.

Simplify or Perish

André Deprit and Bruce Miller
National Institute of Standards and Technology
Gaithersburg, MD 20899, U. S. A.

Abstract

Three suggestions are made for breaking the gridlock which paralyses progress in automating massive symbolic calculations in celestial mechanics:(i) simplifications by canonical transformations, (ii) object oriented programming to endow the algebra of Poisson series with various structures, (iii) massively parallel processing.

Introduction

Old as it may be and out of fashion, celestial mechanics has never ceased challenging computational astronomers to meet its combinatorial complexities.

At the core of the matter is Kepler's equation which cannot be solved explicitly to produce the eccentric anomaly E as a finite symbolic expression in terms of the mean anomaly ℓ. Yet, would it be possible to make do altogether without resorting to expansions in the powers of the eccentricity? Excessive as it may seem at first sight, the question is worth examining. Most recently, the zonal problem in artificial satellite theory has been solved to the third order, without introducing power series in the satellite's eccentricity, not even while averaging over the mean anomaly. Whether the solution of the main problem of lunar theory can be expressed likewise without expanding in powers of the orbital eccentricities of sun and moon is an open question. Should it be answered in the positive, tremendous simplifications would ensue that would make feasible a solution to the gravitational part of the problem accurate to 1 cm., as of late the precision of laser ranging. Pending a resolution of this issue, one could think of carrying the solution in finite terms by successive canonical transformations to a point where development in powers of the eccentricities is likely be of tractable size. This strategy is discussed in Section 1.

Celestial Mechanics **45**: 189–200, 1989.

Complications in the symbolic manipulations of celestial mechanics arise also from the large number of polynomial and trigonometric arguments entering the expressions. An analytical solution for a solar system made of 9 bodies introduces no less than 54 arguments. Current symbolic processors handle these expressions as *flat* Poisson series. Yet much is to be gained from imposing hierarchical structures on the set of Poisson series. In Section 2 we examine how this has been achieved in the framework of abstract programming.

Basic operations in the algebra of Poisson series are of a parallel nature. So far, however, they have been programmed in a serial mode on computers with a serial architecture. But new types of computers have appeared, in which a single instruction broadcast through a network of arithmetic units is executed in parallel on multiple data. Will SIMD machines accelerate development in lunar and planetary theories? Preliminary attempts at reconstructing Cayley's Tables on a massively parallel computer (Section 3) invite one to believe, if one is so inclined, that the new equipment will help break the gridlock which today paralyses progress toward automating celestial mechanics.

Canonical simplifications[1]

What will be said here concerns perturbed Keplerian systems represented by Hamiltonians of the form

$$\mathcal{H} = \mathcal{H}_0 + \mathcal{R}(\boldsymbol{x}, \boldsymbol{X}), \tag{1}$$

where $\mathcal{H}_0 = \frac{1}{2}\|\boldsymbol{X}\|^2 - \mu/\|\boldsymbol{x}\|$. At the heart of General Perturbation Method applied to these Hamiltonians is the Lie derivative

$$L_0 : F \mapsto (F, \mathcal{H}_0) = \nabla_{\boldsymbol{x}} F \cdot \nabla_{\boldsymbol{X}} \mathcal{H}_0 - \nabla_{\boldsymbol{X}} F \cdot \nabla_{\boldsymbol{x}} \mathcal{H}_0.$$

The *kernel* of L_0 is the algebra $\ker(L_0)$ of functions F such that $L_0(F) = 0$; the *image* of L_0 is the set $\mathrm{im}(L_0)$ of functions F such that $F = L_0(G)$ for some $G \in A$. A Delaunay normalization of (1) is a canonical transformation

$$\chi : (\boldsymbol{y}, \boldsymbol{Y}) \mapsto (\boldsymbol{x}, \boldsymbol{X})$$

[1]Professor Sebastián Ferrer (Universidad de Zaragoza) and Professor Carol Williams (University of South Florida) helped considerably in developing the concept of canonical simplification. Joint papers on the topic are in preparation for publication

which maps $\mathcal{H}$ into $\chi^*\mathcal{H}$ in $\ker(L_0)$, thereby making of $\chi^*\mathcal{H}_0$ an integral of the transformed system, and reducing it by one degree of freedom.

The simplest examples of Delaunay transformations arise when $\mathcal{R}$ belongs to the $\ker(L_0)$-algebra A of functions periodic in the mean anomaly. Expansion in Fourier series

$$F = \sum_{j\geq 0}(C_j \cos j\ell + S_j \sin j\ell), \qquad \text{with } C_j, S_j \in \ker(L_0)$$

for any element F in A shows that A is the direct sum

$$A = \ker(L_0|A) \oplus \operatorname{im}(L_0|A). \tag{2}$$

Accordingly $\mathcal{R}$ is normalized by taking

$$\chi^*\mathcal{R} = \langle\mathcal{R}\rangle \tag{3}$$

whereas the normalization χ is built as a Lie transformation whose generator $\mathcal{W}$ is a solution of the partial differential equation

$$L_0(\mathcal{W}) = \mathcal{R} - \langle\mathcal{R}\rangle. \tag{4}$$

However trivial in principle; the scheme is often too costly to be practical. In elementary problems, provided the construction does not go past the first order, general purpose processors like Macsyma, Maple, ScratchPad or Mathematica have proved capable of evaluating the average (3) and of solving the PDE (4). Elsewhere, though, there remains to expand $\mathcal{R}$ in powers of the eccentricity, assuredly a simple operation for current Keplerian processors, but at the cost of a dramatic surge in complexity, especially at the very high orders contemplated in lunar and planetary theories.

We propose to meet this difficulty in stages by decomposing the algebra A in ways less radical than the direct sum (2). Basically we shall be looking for ways of

- identifying a submodule A' of A containing $\mathcal{R}$;
- decomposing A' into a sum of the form $A' = B + C$ where $B \subseteq \operatorname{im}(L_0)$ and $A' \cap \ker(L_0) \subseteq C$, B and C being submodules of A'. The submodule B is chosen so as to make it easy to build the counterimage of any function in B.

In that framework, a ***simplification*** is defined as a canonical transformation $\chi : (\boldsymbol{y}, \boldsymbol{Y}) \mapsto (\boldsymbol{x}, \boldsymbol{X})$ changing the Hamiltonian (1) into a function $\chi^*\mathcal{H} = \mathcal{H}_0(\boldsymbol{y}, \boldsymbol{Y}) + \mathcal{S}(\boldsymbol{y}, \boldsymbol{Y})$ such that $\mathcal{S} \in C$. It aims at embedding the perturbation function into a set of functions where it is hoped normalization will be easier to execute than in the original set.

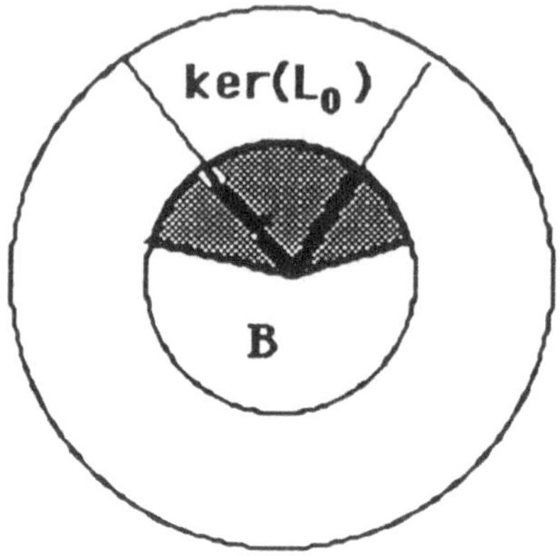

Figure 1: Algebraic effect of a simplification. The outer circle represents the algebra A of periodic functions, the inner circle, a submodule A' of A containing the perturbation; B is the submodule excised from A' by the simplification.

The concept originated in the zonal problem of artificial satellite theory. The perturbation function, which is there the series

$$\mathcal{R} = \frac{\mu}{r} \sum_{n \geq 2} J_n \left(\frac{\alpha}{r}\right)^n P_n \left(\frac{z}{r}\right), \tag{5}$$

belongs to the $\ker(L_0)$-algebra A' of functions of the type

$$F = \frac{1}{r^2} \sum_{j \geq 0} (C_j \cos j\vartheta + S_j \sin j\vartheta) \qquad \text{with } C_j \text{ and } S_j \in \ker(L_0).$$

Choose for B the $\ker(L_0)$-module of Fourier series

$$F = \frac{1}{r^2} \sum_{j > 0} (C_j \cos j\vartheta + S_j \sin j\vartheta) \qquad \text{with } C_j \text{ and } S_j \in \ker(L_0),$$

and for C the $\ker(L_0)$-module of functions γ/r^2 with $\gamma \in \ker(L_0)$. The *elimination of the parallax* converts (5) into $\chi^* \mathcal{R} = \gamma'/r'^2$ where γ' is a series in the zonal coefficients $(J_n)_{n \geq 2}$ over the algebra $\ker(L_0)$, a function no doubt much simpler than (5) in that the argument of latitude ϑ has been removed together with all parallax factors $(\alpha/r)^n$ for $n > 2$.

The elimination of the parallax has been generalized so that only the argument of latitude is removed from (5). The $\ker(L_0)$-algebra A' of functions

$$F = \sum_{i\geq 2} \frac{1}{r^i} \sum_{j\geq 0} (C_{i,j} \cos j\vartheta + S_{i,j} \sin j\vartheta).$$

with $C_{i,j}$ and $S_{i,j}$ in $\ker(L_0)$ is decomposed into the sum of the $\ker(L_0)$-modules B and C made respectively of the functions

$$F = \sum_{i\geq 2} \frac{1}{r^i} \sum_{j>0} (C_{i,j} \cos j\vartheta + S_{i,j} \sin j\vartheta),$$

$$F = \sum_{i\geq 2} \frac{\alpha_i}{r^i} \qquad \text{with} \quad \alpha_i \in \ker(L_0).$$

The PDE $L_0(G) = F$ for $F \in B$ is solved inductively using the identities

$$j\frac{\cos j\vartheta}{r^i} = (i-2)\frac{R}{\Theta}\frac{\sin j\vartheta}{r^{i-1}} + L_0\left(\frac{\sin j\vartheta}{\Theta r^{i-2}}\right),$$

$$j\frac{\sin j\vartheta}{r^i} = -(i-2)\frac{R}{\Theta}\frac{\cos j\vartheta}{r^{i-1}} - L_0\left(\frac{\cos j\vartheta}{\Theta r^{i-2}}\right).$$

Canonical simplifications do for Hamiltonian dynamics what Catastrophe Theory does for modelling physical systems: they strip the perturbation of its non essential features. Take , for instance, the case of functions in the $\ker(L_0)$-algebra of Laurent series

$$F = \sum_{-j\leq i\leq k} \alpha_i r^i + R \sum_{-v\leq u\leq w} \beta_u r^u \qquad \text{with } \alpha_i \text{ and } \beta_u \in \ker(L_0).$$

The *radial simplification* reduces it to

$$F' = \gamma + \frac{\delta}{r^2} + \zeta\frac{R}{r} \qquad \text{with } \gamma,\ \delta,\ \zeta \in \ker(L_0).$$

Indeed, from the basic identity

$$\frac{1}{\mu}L_0(Rr^j) = -ja^{-1}r^{j-1} + (2j-1)r^{j-2} - (j-1)pr^{j-3},$$

there follow, on the one hand, a descending induction,

$$(j+1)a^{-1}r^j = r^{j-2}\left[(2j+1)r - jp\right] - \frac{1}{\mu}L_0(Rr^{j+1}), \qquad (j\geq 1),$$

to remove all strictly positive powers of r, and an ascending induction,

$$(j-2)\frac{p}{r^j} = \frac{1}{r^{j-2}}\left(\frac{2j-5}{r} - \frac{j-3}{a}\right) + \frac{1}{\mu}L_0\left(\frac{R}{r^{j-3}}\right), \qquad (j>2),$$

to remove all negative powers $(r^{-j})_{j\geq 3}$. Other 'superfluous' terms are taken away using the following identities

$$\begin{aligned}
\frac{1}{r} &= \frac{1}{a} + L_0\left(\frac{rR}{\mu}\right), \\
Rr^u &= L_0\left(\frac{r^{u+1}}{u+1}\right) \qquad (u \geq 0), \\
\frac{R}{r^u} &= -L_0\left(\frac{1}{(u-1)r^{u-1}}\right) \qquad (u \geq 2).
\end{aligned}$$

As far as lunar theory is concerned, we signal that double series of the type

$$\mathcal{R} = \sum_{i\geq 0} r^i \sum_{j\geq 0} (C_{i,j} \cos j\vartheta + S_{i,j} \sin \vartheta) \qquad \text{with } C_{i,j} \text{ and } S_{i,j} \in \ker(L_0)$$

can be simplified canonically into Fourier series in the argument of latitude of the simpler type

$$\mathcal{S} = \sum_{j\geq 0} \left[(\gamma_{0,j} + \frac{\gamma_{1,j}}{r} + \frac{\gamma_{2,j}}{r^2}) \cos j\theta + (\sigma_{0,j} + \frac{\sigma_{1,j}}{r} + \frac{\sigma_{2,j}}{r^2}) \sin j\theta \right].$$

The terms in r^{-j} with $j \geq 2$, when they appear, will be removed by the elimination of the parallax. The appropriate Lie transformation is built inductively by descending powers of r along the chain of identities

$$\begin{aligned}
r^i \cos j\theta &= a \left[\frac{2i+1}{i+1} r^{i-1} + \frac{j^2 - i^2}{i(i+1)} p r^{i-2} \right] \cos j\theta \\
&\quad - L_0\left(\frac{a r^i}{i(i+1)\mu} (irR \cos j\theta + j\Theta \sin j\theta) \right), \\
r^i \sin j\theta &= a \left[\frac{2i+1}{i+1} r^{i-1} + \frac{j^2 - i^2}{i(i+1)} p r^{i-2} \right] \sin j\theta \\
&\quad - L_0\left(\frac{a r^i}{i(i+1)\mu} (irR \sin j\theta - j\Theta \cos j\theta) \right).
\end{aligned}$$

Abstract programming

In the overwhelming majority of applications, the expressions to be processed are Poisson series

$$\sum_{(\iota,\kappa)\in I\times K} C_{\iota,\kappa} M[\iota]T[\kappa]$$

where ι and κ are vectors of natural integers ($\in \mathbf{Z}$) of dimension m and n respectively while $M[\iota]$ and $T[\kappa]$ stand for symbolic terms of the type

$$\begin{aligned} M[\iota] &= X_1^{\iota_1} X_2^{\iota_2} \ldots X_m^{\iota_m}, \\ T[\kappa] &= \left\{ \begin{array}{c} \cos \\ \sin \end{array} \right\} (\kappa_1 A_1 + \kappa_2 A_2 + \ldots + \kappa_n A_n). \end{aligned}$$

Poisson series whose coefficients $C_{\iota,\kappa}$ are numbers are said to be *flat.*

Various parameters characterize a Poisson series: the number of its terms, the domain of its coefficients, the number of its polynomial and angular variables, the names of those variables. How and when should these parameters be specified? Ideally, at execution time in the course of an application. It is already done that way in regard to the number of terms when Poisson series are organized internally as linked lists free to grow and shrink in a dynamic way. What about the other parameters? Usually they are built in as macro variables, and the software package is prepared as a bundle of macros intended to generate a source code conforming to the specifications. Either the macro facilities belong to the programming language (MAO in PL/1), or they are added to it—see, for example, how Coffey generalizes Dasenbrock's system using RATFOR—, or they are prepared by the author of the Poisson processor as is the case with TRIGMAN or SASM. In essence, the code is made more general by being made more abstract. A macro code, though, is still not general enough, for it attaches the parameters of a Poisson series not to the series itself but to the compiled code. It precludes, for instance, maneuvers whereby new types of Poisson series are created dynamically at run time.

Polynomial and angle variables, as they appear in a Poisson series, reflect the coordinate map in which the mathematician operates to solve the problem at hand. Often a change of coordinates marks a turning point in the solution, and this in turn should cause a restructuring of the Poisson series involved. A problem may involve initially the eccentricity e and the argument of perigee g; but once it has been recognized that a given Poisson series admits the d'Alembert characteristic in e and g, it becomes more expedient from there on to handle it as a polynomial in $C = e \cos g$ and $S = e \sin g$. Usually one writes off that eventuality by structuring all Poisson series entering the application as expressions in the four symbols e, g, C and S. As a consequence all mathematical advantages that would have been gained at a certain stage of the execution are lost to an overall complication imposed to all objects at all stages. It does not have to be done that way, though, if the programming language puts the characteristics of a Poisson series away from the macro code into the series themselves, in other words, if it treats Poisson series as "objects."

A programming object looks like a record in Pascal, that is, as a template for contents of specific types to be bundled. Additionally, it functions like a portable library of procedures. From that standpoint, multiplying the Poisson series A by the Poisson series B amounts to addressing to the object A a message telling it, if it decides that the operation is legitimate, to create a third object to contain the result of multiplying itself by B. Of course all messages receivable at A have been stored in advance in the library of the object; but they have been coded in an abstract way, independently of the particular context that is the application. Furthermore, the information that A needs from B to execute the multiplication, A gets it not from the compiled code but at run time by addressing the appropriate messages to the object B. Finally, unlike Pascal or Ada records, programming objects can be arranged in hierarchies, somewhat like subsets of a set. As a subcategory of a category $\mathcal{O}$, object $\mathcal{O}'$ owns all the contents appropriated by definition to $\mathcal{O}$ and possibly additional ones. Moreover it inherits from $\mathcal{O}$ the ability to handle all messages one may address to $\mathcal{O}$. Will this new level of abstraction in programming simplify the automation of literal developments in celestial mechanics? The answer is clearly yes.

In our updated version of MAO in Symbolics Common Lisp we have mapped the generic concept of an algebra A over a ring D to an abstract programming object which, in its table of contents, makes reference to another abstract programming object that is the domain of its coefficients. Below the unspecified category of polynomial algebras, we put the subcategories of polynomials in a specified number of variables with specific print names; we have done the same for algebras of Fourier series. At that level of generality, we are able to endow the flat algebra of Poisson series with different structures. Depending on the application and, more interestingly, on the state of the application during execution, we can introduce algebras of polynomials in some letters with coefficients, for instance, in an algebra of Fourier series, this one having in turn for coefficients either polynomials or Fourier series in other variables, and so on. The hierarchy stops eventually at the field of real or rational numbers.

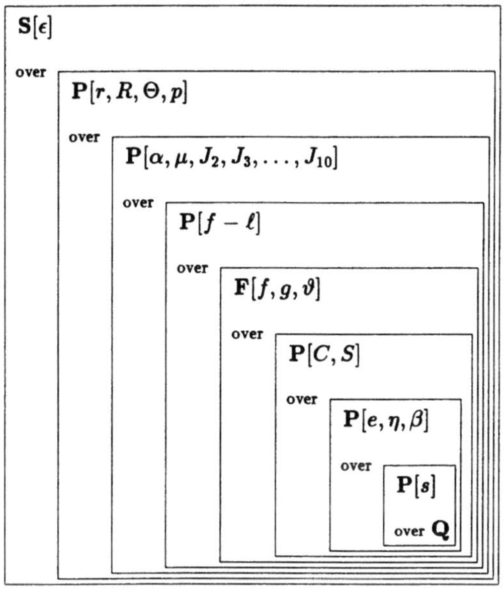

Figure 2: Hierarchy of Poisson series adopted by S. L. Coffey for the zonal problem in artificial satellite theory.

Parallel processing[2]

Boom and bust, this is the basic cycle repeating itself at every step of a calculation involving massive and not so massive algebraic objects. Take the case of two polynomials $F(\boldsymbol{x})$ and $G(\boldsymbol{x})$ in several variables $(x_1, x_2, \ldots, x_n)$, each containing 200 terms; their product $F\,G$ will generate an intermediate result made of 40,000 terms. Yet, after like terms are combined, the result will likely contain no more than 1000 terms.

One can, of course, control the explosion by letting it happen in stages; take every monomial in G, multiply F with it, accumulate the partial product in $F\,G$, and simplify the partial sum. The cost of such serialization in

[2]This research was conducted on the quarter Connection Machine at the Naval Research Laboratory, Washington DC. Etienne Deprit is the author of *MAO!!*; he will report about it in a journal devoted to symbolic or parallel computation.

computing time is very high, at times even prohibitive. But suppose there is somewhere a computer powerful enough to execute all these elementary monomial multiplications at once, sort the immense intermediary result and collapse it into a simplified result in just a few strokes. This dream computer is a reality; it is called a Connection Machine.

Imagine a hypercube $\Pi = \{0,1\}^N$ where $N = 14, 15$ or 16. At each vertex lies a processor. From two operands one bit long plus a flag, the protozoan arithmetic unit produces a one bit result and a carry flag. Processors act like a chorus line; they receive instructions, one at a time but the same for all of them, from a front machine; they execute them in unison, all feet hitting the ground at the same time. The hypercube rests on a gigantic memory made of 2^N columns of 2^{16} bits. To each processor is assigned a column in that matrix. The programmer accesses the memory by rows which he handles as *pvars* or parallel variables. The hypercube comes equipped with a router, a mechanism by which, at the same instant, each processor rings a colleague—as designated by the front machine—to send or receive information. What if several computers are told to ring the same number? Well, collisions happen by design only as, for instance, when the front machine instructs the recipient to sum up numbers brought by the callers. The router is as flexible as any telephone system can be, and just as slow when it is overloaded. On the other hand, engineering and scientific applications tend to establish patterns of communications among processors. To answer this sort of demand, the machine provides another type of connectivity called *grid mechanism.* To the programmer, the grid mechanism appears as made of two ingredients: (1) a decomposition of Π into a product of hypercubes

$$\mathcal{T} = \{0,1\}^{n_1} \times \{0,1\}^{n_2} \times \cdots \{0,1\}^{n_m} \qquad \text{with} \qquad \sum_{1 \le i \le m} n_i = N;$$

(2) a structure of affine space on $\mathcal{T}$ using the module $\mathcal{T}$ itself as the group of translations. On a grid, given a pattern $\boldsymbol{t} \in \mathcal{T}$, a parallel send and a parallel read are implemented respectively as the mappings

$$T(p) \leftarrow S(p + \boldsymbol{t}) \quad \text{and} \quad S(p) \rightarrow T(p + \boldsymbol{t}).$$

Sixty five thousand five hundred thirty six processors may seem like a big ants nest. But think of image processing where the bitmap is applied on Π; a 1024×1024 image would require each processor to be put in charge of managing $2^{20}/2^{16} = 16$ pixels. This way of loading processors is called virtualization. A machine of only 2^{16} physical processors is made to look like one having 2^{20} processors. The concept just emerged, and it is only this past summer that the software gave the possibility of changing virtualization ratios dynamically during execution by making it a characteristic not of the memory but of every individual pvar.

We wanted to test what the Connection Machine could do for massive analytical developments.

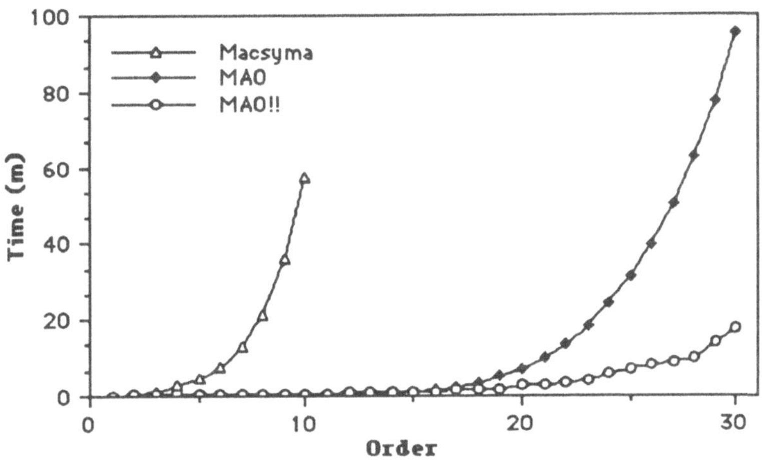

Figure 3: Time taken to develop $(r/a)^4 \cos 5f$ in ascending powers of the eccentricity for each order n in the interval $1 \leq n \leq 30$ (Measurments made by Etienne Deprit).

Cayley's tables provided the test problem. There were several reasons for that selection. The construction starts very small by the number of terms, but grows rapidly with the order. In no time we were able to fill the Connection Machine with series of 200 to 300 terms whose products forced to assess the virtualization mechanism. With no trouble we could program the problem for other computers; actually we moved the problem to a serial Lisp workstation where we used Macsyma, and the special purpose processor, MAO, which Dr. Miller has trained rigorously for speed. Finally it pleased us to work out a new way of solving implicit equations adapted for parallel processing. Ironically we succeeded so well in that direction that the operation itself of solving Kepler's equation could no longer tax the computer resources in the Connection Machine. But, once the Lie transformation settled in the machine, we used it to reproduce and extend the famous Cayley Tables. At last, we were multiplying massive Poisson series, which demanded virtualization ratios of up to 32 in the intermediary results. We turned to the router for some operations, but essentially we treated the machine as a one-dimensional grid so we could materialize polynomials and Fourier series as sequences rather than families of processors.

In all the tests we made, Macsyma proved a very poor runner. For a task that MAO on a Lisp workstation, and MAO!! (pronounced MAO *bang bang*) on the Connection Machine, carried out in a matter of seconds, Macsyma took some 57 minutes. On the whole, as long as the series are small or moderate in size (less than 100 terms), MAO was the champion. Beyond that point, MAO!! turned the undisputed winner.

Conclusions

Endlessly compelling, the grand problems of celestial mechanics keep computational astronomers on the look-out for help from mathematics, software, and hardware. Relief is in sight. How liberating it will then be to step out of a world of flat series and serial manipulations, and into the massively parallel hypercube of array processors now under development!

Acknowledgement

Research leading to this communication was conducted in part at the National Institute of Standards and Technology under a grant from the Applied and Computational Mathematics Program of DARPA, and in part at the Departamento de Física Teórica (Astronomía), Universidad de Zaragoza, with partial support from the Ministerio de Educación.

The Stability of the Lagrangian Point L_4

D.S. Schmidt
Department of Computer Science
University of Cincinnati
Cincinnati, Ohio 45221-0008
USA

ABSTRACT. The Hamiltonian function of the restricted problem of three bodies near the triangular Lagrangian point is normalized through sixth order terms with the help of MACSYMA. The same calculations were done previously with an algebraic processor in order to establish the stability at a critical value of the mass ratio.

1. Introduction

Stability of the Lagrangian point L_4 in the restricted problem of three bodies can only be expected when the mass ratio μ is smaller than Routh's critical value $\mu_1=(1-\sqrt{69}/9)/2 \simeq 0.03852$. Then the linearized Hamiltonian system of differential equations has purely imaginary eigenvalues. Unfortunately the corresponding quadratic terms of the Hamiltonian function are indefinite and Lyapunov's direct method can not be applied. Instead one has to use a theorem of Arnold which invokes the KAM theory to establish stability.

Assume that the Hamiltonian function H is in Birkhoff normal form with respect to the action angle variables $(I_1, I_2, \phi_1, \phi_2)$, that is

$$H = H_2 + H_4 + \cdots + H_{2n} + H^* \tag{1}$$

where H_{2k} are homogeneous polynomials of degree k in I_1 and I_2, $H_2 = \omega_1 I_1 - \omega_2 I_2$ with $\omega_1, \omega_2 > 0$, and H^* contains all higher order terms.

Theorem (Arnold (1961)): The origin of the system of differential equations derived from (1) is stable if for some k, $2 \le k \le n$, $D_{2k} = H_{2k}(\omega_2, \omega_1) \neq 0$.

For a proof see Meyer and Schmidt (1986). The normalization of the Hamiltonian function near L_4 was carried out by Deprit and Deprit-Bartholomé (1967). Excluding the two values $\mu_2 \simeq 0.02429$ and

Celestial Mechanics **45**: 201–206, 1989.

$\mu_3 \simeq 0.01352$ where low order resonance terms appear they calculated

$$D_4 = \frac{36 - 541\,\sigma + 644\,\sigma^2}{16\,(1-4\sigma)\,(4-25\sigma)} \qquad \text{where} \quad \sigma = \omega_1^2\,\omega_2^2 = \frac{27}{4}\,\mu\,(1-\mu).$$

With it they found that $D_4 \neq 0$ except for one value $\mu_c \simeq 0.01094$ where D_4 changes its sign.

Meyer and Schmidt (1986) have investigated this special case. With the help of their algebraic processor POLYPAK they have computed D_6 and shown that its value is different from zero for that special value of μ.

2. Computations with MACSYMA

Coppola and Rand (1987) reported on calculating D_6 with the help of MACSYMA. A closer inspection of their paper reveals that they used the numerical value for μ_c from the beginning. This simplifies the computational task by several orders of magnitude.

Their failure to reproduce D_6 with the help of MACSYMA inspired us to try it ourselves and to find out what are the difficulties in using MACSYMA on a problem that pushes it to the limit. This note reflects on our experience in calculating D_6 with MACSYMA and it describes some of the methods which we used to simplify the calculations. The reader is referred to our previous paper for a description of the method which we used with our algebraic processor POLYPAK which was designed for the efficient manipulation of power series in several variables.

MACSYMA on the other hand is a general purpose computer algebra system, see for example Davenport (1988), and on first sight it appears that it should be the ideal tool for these calculations. The normalization of the Hamiltonian function introduces rational functions in ω_1 and ω_2 and MACSYMA has special facilities to represent such functions efficiently.

But the time consuming aspect of the calculations are the simplification of expressions. It not only involves combining rational functions, removing common factors, etc. but also includes replacements like (sqrt(3))^2 by 3 and ω_1^2 by $1-\omega_2^2$. The reason for the last replacement rule will be apparent later. For now we just want to point out that with MACSYMA it takes a separate step to accomplish these replacements, whereas for the calculations with our own algebraic processor these substitutions can be done as soon as the terms are generated. MACSYMA offers several methods for performing the replacement for ω_1. Any method which reduces ω_1^n for any integer n to terms multiplied with ω_1 raised to the power of 0 or 1 depends on some general pattern matching function and is time consuming. It is more advantageous to insure that at most a quadratic term ω_1^2 can occur and then perform the replacement with a more elementary substitution function of MACSYMA.

The normalization of the Hamiltonian function leads automatically to complex position variables x_1, x_2 and their conjugate momenta y_1, y_2. The time consuming aspects of simplification forced us to reconsider how to represent the homogeneous polynomials

$$\sum c_{ijkl}\, x_1^i\, x_2^j\, y_1^k\, y_2^l \tag{2}$$

of degree 2 through 6. The easiest would have been to represent each polynomial as a single function, but some of the intermediate functions are huge since the coefficients c_{ijkl} can be lengthy rational functions. The simplifier tries to bring all terms on a common denominator and makes matter worse if it succeeds at all.

It was difficult to tell MACSYMA that it should simplify each c_{ijkl} as much as possible but to preserve the structure of (2). We ended up by representing each term of the polynomial (2) as an individual function and we passed it then one by one to the simplifier. Each term of (2) was addressed by an element of a hashed array C[i,j,k,l].

In order to access all terms of a polynomial of degree n we need a loop which generates all nonnegative indices i,j,k,l which satisfy i+j+k+l=n. Such a loop can be generated automatically with the help of the macro facilities of the programming language which is part of MACSYMA. With this it is easy to provide the required operation on polynomials like addition, multiplication, etc.

In effect we used the programming language of MACSYMA to built a primitive algebraic processor. The terms of a polynomial are represented as the elements of a multidimensional sparse array, which in turn is implemented efficiently by MACSYMA with its hashing facilities. The main difference to a typical algebraic processor is that the coefficients of each monomial instead of being simple numbers can now be arbitrary functions and the full power of MACSYMA can be used to manipulate these functions.

3. The four steps of the computation

Step 1: Expand the Hamiltonian function near L_4

$$H = \frac{1}{2}\,(P_1^2 + P_2^2) + Q_2P_1 - Q_1P_2 - \frac{\gamma}{2}\,Q_1 - \frac{\sqrt{3}}{2}\,Q_2 - \frac{1+\gamma}{2\,r_1} - \frac{1-\gamma}{2\,r_2}$$

with $r_1^2 = 1 + \gamma\, Q_1 + \sqrt{3}\, Q_2 + Q_1^2 + Q_2^2$,

$$r_2^2 = 1 - \gamma\, Q_1 + \sqrt{3}\, Q_2 + Q_1^2 + Q_2^2$$

and $\gamma = 1 - 2\,\mu$ through order 6 in the position variables Q_1 and Q_2. This step is easy, especially if one realizes that the expansion of the last two terms in H can be obtained from the expansion of $1/r_1$ by multiplying with γ those terms in the expansion of $1/r_1$ which have an odd power of Q_1.

The last remark also shows how to remove $\sqrt{3}$ from the Hamiltonian. Set $\gamma = 3\sqrt{3}\gamma'$. The homogeneous polynomials of odd degrees in the expansion will continue to have $\sqrt{3}$ as a multiplicative factor but it can be removed by scaling the small parameter which is part of the Lie transformation.

Step 2: The linearized system is derived from the Hamiltonian

$$H_2 = \frac{1}{2}(P_1^2 + P_2^2) + Q_2P_1 - Q_1P_2 - \frac{1}{2}Q_1^2 - \frac{\gamma'}{4}Q_1Q_2 - \frac{5}{8}Q_2^2 .$$

It has the characteristic equation $\lambda^4 + \lambda^2 + (27 - \gamma'^2)/16 = 0$ with the roots $\pm i\omega_1$, $\pm i\omega_2$. The frequencies ω_1 and ω_2 are unambiguously defined by

$$0 < \omega_2 < \sqrt{2}/2 < \omega_1 < 1 \tag{3a}$$

$$\omega_1^2 + \omega_2^2 = 1 \tag{3b}$$

$$\omega_1^2\,\omega_2^2 = (27 - \gamma'^2)/16. \tag{3c}$$

The linear transformation to new complex coordinates leads to

$$H_2 = i\omega_1\, x_1y_1 - i\omega_2\, x_2y_2.$$

The transformation of the higher order terms is the first computational intensive task. Part of the labor comes from the use of (3b) and (3c) to insure that ω_1 and γ' only have exponents 0 or 1 after simplification. The substitution $\omega_1^2 = 1 - \omega_2^2$ destroys the symmetry between ω_1 and ω_2 in the formulas. It is possible to use (3b) as a replacement rule for 1 and one can achieve that the c_{ijkl} are always a ratio of two homogeneous polynomials in ω_1 and ω_2. This approach makes it very difficult for MACSYMA to find and eliminate common factors. Therefore we did the next best thing and set $w = \omega_2/\omega_1$. The replacement rules follow then from (3b)

$$\omega_1^2 = 1\,/\,(1 + w^2) \quad \text{and} \quad \omega_2^2 = w^2/\,(1 + w^2)\,.$$

By scaling some of the variables with ω_1 by setting $\beta = \omega_1\gamma'$ and by stretching the time with the factor ω_1 it is possible to eliminate the direct dependency of the original Hamiltonian function on ω_1 and ω_2. It is then possible to work just with the variables w and β. In order to restrict the exponents of β to 0 and 1 we derive from (3c) the only replacement rule which we still have to use

$$\beta^2 = (27 + 38\, w^2 + 27\, w^4) \;/\; (1 + w^2).$$

Step 3: The Lie transformation of Deprit (1969)is used to bring the Hamiltonian function into its normal form. One of the advantageous of using this systematic procedure is that we can trace the contribution of any term to the normalized terms. In particular one can see that a normalized term at order 6 can only receive contributions from those terms whose exponents satisfy

$$| i - k | + | j - l | \le 3 (6 - i - j - k - l).$$

This holds true for any intermediate terms which may occur within the Lie triangle as well as for terms in the generating function of the transformation. This condition eliminates the calculation of several terms at orders 4 and 5.

Step 4: The previous step was of course computationally the most intensive step. The evaluation of the normalized sixth order terms at the values of the action variables $I_1 = \omega_2$ and $I_2 = \omega_1$ is then a simple task with MACSYMA. The answer written out with w is longer than when it is expressed in terms of the quantity $\sigma = \omega_1^2\,\omega_2^2$ unless one uses the fact that the numerator and denominator each can be written as reciprocal polynomials in w. Since this fact does not appear to be very helpful we give the answer in the form of our previous paper:

$$\begin{aligned} D_6 = (&3105/4 - 1338449/48\,\sigma + 489918305/1728\,\sigma^2 - 7787081027/6912\,\sigma^3 \\ &+ 2052731645/1296\,\sigma^4 + 1629138643/324\,\sigma^5 - 1879982900/81\,\sigma^6 \\ &- 368284375/81\,\sigma^7) \;/\; (\sqrt{\sigma(1-4\sigma)}\,(1-4\sigma)^2\,(4-25\sigma)^3\,(9-100\sigma)) . \end{aligned}$$

The discriminant D_4 is zero for $\sigma = (541 - \sqrt{199945}) / 1288$. This value of σ corresponds to the mass ratio μ_c and we obtain

$$D_6 = - \frac{(1332195623735\,\sqrt{199945} + 585874808946015)}{2594866866020352}\sqrt{760\,\sqrt{199945} - 318424}.$$

Its value is approximately -66.63 and it guarantees the stability of the Lagrangian point L_4 also for μ_c .

4. Acknowledgement

The work was supported by a grant from the Applied and Computational Mathematics Program of DARPA.

5. References

Arnold, V.I. (1961), 'The stability of the equilibrium position of a Hamiltonian system of ordinary differential equations in the general elliptic case', *Soviet Math. Dokl.* 2, 247-249.

Coppola, V.T. and Rand, R.H. (1987), 'Computer algebra implementation of Lie transforms for Hamiltonian systems: Application to the nonlinear stability of L_4', Proceedings of '*Computer Algebraic Integration and solution of Differential Equations*', IBM T. J. Watson Research Center, Nov 19-20, 1987.

Davenport, J.H., Siret, Y. and Tournier, E. (1988), *Computer Algebra*, Academic Press.

Deprit, A. (1969), 'Canonical transformations depending on a small parameter' *Celestial Mech.* 1, 12-30.

Deprit, A. and Deprit-Bartholomé, A. (1967), 'Stability of the Lagrangian triangular points', *Astron. J.* 72, 173-179.

Meyer, K.R. and Schmidt, D.S. (1986), The stability of the Lagrange triangular point and a theorem of Arnold, *J. Diff. Eq.* 62, 222-236.

COMPUTER IMPLEMENTATION OF A NEW APPROACH TO THE IDEAL RESONANCE PROBLEM

R.A. HOWLAND
Department of Aerospace and Mechanical Engineering
University of Notre Dame
Notre Dame, IN 46556

ABSTRACT. A new approach to the librational solution of the Ideal Resonance Problem has been devised--one in which a non-canonical transformation is applied to the classical Hamiltonian to bring it to the form of the simple harmonic oscillator. Although the traditional form of the canonical equations of motion no longer holds, a 'quasi-canonical' form is retained in this single-degree-of-freedom system, with the customary equations being multiplied by a non-constant factor. While this makes the resulting system amenable to traditional transformation techniques, it must then be integrated directly. Singularities of the transformation in the circulation region limit application of the method to the librational region of motion.

Computer-assisted algebra has been used in all three stages of the solution to fourth order of this problem: using a general-purpose FORTRAN program for the quadratic analytical solution of Hamiltonians in action-angle variables, the initial transformation is carried out by direct substitution and the resulting Hamiltonian transformed to eliminate angular variables. The resulting system of differential equations, requiring the expected elliptic functions as part of their solution, is currently in the process of being integrated using the LISP-based REDUCE software, by programming the required recursive rules for elliptic integration.

Basic theory of this approach and the computer implementation of all these techniques is described. Extension to higher order of the solution is also discussed.

1. Introduction

Since the first in a series of articles by Garfinkel (1966), the ideal resonance problem, described by the Hamiltonian

$$- H(x,y) = B(y) + 2\mu^2 A(y) \sin^2 x,$$

has undergone a resurgence in interest. If, as is often done, the functions A(y) and B(y) are expanded about the libration center, y_0, one obtains a series in powers of $p = y - y_0$. In the libration region $p = O(\mu)$, and this ultimately reduces to

$$- H(x,p,\mu) = (\frac{1}{2} B^{(2)} p^{(2)} + 2\mu^2 A^{(0)} \sin^2 x) + O(\mu^3),$$

where superscripts indicate the order of derivative, evaluated at y_0.

In addition to Garfinkel's work, Jupp and Williams have contributed to the literature on the subject. (For a review of recent research on this problem, as well as an extensive list of references, Garfinkel's excellent survey article [1982] is recommended.) In all these cases, a classical approach to the problem has been taken: expansions have been made about the solution of the dominant, 'pendulum', part of Hamiltonian. Unfortunately, this solution entails the introduction of elliptic functions--functions whose derivatives rapidly complicate the expansion process itself--and explicit solution has only been found through second order.

Celestial Mechanics **45**: 207–211, 1989.

The author has implemented a new approach to the librational solution of this problem (1988). The method consists in 'conditioning' the original Hamiltonian system, through a manifestly non-canonical transformation $(x,p) \leftrightarrow (x^*, y^*)$

$$\sin x = \left(\frac{B(2)}{A(0)}\right)^{1/4} y^{*1/2} \sin x^*$$

$$p = 2\mu \left(\frac{A(0)}{B(2)}\right)^{1/4} y^{*1/2} \cos x^* \tag{1}$$

to make it assume a form amenable to analysis in classical action-angle variables. Though the resulting equations of motion are no longer strictly canonical, they do retain some semblance of that form; indeed, the final system becomes

$$\dot{x}^* = \frac{1}{\mu} \cos x\,(x^*,y^*,\mu) \frac{\partial H^*(x^*,y^*,\mu)}{\partial y}$$

$$\dot{y}^* = \frac{-1}{\mu} \cos x\,(x^*,y^*,\mu) \frac{\partial H^*(x^*,y^*,\mu)}{\partial x} \tag{2}$$

in which $H^*(x^*,y^*;\mu) \equiv H(x(x^*,y^*;\mu), p(x^*,y^*;\mu);\mu)$. A subsequent canonical transformation $(x^*,y^*) \leftrightarrow (\xi,\eta)$ then maintains this 'quasi-canonical' form:

$$\dot{\xi} = \frac{1}{\mu} \cos x\,(\xi,\eta;\mu) \frac{\partial H^{**}(\xi,\eta;\mu)}{\partial \eta}$$

$$\dot{\eta} = \frac{-1}{\mu} \cos x\,(\xi,\eta;\mu) \frac{\partial H^{**}(\xi,\eta;\mu)}{\partial \xi} \tag{3}$$

--where $H^{**}(\xi,\eta;\mu) \equiv H^*(x^*(\xi,\eta;\mu), y^*(\xi,\eta;\mu))$ and $\cos x(\xi,\eta;\mu) \equiv \cos x(x^*(\xi,\eta;\mu), y^*(\xi,\eta;\mu))$.

In particular, then, using the latter transformation to eliminate the variable ξ from H^{**}, the equations (3) show that η is rendered constant, while the equation for ξ, of the form

$$\dot{\xi} = \pm\frac{1}{\mu} \sqrt{1-\sin^2 x(\xi\eta;\mu)}\ \frac{\partial H^{**}(-,\xi;\mu)}{\partial y}$$

$$= \pm\frac{1}{\mu} \sqrt{1-\sqrt{\frac{B(2)}{A(0)}}\eta \sin^2\xi + \mu h(\xi,\eta;\mu)}\ \frac{\partial H^{**}(-,\eta;\mu)}{\partial \eta}, \tag{4}$$

can be integrated through a separation of variables to give $\xi(t)$ implicitly in a 'Kepler equation'. The sequence of transformations $(x,p) \leftrightarrow (x^*,y^*) \leftrightarrow (\xi,\eta)$ then yields the solution for the original variables.

The equations (3) are obviously a generalization of the traditional Hamiltonian equations of motion, in which the cosine factor reduces to a constant. Though a final explicit quadrature is necessary in this approach, it does forestall the introduction of the elliptic functions characterizing the ideal resonance problem until this integration; these appear only in the solution for $\xi(t)$, while the transformations $(x,p)\leftrightarrow(x^*,y^*)\leftrightarrow(\xi,\eta)$ are all purely trigonometric. The transformed Hamiltonian can thus be found to arbitrary order without taking the derivatives of elliptic functions, required when the simple pendulum is used as the 'unperturbed' solution.

The above approach has been implemented through the use of computer-assisted algebra at all stages: the substitution (1) and transformation elimination of the angular variable have been effected using the author's general-purpose FORTRAN package, QUAD; the subsequent integration of the time equation, (4), is in the process of being carried out using the LISP-based REDUCE software. A brief description of each package and its application to the present problem are outlined in what follows.

2. Transformation of the Equation of Motion

Since the proposal by Kolmogorov (1954) of an approach to perturbation analysis conjectured to give mathematically convergent solutions, and subsequent proof of this convergence by Moser (1962) and Arnol'd (1963), the 'KAM' method has been the paradigm for 'superconvergent' techniques. The basic approach, in essence, relies on continual, complete retransformation of the Hamiltonian to effect quadratic *formal* convergence, reminiscent of Newton's method for finding roots; *mathematical* convergence, in turn, requires that the original frequencies in the problem satisfy an 'irrationality condition' bounding all possible linear combinations of these frequencies away from zero. The original formulation was cast in terms of classical mixed-variable canonical transformations, with the attendant problems of inversion exacerbated by the need to go through a sequence of [complete] transformations to the high orders of which the approach was capable.

The author (1977, 1979) has implemented this basic retransformation philosophy utilizing Lie transformations (Deprit, 1969), overcoming the problems of inversion. In the event the Hamiltonian is expressed in action-angle variables, the quadratic approach has the advantage of giving perturbed frequencies of the system to twice the order of explicit solution (transformation) at the last step, avoiding the mixed-secular error resulting from errors in the frequencies over time intervals orders of magnitude greater than possible in a 'linear' approach. The technique was subsequently implemented in a general-purpose FORTRAN package, QUAD, for the quadratic solution of Hamiltonians in action-angle variables through the elimination of the angles (Howland, 1986); this, in turn, relied heavily on a package of subroutines, SAP, written by Richardson (1980) for the algebraic manipulation of Poisson series.

Among the features of QUAD is a programmed entry into the subroutine which reads in the starting Hamiltonian, to allow an arbitrary initial manipulation of that function. It is this feature which effects the transformation (1) through the use of a SAP subroutine directly substituting the variables into the Hamiltonian.

The program then automatically carries out the transformation eliminating angular variables. Because of the degeneracy inherent in the initially-transformed Hamiltonian (2), a first transformation will eliminate the angle to fourth order. The results can be found in the author's 1988 paper.

3. Integration of the Kepler Equation

As explained in the introduction, η is now known to be constant, but there remains the integration of the time equation (4). QUAD normally determines only the frequencies to maximum order, leaving the variables' explicit transformation at the order of the previous transformation. The square root factoring this equation, however, depends on the [transformed] sin $x(\xi,\eta;\mu)$, which thus must be known. Again, an entry point in QUAD at the end of each transformation step allows determination of this last transformation's generating function. Now the form of the denominator can be found: assuming that

$$\left| \frac{\mu h(\xi,\eta;\mu)}{1-\sqrt{\frac{B^{(2)}}{A^{(0)}} h \sin^2 x}} \right| < 1$$

--it is this condition ultimately limiting the solution to the librational region--the radical can be expressed through a binominal expansion[1] and the resulting series integrated term by term. A simple inductive argument shows that the Hamiltonian always involves only cosine terms, so solution to the equation (4) then entails finding terms of the form

$$I[n,m] \equiv \int \frac{\cos n\phi \, d\phi}{(1-k^2 \sin^2\phi)^{m/2}} \tag{5}$$

in which m is always odd; thus it suffices to be able to evaluate such integrals. This is expedited by the recognition that an integration by parts of (5) yields a recurrence relation

$$I[n,m] = \frac{4}{(m-2)k^2}\left((n-2)\, I[n-2,m-2] - \frac{\sin(n-2)\phi}{(1-k^2 \sin^2\phi)^{(m-2)/2}}\right) + I[n-4,m] \tag{6}$$

The even nature of the integrand in (5) and use of (6) ultimately reduce the problem to finding I[0,m] and I[1,m] for [odd] positive or negative m (the cases being treated separately). These [four] integrals can be expressed using the elliptic functions nd and dn -- functions whose integrals possess their own recursion relations--and more elementary trigonometric functions. (For details, see Howland, 1988.)

The recursive nature, both of these integrals and of the relation (6), makes this integration a natural candidate for automated solution through a recursive language--in this case, LISP, as implemented through the REDUCE algebraic software. By programming these relations and [four] particular values for each of the integrals, the I[n,m] above can be found for all values of m and n.

---

[1] A call is made to another SAP routine effecting such binomial expansion at the above-mentioned entry point.

4. Extension of the Approach to Higher Order

The above procedure, though only described (and, at this point, implemented) through one transformation step, can be carried out to arbitrary order: the new Hamiltonian is continually of a form to make the terms to be integrated remain as in (5). Thus the same REDUCE program used to find the above solution can be utilized in the higher-order one. As an issue of convenience--even practicality--however, it becomes useful for the output of QUAD to be able to be input directly into the LISP routines, and work to format the former program's results automatically in a form suitable for REDUCE is in initial stages.

References

Arnol'd, V.I.: 1963, *Russ. Math. Surv.* **18**, 9.

Deprit, A.: 1969, *Celes. Mech.* **1**, 12.

Garfinkel, B.: 1966, *Astron. J.* **71**, 657.

Garfinkel, B. : 1982, *Celes. Mech.* **28**, 275.

Howland, R.A.: 1977, *Celes. Mech.* **15**, 327.

Howland, R.A.: 1979, *Celes. Mech.* **19**, 95.

Howland, R. A. and Richardson, D.L.: 1986, *J. Comp. Phys.* **67**, 19.

Howland, R.A.: 1988, to appear in *Celes. Mech.*

Kolmogorov, A.N.: 1954, *Dok. Ak. Nauk. SSSR* **98**, 527.

Moser, J.: 1962, *Nach. Akad. Wiss. Gottingen Math. Phys.* **1**, 1.

Richardson, D.L.: 1980, Publ. Dept. of Aerospace Engineering (U. of Cincinnati).

A LABORATORY FOR GRAVITATIONAL SCATTERING EXPERIMENTS

PIET HUT
Institute for Advanced Study
Princeton, NJ 08540, U.S.A.

ABSTRACT. An accurate macroscopic description of the evolution of star clusters, galaxies, and clusters of galaxies should be based on a detailed knowledge of the underlying microscopic processes. Examples of such processes are the interactions between single stars and binaries, in the context of a star cluster, or between whole galaxies, in the context of a cluster of galaxies. A statistical description of these processes is most conveniently given in terms of cross sections and reaction rates, just as in atomic or nuclear physics. The main difference is that in gravitational physics our particle accelerator is replaced by a computer program, which plays the role of star accelerator.

Work in progress towards realization of a laboratory for gravitational scattering is summarized, by concentrating on the three levels of software needed to build such a laboratory. First, one needs a program which computes the actual particle orbits (the accelerator). Second, one needs a program which monitors the progress of individual scattering experiments, and decides automatically when to halt computation (the servo mechanisms which guide and control the accelerator). Third, one would like to have an intelligent program which takes a physics question, such as "what is the reaction rate for the occurrence of hierarchical scattering for a binary with a binding energy of 10 kT in a heat bath with a Maxwellian distribution of single star velocities?", and translates that question in an agenda of scattering experiments to be carried out (the laboratory assistant).

1. Introduction

With the steady increase in hardware speed, two different challenges are posed to the theoretical astrophysicist using computers. The first challenge, with which everyone is familiar, is to develop new algorithms and implementations thereof which allow problems to be addressed which are sufficiently complicated that previous generations of computers were not powerful enough to deal with them. The second challenge, which has generally received less attention in the literature, is to repeat calculations which were already possible on smaller computers, but now in a systematic parameter study. Sometimes this second approach is relatively straightforward, for example if there is a clear *a priori* set of parameter values through which one wants to step in a regular grid-like way. But often the most desirable way of repeating similar calculations is much less predictable, and in many cases the parameters to be chosen are affected by the outcome of previous calculations in the series.

It is the dynamic feedback of the results of some of the previous calculations on the choice of initial conditions of some of the later calculations, which makes a general parameter search an algorithmic challenge. I will concentrate in this paper on an astrophysical application with which I am familiar: gravitational scattering

Celestial Mechanics **45**: 213–218, 1989.

experiments. In this case the complexity of the guidance of the parameter study can vastly outweigh the complexity of the original program which is capable of performing one calculation for a given set of parameters.

2. Gravitational Scattering

Scattering experiments play an important role in many areas of microscopic physics. In most systems the interactions of individual constituents display a bewildering variety of behavior when investigated in detail. In general, such information is not relevant for a global, thermodynamic description of the system. What we are mainly interested in can often be expressed by a few numbers, such as the heat capacity and the heat conductivity. A convenient way to arrive at such final bulk numbers is a three-step approach: (1) one starts by measuring a few relevant numbers at the end of each individual scattering experiment, where these numbers are the values of some quantities which are judiciously chosen so as to shed maximum light on the property of the system one is after; (2) the next step involves a statistical averaging (ensuring a proper weighting procedure) of all individual results in similar runs in order to express all these results in the form of a few cross sections and reaction rates; (3) the final step is to start with the statistical nature of the individual constituents, as characterized by the cross sections and reaction rates, and to average their behavior over the system as a whole, by using the known statistical distributions of properties of the constituents.

An astrophysical example of an application of microscopic scattering experiments is given by the low-energy nuclear physics experiments which are performed to determine the cross sections and reaction rates which are used as input in stellar evolution calculations. In this case, the first two steps of the classification given above are performed in the laboratory, and the third one on a computer in a stellar evolution code. Other astrophysical examples, which are examples of macroscopic scattering experiments, are sometimes purely computational by necessity, whenever the individual experiments are unpractical. The subject of gravitational scattering falls in this category, since encounters between single stars and double stars, or for that matter between whole galaxies, are not easy to orchestrate in the real world.

A specific example of an application of gravitational scattering in astrophysics is based on the gravitational three-body problem. After measuring the outcome of orbit calculations for individual encounters between single stars and binary stars, it is possible to produce statistical expressions, in a given environment, for the average rate of increase of binary binding energy and the resultant rate of heating of the stellar environment. In the three-step classification given above, the first step is relatively simple when we limit ourselves to gravitationally interacting point masses. The third step consists of writing a computer code which can follow the evolution of a star cluster in a statistical way, for example in the Fokker-Planck approximation. After core collapse, a star cluster can begin to re-expand when binaries are formed dynamically and begin to give off excess heat to the environment when their binding energy increases on average. A major complication arises at the second step: how do we insure a proper sampling of initial conditions, as well as a proper statistical interpretation of the final results of the individual experiments? We will come back to this interesting question in §5, after having set up the stage in the form of a laboratory. Another example of gravitational scattering is formed by encounters of galaxies, discussed in §6. See Heggie (1988) for a general discussion of gravitational scattering.

3. A Computational Laboratory

Gravitational scattering experiments can be performed on a computer in a variety of ways. The simplest approach is to write a computer code which is restricted to solving the equations of motion and propagating the particles, while giving periodic output of the intermediate results. A human controller can then study the successive output results and determine at which point the calculation has run its course and should be terminated. The halting criterion can be relatively simple, such as the separation of two galaxies beyond a certain distance, in the case of a high-speed encounter between two galaxies where it is *a priori* known that the two galaxies will not stick together. A more complicated criterion is necessary in the case that the encounter speed is low enough to allow the possibility (but not the certainty!) of a merging of the two galaxies. Not only is a decision needed as to which of the two outcomes has been established (merging or escape), also there is a lot of freedom in determining a halting criterion in the case of a merger. It is sensible to continue the run as long as the merger remnant still shows clear signs of evolution, of not yet having settled down in its new form. However, the translation of this qualitative remark into hard objective criteria is not so simple, and certainly not unique.

This brings us to a second type of approach to performing gravitational scattering experiments on a computer. We can extend the task of the computer, by writing a second code which controls the action of the code used for getting the computational crunch work done. Again, depending on the subtleties involved in deciding when to end a calculation, this second-level code can be short and simple, or it can be longer and more complicated than the first-level code, for example if something like pattern recognition is involved in classifying the status of the system at a given time.

The task of the human researcher is obviously greatly simplified with the halting decisions for each experiment being taken care of by the computer. What remains to be done by hand, though, is to chose a set of initial conditions for each experiment which will be carried out. This third-level type of activity can be automated as well, as will be discussed in the remainder of this paper.

4. A Laboratory Assistant

The central problem in automatizing the setup of experiments is formed by the much more extended amount of knowledge which needs to be built into the computer program. On the first level, the program should have a working understanding only of the equations of motion. On the second level, not only the mechanics but also the aim of each experiment should be coded into the program. For example, the question what exactly we want to measure to what (quantitative or qualitative) accuracy at the end of a calculation can cause the structure of the second-level program to be vastly different for different types of experiments, even when the first level program remains completely unchanged.

At the third level, we try to automatize not only the halting but also the starting-up and the choice of start-up parameters of each experiment. This requires a much wider type of knowledge than that used on the second level. In laboratory terms, a machine to perform the experiments is often modified in such a way that it can also signal the end of an experiment, and in addition that it can do some of the early data reduction and interpretation as well. However, the choice of a complicated suite of experiments is often left to a human laboratory assistant, rather than a computer.

One important distinction to make at this level is that between pilot studies and production runs. When we start a new series of experiments, it may well be that we do not know the precise part of parameter space which we will want to search systematically. Instead, we may want to do a few "shots in the dark", to get a preliminary feeling for the relationship between the choice of initial conditions and the type of outcome of the experiments. Once we have a clear enough picture, we can start to carry out production runs, in which we systematically explore the parameter range which seemed appropriate as a result of the pilot studies. In general it is a good idea to use these production runs also to keep checking the appropriateness of the calculations, as was done first in the pilot studies.

Of these two types of third-level activities, perhaps the pilot studies are the most complex from a computer science point of view (as well as from a coding point of view). The reason is that at this point most is demanded in terms of imagination and creativity, because the outcome of experiments is completely new and different parameter regimes are often unknown until the pilot experiments are actually performed. What is needed goes beyond the more familiar rule-based expert systems, in that some of the rules of the game have to be determined empirically, while performing pilot experiments.

5. A Small-Scale Example: Three-Body Scattering

The ideas presented above can be illustrated by a specific example which I happen to be working on, although it has to be kept in mind that the underlying ideas have a general validity for any type of computer use where massive parameter studies are performed in as yet unexplored parameter domains.

When a single star encounters a tight binary star in a globular cluster, the resulting interaction will on average leave behind a binary with a higher binding energy, and a single star therefore will emerge at higher speed than the ingoing star. Each example of a simulation of such a gravitational three-body scattering process needs nine parameters to uniquely describe the initial conditions. A grid search in a nine-dimensional parameter space being far too time-consuming, a Monte Carlo sampling of this huge parameter space is called for (see Hut 1985 for a review and earlier references). Most parameters can be chosen at random from a compact domain of values, according to a well-defined statistical weighting procedure which can be determined *a priori*. However, there is one exception, in the form of the impact parameter, which introduces a serious complication.

The impact parameter for a scattering experiment is defined as the off-set from a head-on collision, measured at infinity, *i.e.* asymptotically far away from the target. In principle, allowed values lay between zero and infinity, although in many cases we can limit the domain of interest to a compact one. For example, if we want to study exchange of energy between a single star and a binary, and if we are interested only in amounts of energy exchange larger than 10^{-4}, then we can use approximation techniques to put a firm upper limit on the impact parameter values for which such energy exchanges are at all possible.

In practice, however, theoretical considerations often fence off a parameter domain which is much larger than the actual parameter domain in which the processes of interest happen to take place. An indiscriminate Monte Carlo parameter sampling in too large a domain can lead to terrible inefficiency. At the same time, an underestimate of the parameter domain can lead to incorrect conclusions, by leaving out a part of parameter space which ought to be sampled as well.

In this situation, a pilot study of a large area is called for. The initial overestimate for the size of the area can be based on theoretical considerations, or on previous experience in nearby parameter domains, or simply on an arbitrary shot in the dark. While the pilot calculations are in progress, it will soon become clear whether the initial guess is anywhere close to being satisfactory. If so, the pilot study is used to sharpen the original estimate a bit, and production runs can begin. If not, the type of gross deviations are analyzed, and based on the conclusion of that analysis a different guess is made for a new series of pilot experiments.

In the production phase, it is wise to define a substantial buffer zone along all the edges of the parameter space handed down from the analysis of the pilot experiments. By performing experiments in the buffer zone simultaneously with those in the inner parameter zone, it is possible to check whether the pilot analysis has really been accurate. It may occur that an experiment in the buffer zone shows an outcome of the type which the investigator is interested in, and therefore should have only occurred in the inner zone. At that point the inner zone should be enlarged to encompass the unexpected parameter combination, and the buffer zone should be enlarged around that area as well. In this way, dynamical feedback will guarantee accuracy of the final results while the data gathering is in progress.

6. A Large-Scale Example: Galaxy-Galaxy Scattering

Encounters between galaxies have been studied in computer simulations, beginning with the pioneering studies by Toomre & Toomre (1972), in which the galactic disks were represented by test particles. Only recently have galaxy collisions been modeled in a reasonably realistic way. Barnes (1988) has modeled the encounter between disk galaxies, each containing three components in the form of a bulge, a disk and a halo, each of which were represented by thousands of particles, all mutually interacting. In these calculations he showed how the merger remnants of a collision between spiral galaxies indeed look very similar to normal elliptical galaxies. Calculations of this type require hours of supercomputer time, and therefore it is not yet feasible to perform very many of these types of experiment.

Before long, however, a detailed parameter study of galaxy-galaxy scattering will become possible. An increase in hardware speed of an order of magnitude, together with increased availability of computer time through supercomputer centers, will enable us to make a systematic survey of many different types of encounters. On the software level, the set-up, orchestration and analysis of such a survey will pose a number of new and interesting problems, close to the areas of pattern recognition and artificial intelligence in general.

If a Gigaflop computer would be available full-time for a project to investigate scattering processes between galaxies, automatization of the runs would be necessary to some extent, since individual runs could be carried out day and night at a speed of several encounters per hour (depending on the degree of accuracy required and the complexity of the encounters). And with the rapid increase in speed and parallellism of computers, such requirements will be felt before long. Since it will take years of work to build a software environment in which to realize the automatization of galaxy-galaxy scattering experiments, the time seems ripe to start laying some of the foundations for this type of software development.

Our first step has been to construct a toolbox, in the form of an environment with a large set of moderate-size programs which perform individual tasks, involving manipulations of N-body systems. Examples are the generation of initial conditions, the selection of subsystems, the addition, translation and rotation of systems, and the representation of systems in the form of various two-dimensional projections of the full phase space information. An example of the latter is given by a program which allows the user to make a two-dimensional projection of the positions of all particles, as a plot of the "sky". The user can then put a "slit" in an arbitrary position and orientation on the "galaxy" and obtain velocity information along the one-dimensional line define by the slit, just as in observational spectroscopy.

This toolbox is known under the name NEMO†. The foundations for this toolbox were laid in 1986-'87 by Josh Barnes, Peter Teuben and myself, and since then it has been growing steadily. In principle we are happy to share our NEMO environment with other interested astrophysicists, although we currently face a serious problem in that we are severely limited in people power for maintenance and documentation, two essential ingredients in any publicly available non-trivial body of software. Since NEMO forms a computational observatory, we need the equivalent of engineers and night assistants, in the form of software engineers and software archivists, before we can be of service to the community of stellar dynamicists at large. As soon as funding would become available for the NEMO project, we will be happy to make our software environment generally available.

References.

Barnes, J., 1988. *Astrophys. J.* **331**, p. 699.

Heggie, D.C. 1988, in *The Few-Body Problem*, I.A.U. Colloq. 96, ed. M.J. Valtonen (Kluwer Acad. Publ.), p. 213.

Hut P. 1985, in *Dynamic of Star Clusters, I.A.U. Symposium 113,* eds. J. Goodman and P. Hut (Dordrecht: Reidel), p. 231.

Toomre, A. & Toomre, J., 1972. *Astrophys. J.*, **178**, 623.

† The name NEMO was chosen because it grew atthe Institute for Advanced Study as a set of shared software, for which we created a special directory, owned by nobody in particular. However, the name nobody being a reserved word already, we chose the Latin translation instead.

PROBLEM ORIENTED LANGUAGE FOR EPHEMERIS ASTRONOMY AND ITS REALISATION IN THE SYSTEM ERA

G.A.Krasinsky, F.A.Novikov, V.I.Scripnichenko
Institute for Applied Astronomy, Leningrad, USSR

1. GENERAL OUTLINE OF THE SYSTEM

Practically any astronomical observational program needs ephemeris calculations either for prediction of the observables or for the aims of slewing. Sometimes these calculations are almost trivial and printed Astronomical Ephemerides may be useful. But as a rule that is not the case and observers are obliged to call for help from specialists in various branches of ephemeris astronomy. In many cases the ephemeris software is a real nuisance if one analyses the standard type acknowledgements in astronomical papers: "we are indebted to Dr. X who has provided us with the ephemerides of Galilean satellites" or "our thanks to Dr. Y for calculations of the lunar occultations". One of the aims of project ERA (Ephemerides for Research in Astronomy) was to satisfy the interests of users inexperienced in ephemeris astronomy. However that was not the single motivation. The system ERA was designed for professional use to facilitate solving the most complicated problems of ephemeris astronomy (calculation of the observables, comparison of the dynamical theories with observations, scheduling observational activities, prediction of astronomical phenomena, and so on). A technology for scientific work in the field of the ephemeris astronomy was worked out. Moreover running time operations were included in the system, because one of the intentions was to make it appropriate as a kind of operational system for large astronomical instruments, or for instrumental complexes (such as VLBI nets).

The core of the system is a problem-oriented language SLON which controls a developed applied program package. According to our experience, any observational program may be easily formulated in terms of the language. After translation, a resulting SLON-program is executed making calls to the subroutines from the applied program package. The base units of the language are tables (supplied by names and kept as files) of observations (real or synthesied). In the language there are well developed means to manipulate the tables, combine, modify and keep them.

Thus, the system consists of three parts:

1. SLON-processor.
2. File management system.
3. Applied program package.

The user need not know the structure of the applied program package ; he interacts with the package only in terms of the problem-oriented language SLON. The system is intelligent in the sense that there is no predescribed set of tasks, but any problem formulated in terms of the language may be solved.

The accuracy of the modelling in the system corresponds to the level of demands from the modern high precision observations (such as VLBI observations). The system ERA includes a

Celestial Mechanics **45**: 219–229, 1989.

specialised database for coordinates of the Earth pole and for differences TAI-UT1. These data are refreshed one time in a year after the BIH Annual Report is issued. As a standard for lunar and planet ephemerides the tape with DE 200 Chebyshev polynomials is in use, but the planetary theory [5] may be chosen as an option. The DE 200 tape was restructured and enriched by adding polynomials for lunar physical librations and for differences TAI-TDB due to General Relativity. For dynamical theories of other celestial bodies there are utilities for calculating Chebyshev polynomials according to the interface with the system ERA.

As the system is comparatively large and rather sophisticated, from the very beginning attention was paid to the principles of the system supporting, mobility, discipline of programming, documentation and testing. In order to facilitate administrating the system and satisfy diverse demands from users, a special language for description of the desired configuration of the system was realised (LAC - Language for Configuration). The output of the LAC-program is a database, which contains full information on all objects and constructions of the system ERA. This approach is quite effective in unifying the work of many people participating in the project.

Structural realisation of the system made possible the deep segmentation and as a result restrictions from computer memory are comparatively weak. The running version of the system consists of more than 20000 lines of Pascal and Fortran text and needs 128 Kbytes of core memory (for Soviet BESM-6 computer). This version of the system ERA was at work in summer 1986 and now it is thorougly tested in all its subsystems. The testing is a continuing process because in many cases a special scientific research presents new demands. The correctness of the ephemeris predictions is tested by discussion of observations from very different observational programs - from solar and lunar eclipses to VLBI observations of IRIS program and space probes of VEGA mission (3).

The ideas of the system evolved during 1983-1985 and at the end of this period they were fully crystallized and fixed. There are three programming efforts which have influenced the system. The main concept of the language has been developed from the ideas of the applied program package MAMONT [1]. The technology and style of the system programming were strongly influenced by works of S.Lavrov and his collaborators on automatic program synthesis [2,3]. The file management system has its origin in [4].

Apart from the authors many people took part in the project in various stages. We thank our colleagues I.Agamirzian, O.Breslav, E.Chlibov, I.Vasiliev, T.Ivanova, E.Chotimskaija, N.Krivova, E.Aleschkina, M.Eliseeva, E.Iagudina, and S.Tarasevitch for their help.

2. THE LANGUAGE SLON: INFORMAL DESCRIPTION.

2.1. Variables.

The user of the system ERA formulates his observational

program in terms of more or less convenient astronomical notation. For instance, he must choose OBJECT from a set of permissible values (SUN, MOON, MERCURY,...), indicate TYPE_OF_OBSERVATION from (SPHERICAL, RANGING, INTERFEROMETRY,...) and so on. Any variable is characterized by its type (in the sense of the type conception of program languages). The types of the SLON variables may be: 1. Integer, 2. Real, 3. Double precision, 4. String, 5. Double string, 6. Literal (for instance, the type OBJECT) and 7. Boolean.

SLON-translator makes strong type checking. The most usable variables for the running version of the system are:

1. Double precision: DATE, APPROXIMATION
2. Real: LONGITUDE, LATITUDE, ALTITUDE, PX, PY, X__POLE, Y__POLE, UT1_TAI, TAI_UTC, ALPHA, DELTA, MU_ALPHA, MU_DELTA, PARALLAX, X_CRATE, Y_CRATE, Z_CRATE, LIMB_R, LIMB_F, EPOCH, EQUINOXY, REFRACTION, CX, CY, O_CX, O_CY, DX, DY, STEP_OF_ITERATIONS, EQUINOX_OF_CATALOG, T, P, X, Y, Z
3. Integer: SUBTYPE, N_OF_ITERATION, N_OF_STAR, N, PARAMETER_FIXED,
4. Literal: TYPE_OF_OBSERVATION, COORDINATES, OBSERVER, OBJECT, TIME_SCALE, SYSTEM, CHANGING
5. Double string: OBSERVATORY, NAME_OF_STAR, COMMENT
6. Boolean: FLAG

String-type variables are presented in SLON-program as an arbitrary text bracketed by the symbol " (for instance, as "WASHINGTON"). The text is cut from right side to 6 (12) symbols for single (double) strings. The values of boolean variables are the symbols YES and NO.

The values of literal type variables are given by the following list:

TYPE_OF_OBSERVATIONS:
SPHERICAL, RANGING, TIMING, INTERFEROMETRY, DIFFERENTIAL, ANGULAR_DISTANCE, PHASE, ECLIPSES

COORDINATES:
EQUATORIAL, HORIZONTAL, HOUR_ANGLE, ECLIPTICAL, GALACTICAL

TIME_SCALE:
UTC, ET, UT1, TAI

SYSTEM:
FK4, FK5, FK4_FK5

OBJECT:
SUN, MOON, MERCURY, VENUS,MARS, JUPITER,SATURN, URANUS, NEPTUNE, PLUTO, STAR, PLANET VENUS_SATELLIT LUNAR_SATE LLI TE, EARTH_SATELLITE, MARS_SATELLITE, PHOBOS, DEIMOS, IO, EUROPE, GANIMED, CRATE, PROBE

OBSERVER:
TOPOCENTRIC, BARYCENTRIC

CHANGING:
INCREASE, DECREASE

To complete this description it is necessary to include into the list of the values for the type OBSERVER the set SUN_CENTERED,

MOON_CENTERED and so on (any OBJECT may be used as a prefix apart of STAR). Similarly, for the type COORDINATES the values SUN_CENTRIC, MOON_CENTRIC and so on have to be added.

In many cases in the process of observing, more than one object or observer are involved. To describe such complex observations two additional series of variables are constructed adding the postfix 1 or 2 to any variable. For instance, there exist variables DATE_1, DATE_2 and OBJECT_1, OBJECT_2.

2.2. Semantics of the variables

Any observation is determined by assigning concrete values to the SLON-variables. Corresponding observables may be calculated by a special operation COMPUTE with results assigned to the variables CX, O_CX (for the first parameter observed) and to CY, O_CY (for the second one). For correct calculation of the residials O_CX, O_CY, observed values are to be placed into OX, OY. In order to indicate that some variables are to be computed, before application of COMPUTE a special symbol * must be assigned to them. If symbol * is assigned to another variable than CX, CY, O_CX, O_CY, then an indirect ephemeris task arises. The system checks whether the task is meaningful and solves it by iterations. Variables STEP_OF_ITERATIONS, N_OF_ITERATIONS, APPROXIMATION, CHANGING rule the process of iteration and the variable PARAMETER_FIXED indicates either CX or CY has the desirable constant value of the observable for the indirect task.

Meaning of the observables is determined mainly by values of the variables TYPE_OF_OBSERVATIONS and SUBTYPE. In brief the meaning is:

SPHERICAL. Calculation of the longitude and latitude in the coordinate frame determined by the variable COORDINATES. If SUBTYPE is equal to 1 then it is the latitude of the object that is calculated else - the colatitude. The system of reductions and values of the astronomical constants are determined by the variable SYSTEM. By a default convention the calculated coordinates are apparent ones but, if EQUINOXY is not equal to zero, they are the mean coordinates for the date of EQUINOXY.

ANGULAR_DISTANCE. Calculation of the angular distance and positional angle between OBJECT and OBJECT_1. In dependency on SUBTYPE not only the angular distance D between the centers but a number of linear combinations between the distance and visible radii r and r1 of the celestial bodies are calculated according to the table:

SUBTYPE	observable
1	D
2	D - r1
3	D - r
4	D - r -r1
5	D + r -r1
6	D + R +r1

DIFFERENTIAL. Calculation of spherical coordinate differences of the bodies OBJECT_1 and OBJECT.

RANGING. Calculation of the time delay and Doppler shift. If time of observation corresponds to reception of the signal, then SYBTYPE is equal to 1, if it is a time of shooting then to 2.

INTERFEROMETRY. Calculation of the time delay and fringe frequency if SUBTYPE is equal to 1 and the space frequencies otherwise.

TIMING. Calculation of the time correction to the observed arrival time (i.e. reduction to the barycenter of Solar system). The second observable is the time derivative of the first one.

PHASE. Calculation of the angle D between directions OBJECT-OBSERVER and OBJECT-OBJECT_1. The observables are linear combinations of D, the apparent radius r of the body OBJECT_1, and r1 of the Earth (for the observer on the body OBJECT).They are calculated according to the value of SUBTYPE by the same way as for the value ANGULAR_DISTANCE.

ECLIPSIS. Calculation of angular distances between directions OBJECT-OBJECT_1 and OBJECT-OBJECT_2.

If the value of OBSERVER is TOPOCENTRIC, then coordinates of the observatory may be presented either as the geodetic LONGITUDE, LATITUDE and ALTITUDE or as the geocentric LONGITUDE, equatorial projection PX, polar projection PY., The three later variables having the higher priority.

The system automatically makes all reductions of the coordinates and times due to the Earth rotation irregularities according to BIH data and in correspondence with the value of TIME_SCALE. If BIH data are not yet available, then the user may introduce his own data by assigning "by hand" the numerical values to the variables X_POLE, Y__POLE, TAI_UTC, UT1_TAI.

Star coordinates are described by a set of parameters; ALPHA, DELTA, MU_ALPHA, MU_DELTA, PARALLAX, EPOCH, EQUINOXY_OF_CATALOG, RADIAL_VELOCITY.

It is possible to calculate observables for any points on the surface of the object. The user has to assign the values to polar coordinates LIMB_R and LIMB_F of the point on the limb. The radial coordinate LIMB_R must be presented in units of the apparent radius of the planet and the polar coordinate LIMB_F is supposed to be in the coordinate frame determined by the variable COORDINATES.

If one would like to calculate observables for the point on the surface of the planet, then numerical values are to be assigned to the variables X_CRATE, Y_CRATE and Z_CRATE. By this way the coordinates of the point are fixed in the standard coordinate frame with its origin in the barycenter of the central planet.

No need to assign values to all the variables. It is quite essential that there exists a default convention supported by LAC description of the system. The default values of the literal type variables are the first ones in the corresponding lists given above.

2.3. Tables and presentation of variables.

Due to the default convention it is possible to consider that after assigning the values to any subset of variables some observation is determined and thus an "elementary" ephemeris task arises. The most characteristic feature of the language is the advanced means for describing a flow of the elementary tasks which in the whole may correspond to a rather complicated observational program. The simplest way to describe such a flow is to introduce a table which contains the values of the variables for each task in every row. These values will be referred to as a "Tuple." The head of the table determines the names of the corresponding SLON-variables. This table may be supplied by a name. For instance, the SLON operator

```
EXAMPLE  :=//OBJECT,       DATE,  CX /
                SUN, 2442000.5D,   *  /
               MOON,     -      ,  *  //;
```

determines a table EXAMPLE which is kept for one run of the program (the postfix "D" marks the double precision numbers). In order to save the table, the symbol of assigning <= ("arrow") must be used instead of the symbol :=. The symbol "*" indicates that the variable CX is unknown; the symbol "-" stands for the default values (for DATE that is JD 24442000.5).

The internal presentation of the variable as a rule differs from the external one (in the SLON-program presentation) and the user may choose a system of units which is the most adequate to his problem. For instance, if the calendar dates are preferable, then the table EXAMPLE may be presented in the form:

```
EXAMPLE :=//OBJECT, CALEND (DATE), CX/
               SUN,   19720101.0D, * /
              MOON,        -     , * //;
```

2.4. Algebra of tables

Any table may be presented in SLON-program by three ways:

1. By explicit denotations of all its tuples (see 2.3).

2. By reference to the name of a table determined earlier.

3. By a program which consequently gets all tuples according to a "table expression".

Table expressions are constructed by special operations "addition" and "multiplication", which operate tables as their arguments. Thus, an algebra arises and SLON_operators have the form of table polynomials. Let us describe this algebra.

1. Addition of tables. The sum A+B of the table A (with n tuples) and B (with m tuples) is a table constructed according to the rules:
a) Headline of the sum A+B consists both of names of variables in A and B.
b) The tuples of the sum are n tuples of A and m tuples of B. Missing values are set by the default convention.

2. Multiplication of tables. The table A*B is a result of multiplying the table A and B. The operation is meaningful if no common variables are present in A and B. Just as for addition, the headline unites names of variables in A and B. Tuples of the product (in number n*m) are obtained by consequent joining every tuple of A with all tuples of B. For instance, the table EXAMPLE may be presented as a sum:

```
EXAMPLE:=//DATE,OBJECT,CX/ 2442000.5,   SUN, *//
         +
         //DATE,OBJECT,CX/ 2442000.5,MOON, *//;
```

or as a product:

```
EXAMPLE:=//DATE, CX /2442000.5, * //
         *
         //OBJECT/SUN/MOON//;
```

From this example, it is clear that the operation "multiplication" makes it possible to factorize the tables with regular structure and to save the place and efforts while determining the tables.

3. Iteration of a table. Instead of adding the table several times, one may use the operation of "iteration" (multiplying the table by an integer constant). For instance: A+A+A is equivalent to 3A.

2.5. Actors

Any table may be accompanied by "actors" which work on every tuple and either transform the values of the variables or simply have some side effects (for instance, they may print the values or calculate observables). For calculation of observables the actor COMPUTE which incorporates all subroutines from the applied program package was realized. For output of results the actors PRINT or SCREEN_OUTPUT may be used. Let us consider a SLON operator which refers to the table EXAMPLE determined above:

```
:= EXAMPLE COMPUTE PRINT( CALEND(DATA), OBJECT, HMS(CX) );
```

In this case the resulting table is rewritten not to another one but "into air". Result of the work of the operator is a printed listing which consists of the headline of the table EXAMPLE and its tuples. The actor COMPUTE has assigned to the variable CX the value calculated according to the meaning of the ephemeris task. Namely, it is the apparent right ascensions CX of the Sun and the Moon for JD 2442000.5 that is calculated (by the default convention the TYPE_OF_OBSERVATION is SPHERICAL, the COORDINATES are EQUATORIAL and so on).

This example demonstrates that the function-transformers may be applied to parameters of actors. The interpretation in this case depends on the concrete actor. For PRINT these functions transform internal representation of the variables into an external one according to the chosen system of units (calendar data for DATE and hours, minutes, seconds for right ascension CX).

There are four embedded (built-in) actors of projection and selection having the form:

ONLY (list of the variables)
(rejection of columns of the table, which are not in the list).
EXCEPT (list of the variables)
(rejection of columns of the table from the list).
IF (boolean expression)
(rejection of tuples for the value NO of the boolean expression).
IF_NOT (boolean expression)
(rejection of tuples for the value YES of the boolean expression).

Boolean expressions are constructed in the usual way from variables and their values by means of logical relations ">" , "<", "=" , "<>", "OR", "NOT", "AND". As operands of the boolean expressions, the algebraic expressions involving SLON variables are allowed. Moreover, all sets of elementary functions as well as the function-transformers mentioned above may be used. For instance

```
:=EXAMPLE IF( DATE > CALEND(19720101.0) AND OBJECT=MOON);
```

Besides the one-argument elementary functions and function transformers, there exist "function-constants" without any arguments. They give the values of the most usable constants (value PI, light interval TAU, astronomical unit AU and so on). One such function of special interest is the function "I" which gives the running number of the corresponding tuple.

Users may construct their own actors and include them in the system of LAC program. For any actors in addition to the four embedded ones, it is possible to organize conditional calls by the indicator ON with a boolean expression. For instance:

```
:=EXAMPLE COMPUTE ON (OBJECT=SUN);
```

In order to call actors only on the last tuple of the table, one may use the indicator ON_END - all actors after this indicator work only at the end of the table.

Now we give the list of the actors which were realised by May 1988. Parameters of the actors are either a list of names of variables or expressions (arithmetical or logical). Strong type checking is guaranteed. If the list of names is not a single parameter, then this list must be bracketed.

Table 2. List of the actors

Actors and their parameters	semantics
COMPUTE	Calculation of the observables
PRINT(<list>)	Printing
SCREEN_INPUT (<list>)	Input from the screen for the interactive work
SCREEN_OUTPUT(<list>)	Output from the screen
CHANGE ([<list>], <any>)	Assigning the value of the second parameter to the variables from the list
PLOT (<integer>, [<list>], <double>, <double>, <double>, <double>, <string>)	The first parameter - number of the plot, the first two elements in the list -the variables to be plotted (abscissa and ordinate), the four double variables - the boundaries; the string is a symbol for drawning.
PLOT_OUTPUT (<integer>)	Output of the picture formed by consequent application of PLOT. The integer is a number of the plot.
SAVE (<list>)	Saving one tuple in an internal memory.
RESTORE (<list>)	Restoring from the internal results which have been saved by SAVE
PARTIALS (<double-string>)	Calculation of partials and saving them in the file with name given by the string parameter.
PARTIAL_OUTPUT	Output of the partials.
SURFACE(<list>)	Calculation of the height on the apparent limb of celestial bodies with polar LIMB_F and LIMB_R coordinates. (Mars, Venus and lunar marginal zone)
TIME (<list>)	Running time (Julian date) from a timer is assigned to the first element of the list
PAUSE (<real>)	Pause in sec. in real time calculations

2.6. Loops.

Making use of the iterators of tables, the actor CHANGE and the function "I" (which gives the sequence number of the tuple) one may construct the table with arbitrary complex cyclic structure. For instance, the operator

```
DATA:=(100//DATE/2444000.5//)
      CHANGE( [DATE], DATE+(I-1)*2 ) ;
```

has the result that the table DATA, which consists of 100 tuples, the values of DATE increase with the step 2.

3. TESTING.

There is a set of tests for every subsystem of ERA. Here we present the results of testing the most important applications of actor COMPUTE (calculation of observables). The list of the tests for COMPUTE consists of:

1. Ranging of Mars (Haystack, 1971)
2. Meridian observations of the Moon (USNO 1971)
3. Occultation of Eps Geminorum by Mars (Lowell obs. 1976)
4. Transits of Mercury (1972). Isochrones for the first contact.
5. Earth pole positions and differences UT1-TAI
6. Differential VLBI observations of VEGA-1 space probe (DSN observations).)

7 Observations of Phobos by MARINER space probe (1972)

In SLON-program comments are marked by a pair of symbols "--". Due to lack of space, plots of results are not given here. Note that listings generated by the actor PRINT have the form of SLON tables. Thus the listings may be used while constructing new SLON program.

Note that the SLON-processor is an interpreter, the operators are interpreted and executed consequently and results are printed before translation of the next operator. Test 1. is given as an example.

3.1. Ranging of Mars (Haystack, 1971)

```
OBSERVATIONS:= (          //DATE     ,  MKS(OX)   /
                2441057.479444623D, 1031949744/
                2441061.948874950D,   993299697//
                                    *
  // HSEC (LONGITUDE), RE(PX),     RE(PY),    O_CX, OBJECT,
          TIME_SCALE,   TYPE_OF_OBSERVATIONS,  SUBTYPE /
          17157.28, 0.736972, 0.6736913,     *  ,   MARS,
                ET,                RANGING,         2 //)
          COMPUTE  PRINT( CALEND (DATE), MKS (O_CX) ) .
```

```
        //CALEND(DATE),              MKS(CX),        MKS(O_CX)/
--------------------------------------------------------------
    19710415.23302401D,       1031948752.844,        -8.844  /
    19710420.10462279D,        993299697.828,        -0.425  //
--
--   This listing is a result of work of PRINT
```

References

1. Krasinsky G.A. - System for automatisation of ephemeris calculations MAMONT-1981. -Algorithms of Celest. Mech., 39, pp. 3-32, 1982 (in Russian)

2. Babaev I.O., Novikov F.A. Petrushina T.I. - Language DESCARTES of the system SPORA - Applied informatics, 1, pp. 35-73, 1981 (in Russian)-

3. Agamirzian I.R. "STEP" - a computer design technological support system. Analyzer development tools. Algorithms of Celest. Mec., 46, pp. 3-50, 1980 (in Russian).

4. Shripnichenko V.I.- Dynamic exchange with external storage in UPP system. - Algorithms of Celest. Mech., 36, pp. 3-30, 1981 (in Russian)

5. Krasinsky, G. A., Saramonova E. Ju., Sveshnikov M.L., Sveshnikov, E.S. - Universal time, lunar tidal deceleration and relativistic effects from observations of transits, eclipses and occultastions of XYIII-XX centuries. - Astron. Astropoh., 145 pp. 1985.

NUMERICAL INTEGRATION METHODS IN DYNAMICAL ASTRONOMY

HIROSHI KINOSHITA and HIROSHI NAKAI
National Astronomical Observatory
Osawa 2-21-1, Mitaka, Tokyo, Japan

Abstract. We review what kinds of numerical integrators are used by astronomers in the field of dynamical astronomy and to what problems they are applied . This review is based on the questionaires distributed mainly to the members of IAU Commission 7(Celestial Mechanics). Because of the restriction to the Commission 7 members,the answers are mainly from astronomers in the solar system dynamics and problems mentioned in the answers are also related to celestial bodies in the solar system. Other than above, two questions,how to check the precision or accuracy of numerically integrated results and how to treat a close approach, are also surveyed . The problem of the suitable choice of a numerical integrator from various numerical integrators is out of the scope of this review,and it depends strongly on the dynamical nature of a particular dynamical system and the required accuracy.

1. Distribution of Questionaires

The questionaires were sent mainly to members of IAU Commission 7 (Celestial Mechanics),who are picked from an Alphabetical List of Members(1986) published by IAU. The distribution of questionaires is summarized in Table 1 and the number of replies are also listed.

TABLE 1. Number of distribution of queationaires to members of Commission 7(Celestial Mechanics) and number of replies

	Number of distribution	Number of Replies
Argentina	2	1
Brazil	5	2
Colombia	1	0
U.S.A.	47	26
Australia	1	1
China	19	5
India	1	1
Japan	11	9
Korea	1	1
Egypt	5	0
Austria	2	1
Belgium	4	0

Celestial Mechanics **45**: 231–244, 1989.

Denmark	1	0
Finland	2	1
France	26	9
Germany,F.R.	4	3
Greece	10	5
Ireland	1	1
Italy	5	2
Norway	1	0
Portugal	1	1
Spain	5	0
Sweeden	1	1
United Kingdom	14	5
Bulgaria	10	0
Czechoslovakia	4	2
Hungary	1	0
Poland	3	3
Rumania	1	1
U.S.S.R.	29	4
Yugoslavia	3	3
Sum	212	88

Some replies are from non-commission members,since I asked commission members to pass the questionaires to their colleagues. About 40 percent replies are quite good,since the majority of commission members seem to be theoreticians who do not use a computer themselves.

2. Numerical Integration Methods

Recently a number of new numerical integration methods have been developed for a system of ordinary differential equations. Some of them are based on an extension of traditional ideas which assume calculation by hand,and become feasible with use of computers. They are divided into two groups:
a)multistep methods,
b)one step methods,
   1)Runge-Kutta type methods,
   2)Taylor type methods,
   3)Extrapolation methods.
At first we explain briefly in Sections(2-1,2,3,4) the principles of these methods and some packages based on these methods,which are mentioned in the answers to the questionaires. Then we show in Section(3-1) what kind of integrators are widely used in the field of dynamical astronomy.

2.1. MULTISTEP METHODS

Consider the initial value problem for first-order differential

$$\frac{dx}{dt} = f(x,t), \quad x(t_0)=\xi \tag{1}$$

where x and f are vectors. We introduce the following notations,

$$x_n=t_0+nh, \qquad f_n=f(x_n,t_n),$$

where x_n is an approximate solution to $x(t_n)$, and h is called the steplength or stepsize. A general linear multistep method or a linear k-step method can be

$$\sum_{j=0}^{k}\alpha_j x_{n+j} = h\sum_{j=0}^{k}\beta_j f_{n+j}, \tag{2}$$

where α_j and β_j are constants which define the integrator (2) with step number k, and $\alpha_k=1$. The method (2) is called an explicit integrator or a predictor if $\beta_k=0$, and an implicit integrator or a corrector if $\beta_k \neq 0$. The implicit integrators in general need more computational time than the explicit integrators, since the equation is implicit with respect to the unknown x_{n+1} and the approximate solution x_{n+1} is obtained by an iterative procedure. Even if the implicit methods are time-consuming, the difference between the approximate solutions by the predictor and corrector can be used for the control of changes of stepnumber k and stepsize h. A method of the form(2) is called of order p, if the local truncation error is of order h^{p+1}.

The typical integrators (2) are Adams-Bashforth methods(explicit) and Adams-Moulton(implicit), in which $\alpha_{k-1}=-1, \alpha_{k-j}=0, j=1,2,\cdots,k-2$. Sometimes both methods are simply called Adams type integrators and their order is equal to the number of evaluating f.

There are also linear multistep methods for second-order differential equations

$$\frac{d^2x}{dt^2} = f(x,t), x(t_0)=\xi, \frac{dx}{dt}(t_0) = \eta. \tag{3}$$

A general linear multistep method for (3) can be written

$$\sum_{j=0}^{k}\alpha_j x_{n+j} = h^2\sum_{j=0}^{k}\beta_j f_{n+j}. \tag{4}$$

Popular integrators (4) are Stormer methods(explicit) and Cowell methods(implicit), where $\alpha_{k-1}=-2, \alpha_{k-2}=1, \alpha_j=0, j=1,2,\ldots,k-3$, and the order of both Stormer and Cowell methods is equal to the number of evaluating f. Explicit expressions α_j, β_j in (2) and (4) are given in various textbooks(ex. Henrici(1961) and Lambert(1973)). The Stormer method has been given several different names such as Gauss- Jackson method , the second sum method. Other than the conventional methods mentioned above , other types of families of multistep integrators are described in various textbooks(see references). Lundberg(1981) proposed the integrators which use the velocity with the positions and accellerations for second-order differential equations. Huang and Valtonen(1988) proposed a new family of linear multistep integrators which use higher derivatives ($x^{(l)}, l \geq 3$), which are called Oberchkoff methods(see Chapter 7 of Lambert(1973)). Feagin(1976) proposed a new class of linear multistep methods which are generalizations of the Stormer and Cowell methods.

Software packages based on multistep methods are

DVDQ - variable-order, variable-step method by Krogh(1970) for a system of first- and second-order ordinary equations,

DIVA - an extension of DVDQ, step changing algorithm mathematically like that in EPISODE, available from Krogh,

ODE - variable-order, variable-step method by Shampine and Goldon(1975)

for a system of first-order differential equations,

ABFS - fixed-order,fixed-step Adams method by Lundverg(1981) for a system of first-order differential equations,

KSGFS - an extension of the work by Shampine and Goldon(1975) for a system of second-order differential equations,(?)available from Lundberg,

Schubart and Stumpf(1966)
- fixed-order,fixed-step for a system of second-order differential equations(predictor-Stormer,corrector-Cowell),

EPISODE-variable-order,variable-step Adams-Moulton(implicit)method by Hindmarsh and Byrne(1975) for a system of first-order differential equations.

2.2. RUNGE-KUTTA

A general Runge-Kutta method for first-order differential equation (1) can be written

$$x_1=x_0+h\sum_{i=1}^{s}b_i f_i, \tag{5-1}$$

where

$$f_i=f(x_0+h\sum_{j=1}^{i-1}a_{ij}f_j,\ t_0+c_ih),\ i=1,2,\dots,s. \tag{5-2}$$

The parameters a_{ij}, b_i, c_i, and s(the number of stages) governs the properties of the integrators(5)(e.g. order). The difference between the solution $x(t_0+h)$ and the approximate solution x_1 is of order h^{p+1},where p is called order of (5) and is less than s if$s\geq 5$. Fehlberg(1968,1969) derived formulae that have an ability to control the stepsize. With use of same f_i(5-2),another approximate solution is given by

$$x_1^*=x_0+h\sum_{i=1}^{s^*}b_i^* f_i, \tag{5-3}$$

The order of (5-3) is p-1 and $s^*<s$. The difference$(x_1-x_1^*)$ represents an approximation of the leading local truncation error,which can be used to estimate a new stepsize. Fehlberg's integrators are sometimes called embedded Runge-Kutta methods or Runge-Kutta-Fehlberg methods,simply RKF methods. Runge-Kutta type integrators for a system of second-order differential equations are called Runge-Kutta-Nystrom methods or RKN methods. Fehlberg(1972) also derived embedded Runge-Kutta-Nystrom methods ,which are called Runge-Kutta- Nystrom-Fehlberg methods or simply RKNF methods. Dormand and Prince(1978,1988) have developed another family of RK and RKN methods. Everhard developed another family of implicit RK and RKN. In Everhart methods,the spacings for substeps c_ih are the same as those for Gauss-Radau and Gauss-Lobatto quadratures,which are numerical integration methods of definite integrals with unequal stepsizes.

Popular software packages based on Runge-Kutta methods are

RKFp(p+1),RKNFp(p+1) - fixed-order,variable-step Runge-Kutta method by Fehlberg(1968,1974) for a system of first- and second-order differential equations,

RADAU - fixed-order, variable-step Runge-Kutta method by Everhart(1974) for a system of second-order differential equations.

2.3. TAYLOR-SERIES METHOD

A Taylor-series method for a system of first-order differential equations(1) has a simple form

$$x_1 = x_0 + hx_0^{(1)} + \frac{h^2}{2}x_0^{(2)} + \cdots + \frac{h^r}{r!}x_0^{(r)}, \tag{6}$$

where $x_0^{(i)}$ is i-th derivative of x_0 with respect to time. This method is also called time series method. Even if the formula(1) itself is simple, the calculation of higher derivatives is not simple:

$$x^{(1)} = f,$$

$$x^{(2)} = f_t + f_x f,$$

$$x^{(3)} = f_{tt} + 2f_{tx}f + f_{xx}f^2 + f_x^2 f + f_x f_t.$$

It is immediately evident that the easy calculation of the higher derivatives or availability of a recurrence relation for them is an essential requirement. Steffensen (1957) constructed practical recurrence formulas of both the restricted three-body problem and the general three-body problem, and, then, Moran (1973), Moran, Roy, and Black (1973), Emslie and Walker (1979) derived other forms of recurrence formulas for the restricted problem. Broucke(1971) and Black(1973) gave recurrence formulas for the general n-body problem, which are based on the idea of Steffensen (1957). Kinoshita(1977)applied a Taylor-series method to check the accuracy of the analytical third-order solution of an artificial satellite theory with J_2, J_3, J_4 terms as the geopotential.

A series of programs(NBODY) developed by Aarseth(1972,1985) adopt the idea of Taylor series expansion in the elementary process of numerical integration. His programs are widely used in the simulations of a many-body problem like a globular cluster.

Instead of developing the solution into Taylor series, Grobner(1967) derived an integration formula which expands the solution into Lie series. A recurrence relation for higher Lie derivatives is necessary for accurate integrations by Lie series, like in the Taylor series method. Recently Hanslmeier and Dvorak(1984) applied Grobner's idea to the n-body problem and derived recurrence formulas for higher Lie derivatives. They applied their method to planetary motions(Sun,Venus,...,Pluto) and found that 1)Lie-series method is relatively fast(about factor 3-10 times faster in comparison with the n-body program by Schubart and Stumpf (1966) and 2)because larger steps(e.g. stepsize 135d for Jupiter)can be used, round-off errors are smaller. The serious problem in Taylor-type methods is to find specific recurrence formulas for a chosen equation of motions, and it is not easy to write a program without errors compared with other general purpose integrators.

2.4. EXTRAPOLATION METHODS

Let us consider the solution for a substepsize h at some fixed point $t+H$ to the form

$$x(t+H,h)=x(t+H) + \sum_{i=1}^{m}\tau_i h^i + O(h^{m+1}). \tag{7}$$

What we want to know is $x(t+H)$ as $h\to 0$. We can eliminate $\tau_1,\ldots,\tau_m$ with use of $x(t+H,h)$ for $h=H,h_1,\ldots,h_m (H>h_1>\ldots>h_m>0)$ and express $x(t+H)$ in the following form

$$x(t+H)=\sum_{i=0}^{m} c_{i,m} x(t+H,h_i) + O(H^{m+1}). \tag{8}$$

The conditions under which an expansion of the form (7) exists have been investigated by Gragg(1963,1965) and Stetter(1970). Gragg(1963,1965) found the asymptotic expansion

$$x(t+H,h) = x(t+H) + \sum_{i=0}^{m}\tau_i h^{2i} + O(h^{2(m+1)}, \tag{9}$$

with the modified mid-point method for evaluating $x(t+H,h)$. See Chapter 6 of Lambert(1973) or Chapter 6 of Gear(1971) for more information on extrapolation methods. According to Fukushima (1988) who compared the extrapolation method with RKF, DVDQ, and the Adams-Moulton method, the extrapolation method is suitable for the problem requiring the highest accuracy.

One of the popular packages based on extrapolation methods is

DIFSYS - a rational extrapolation method with variable-stepsize by Burlirsch and Stoer(1966) for a system of first- and second-order differential equations. This program was written originaly in Algol and the FORTRAN version is listed in the textbook of Gear(1971).

2.5. CANONICAL INTEGRATORS

Differential equations widely used in dynamical astronomy are derived from a Hamiltonian. The exact solution of such a system of differential equations is considered to be a canonical transformation (a symplectic transformation) from the initial conditions. All integrators mentioned in this section are not canonical. A canonical transformation of a Hamiltonian system is an area preserving map. An integrator, however, that is not canonical is not area preserving , and so the integrated system will be spuriouly damped or excited. For the study of a chaotic behaviour of a dynamical system with use of numerical integration, canonical integrators will be powerful and desirable. Rcently canonical integrators are proposed by Ruth (1983), Channell (1983), Menyuk (1984), Aizu (1985), Fen (1987), and Neri (1987).

The Wisdom mapping (1982) (someone calls it Chrikov-Wisdom mapping) is a canonical transformation.

3. Results of Questionaires

3.1. "WHAT KIND OF NUMERICAL INTEGRATOR DO YOU USE?"

The answer to this question is summarized in Table 2. To my surprise there are many confusions on naming numerical integration methods, including myself, because the identical method is called in different ways. In section 2, we tried to avoid confusions on names by giving other names in addition to the commonly used name.

TABLE 2. Integrators mentioned in the answers.
(Cowell* means Adams-Bashforth, Adams-Moulton, Stormer, and Cowell methods.)

Type	number(%)	type	number(%)
Taylor	20(12)	Taylor series	13(65)
		Lie series	5(25)
		Chebyshev series	2(10)
Extrapolation	24(15)		
Runge-Kutta	66(40)	RK(4)	25(38)
		packages	
		RADAU	19(29)
		Fehlberg	16(24)
		RK-Dormand	3(5)
		other	3(5)
Multistep	54(33)	Cowell*	34(63)
		packages	
		Schubart-Stumpf	9(17)
		DVDQ	3(6)
		ODE	3(6)
		Episode	1(2)
		other	4(7)

From Table 2 we see popular integrators are Runge-Kutta type integrators and multistep methods, and fourth-order Runge-Kutta and Cowell* integrators are widely used. Here Cowell* integrators are defined as Adams-Bashforth (predictor), Adams-Moulton (corrector) for first-order differential equations, and Stormer(predictor), Cowell (corrector) for second-order differential equations.

Runge-Kutta-Gill(RKG) method is one of fourth-order Runge-Kutta methods, which reduces the round-off errors. RKG method assumes the use of a fixed point arithemetic. When we use a floating point arithmetic, RKG is not superior to the traditional fourth-order Runge-Kutta methods.

Most users of multistep integrators apply correctors in their integrations. Some who seem to investigate the global behaviour of planetary motions do not apply correctors. When the dynamical behaiviour is regular (not chaotic, no appearance of close-approach, small eccentricity), it is not necessary to change stepsize. When the time span for integration is quite long in addition to the regularity, we cannot expect the improvement of the approximate solution with use of a corrector, because the global accumulated error due to round-off errors is larger than that due to truncation errors for very long time integration (ex. Cohen, Hubbard, and Oesterwinter 1973, Kinoshita 1984). By omission of a corrector, we can save computer time.

One step methods are self-starting, because they use only initial values. On the other hand, multistep methods need several starting values, which should be calculated at the beginning. This difference, whether starting values are necessary or not, is not an important factor in the criterion of the choice between one-step or multistep methods, since the calculation of the starting values does not take computing time compared with long time span integrals. Ease in changing stepsize is also not an important criterion in the present computer age. It is, however, quite difficult to give criteria for the selection among these various integrators from the standpoints of efficiency (or speed) and accuracy, since they depend mainly on the dynamical nature of a particular dynamical system and the required accuracy. At least we can say any package mentioned above can be used, if the time span of integration is not so long compared with the fundamental periods of the chosen dynamical system and the required accuracy is not high. The merits and demerits of traditional integrators are discussed in textbooks(see reference). As for the newly proposed integrators, the proposer discusses the merits of his own method against other integrators in his paper. Many papers have discussed the relative merits of various integrators, and some of them are

Fox(1984) :choose 3 methods from multistep integrators,4 methods from Runge-Kutta integrators,one from Taylor series integrators,one from extrapolation methods,apply them to the two-body problem,and compare their speed and accuracy,

Lundberg(1981):choose 7 methods from multistep integrators and 2 methods from Runge-Kutta integrators,apply them to differential equations with periodic solutions(harmonic oscilator,two-body problem,Euler rigid-body problem,restricted three-body problem) and two artificial satellite problems with two force models.and compare their merits and demerits,

Parp,Innanen,and Patrick(1978,1980):choose DVDQ,RADAU,RKF6(7),DIFSYS, EPISODE,apply them to stellar orbits in three model galaxies, and compare their relative merits on accuracy,number of function calls,and computer time,

Kalbouridis(1988):choose several fourth,fifth,and sixth order Runge Kutta integrators(explicit),apply them to the motion of a charged particle in the electromagnetic field of a magnetic binary,and compared against each other regarding cost and accuracy.

3.2."HOW DO YOU CHECK THE PRECISION OR ACCURACY OF NUMERICALLY INTEGRATED RESULTS?"

The conservation of the energy integral is frequently used, but it is not a good criterion for the error of positions. According to the analysis of accumulation of truncation (or discretization) error in the two-body problem (Kinoshita 1968), the error of the semi-major axis (equivalent to the energy) changes proportional to time and the mean motion also changes proportional to time through the Kepler's third law, and, therefore, the global error of the longitude increases proportional to the square of time.

Instead of use of integrals,Szebehely and Bettis(1970) and Huang and Innanen(1983) suggested to use the integral invariant relations. Huang and Innanen(1983) concluded on the invariant relation "This technique is a good check in the stable and order(regular) regions,but is not an ideal one in the unstable and stochastic regions".

Another popular method to check the accuracy is to check the reversability of solutions. If the integrator has a time-reversal nature,this check does not give any information on the accumulation of truncation errors. Integrators discussed in Section 2 are not time-reversal.

TABLE 3. Methods of reliability check of numerically integrated results

	number(%)
conservation of integrals	45(33)
reverse of time	43(32)
change of stepsize	12(9)
use of two different methods	9(7)
higher order methods,change of tolerence	7(5)
use of higher precision arithmetic	6(4)
comparison with soluble problems	5(4)
comparison with observation	2(2)
repeat integration	2(2)
integral invariant relation	1(1)

Another main source of errors in numerical integrations is the round-off error. Brouwer(1937) derived his famous law that the accumulated round-off error in the longitude after a large number N of steps is proportional to $N^{3/2}$ under the assumption the round-off error is statistically even. According to Milani(1988),"What happens when the rounding off is not statistically even has been studied by Fabri and Penco(1984); they have found that the accumulated round-off error in the longitude can be described by a normal distribution:

$$\Delta\lambda \simeq cnhN^2\varepsilon + N(0, dN^{3/2}\varepsilon), \tag{10}$$

where ε is the last stored digit in the used computer,and the constant c and d can be estimated if the details are known on how the processor really handles floating operations ". The round-off error in floating point operations of a majority of computers currently used does not seem to be statistically even. In the present situation,both truncation and round-off errors are considered to be proportional to the square of the overall integrated time span. Milani(1988) also discussed in detail other error sources which become important for the long time span integral and his investigation is based on his 100 million years integration of the outer five planets.

Here we should mention a new idea by Sussman and Wisdom (1988) to reduce the accumulated errors in position. The error in energy of the system produces the error of the longitude , which increases proportional to the square of time, and does depend on the chosen step size. They found the special stepsizes drastically reduce the slope of the energy error, by roughly three orders of magnitude in their 845 million years integration of outer planets (1988). The special step size does depend on the differential equations of a particular problem, a chosen numerical integrator, and the computer hardware used.

3.3. "IN WHAT PROBLEM DO YOU USE NUMERICAL INTEGRATION?"

The answer to this question is shown in Table 4.

Table 4. Subjects of researches

	number (%)
Asteroid	41(19)
Comet	25(12)
Planet	26(12)
Natural satellite	18(8)
Artificial satellite	29(14)
Nutation	2(1)
Periodic orbit	20(9)
Few-body problem	30(14)
Many-body problem	9(4)
Galaxy dynamics	4(2)
Globular cluster	2(1)
Astrophysics	4(2)
other	4(2)

Table 4 shows that the half of subjects appeared in the answers are related to the solar system dynamics, partly because the questionaires were mainly distrubuted to the members of Commission 7 (Celestial Mechanics).

3.4. "HOW DO YOU MANAGE A CLOSE APPROACH?"

The answer to this question is summarized in table 5.

TABLE 5 . Treatments of a close approach

	number (%)
change of stepsize	57(64)
regularization	22(25)
change of coordinate system	5(6)
change of potential	2(2)
stop calculation	2(2)
change of order of integrator	1(1)

A majority of software packages mentioned in this review automatically detect a close approach and change the stepsize. When the method of the stepsize does not work,we have to regularize the equations of motion if we want to know the behaviour after the close approach. Kustanheimo-Stiefel transformation (KS-transformation) is well known for regularization in three-dimensional space(see the textbook by Stiefel and Scheifele 1971). In the general N-body system close approaches between more than two bodies are common and frequent changes of the regularized pairs are necessary. Aarseth and Zare(1974) developed a three-body regularization method in which two of the possible three pair-collisions are regularized, while Heggie(1974) found a way to regularize simultaneously all the possible pair collisions. Heggie(1974) also generalized the method to the N-body system,and Mikkola(1985) derived globally regularized equations of motion for the N-body problem which are based on the method of Heggie(1974) and much easier to handle. There exist many papers on regularization in the field of stellar dynamics. A summary of recent work on the regularization and integration methods is contained in a review by Aarseth(1988). His review also discusses stabilization methods which are based on some energy constraints, time smoothing or the variation of parameters. For example he recommends a stabilization of perturbed Keplerian motion introduced by Baumgarte(1972) for a dynamical system where the binding energy of the dominant two-body motion changes slowly with time.

Numerical integration becomes more important and vital to the study of dynamical astronomy, as not only the time scale but also the number of bodies in the dynamical system increase. For the study of regular motions of a dynamical system we can use any integrator mentioned in this review and choose a high-order integrator if high precision is necessary. For the study of chaotic behaviors which appear in long time scale and are caused by close approaches, canonical integrators and regularization of the system will be desirable.

I would like to thank people who kindly responded the questionaire on which this review was based.

References

Aarseth,S.J.,1972,Numerical Experiments on the N-Body Problem,*in Gravitational N-Body Problem,ed.by Lecar,D.Reidel*,29-43.
Aarseth,S.j.,1985,Direct Methods for N-Body Simulations,*in Multiple Time Scales,ed.by Brockbill and Cohen,Academic Press*,377-418.
Aarseth,S.J.,1988,Integration Methods for Small N-body System,*in the Few Body Problem,ed. by Valtonen,Kluwer Academic Publisher*,287-307.
Aarseth,S.J.and Zare,K.,1974,A Regularization of the Three-Body Problem, *Celest.Mech.*,**10**,185-205.
Aizu,K.,1985,Canonical Transformation Invariance and Linear Multistep Formula for Integration of Hamiltonian Systems,*J.Comput.Phys.*, **58**,270-274.
Baumgarte,J.,1972,Numerical Stabilization of the Differential Equations of Keplerian Motion,*Celest.Mech.*,**5**,490-501.
Baumgarte,J.,1976,Stabilization by Making Use of a Generalized Hamiltonian Formalism,*in Dynamics of Planets and Satellites and Theories of Their Motions,ed.by Szebehely,D.Reidel Pub.Co., Dordrecht-Holland*,149-156.
Black,W.,1973,Practical Considerations in the Numerical Solution of the N-Body Problem by Taylor Series Methods,*Celest.Mech.*,**8**,357-370.
Broucke,R.,1971,Solution of the N-Body Problem with Recurrent Power Series,*Celest.Mech.*,**4**,110-115.
Brouwer,D.,1937,On the Accumulation of Errors in Numerical Integration, *Astron.J.*,**46**,149-153.
Burlirsch,R. and Stoer,J.,1966,Numerical Treatment of Ordinary Differential Equations by Extrapolation Methods,*Num.Math.*,**8**,1-13.

Channell,P.J.,1983,Symplectic Integration Algorithms,*Los Alamos National Laboratory internal report AT-6:ATN-83-9*.
Cohen,C.T.,Hubbard,E.C.,and Oesterwinter,C.,1973,Elements of the Outer Planets for One Million Years,*Astronomical Papaers of the American Ephemeris*,**22**,pt.1.
Dormand,J.R.and Prince,P.J.,1978,New Runge-Kutta Algorithms for Numerical Simulation in Dynamical Astronomy,*Celest.Mech.*,**18**,223-232.
Dormand,J.R.,El-Mikhawy,M.E.A.,and Prince,P.J.,1987,*IMAJ.Numerical Analysis(in press)*,*(Teeside Polytechnic Mathematical Report, TPMR 86-2)*.
Emslie,A.K. and Walker,I.W.,1973,Studies in the Application of Recurrence Relations to Special Perturbation Methods II,*Celest. Mech.*,**19**,147-162.
Everhart,E.,1974,Implicit Single-Sequence Methods for Integrating Orbits,*Celest.Mech.*,**10**,35-55.(Physics/Astronomy Tech.Rep.,University of Denver:Listing and Exploration of Program Radau).
Feagin,T.,1976,Multistep Methods of Numerical Integration Using Back-Corrections,*Celest.Mech.*,**13**,111-120.
Fehlberg,E.,1968,Classical Fifth-,Sixth-,Seventh-,and Eighth-order Runge-Kutta Formulas with Stepsize Control,*NASA TR R-287*.
Fehlberg,E.,1974,Classical Fifth-,Sixth-,Seventh-,and Eighth-order Runge-Kutta-Nystrom Formulas with Stepsize Control for General Second-Order Differential Equations,*NASA TR R-432*.
Fen,K.and Qin,M-Z.,1987,The Symplectic Methods for the Computation of Hamiltonian Equations,*in Numerical Methods for Partial Differential Equations,ed.by Zhu and Guo*.
Fox,K.,1984,Numerical Integration of the Equations of Motion of Celestial Mechanics,*Celest.Mech.*,**33**,127-142.
Fukushima,T.,1988,Tests of the Extrapolation Method for the Numerical Integration of the Keplerian Motion,*in this issue*.
Gragg,W.,1963,Repeated Extrapolation to the Limit in the Numerical Solution of Ordinary Differential Equations,*Doctoral Dissertation, University of California,Los Angels*,pp.103.
Gragg,W.,1965,On Extrapolation Algorithms for Ordinary Initial Value Problems,*SINUM*,**2**,384-403.
Grobner,W.,1967,Die Lie-Reihen und ihre Anwendungen,*VEB Deutscher Verlar der Wissenschaften,Berlin*.
Hanslmeier,A. and Dvorak,R.,1984,Numerical Integration with Lie-Series, *Astron.Astrophys.*,**132**,203-207.
Hanslmeier,A.,1984,Application of Lie-Series to Regularized Problem in Celestial Mechanics,*Celest.Mech.*,**34**,135-143.
Heggie,D.,1974,A Global Regularization of the Gravitational N-Body Problem,*Celest.Mech.*,**10**,217-241.
Hindmarsh,A.C.and Byrne,G.D.,1975,EPISODE,*UCID-30112*,Lawrence Livermore Laboratory.
House,F.,Weis,G.,and Wiegant,R.,1978,Numerical Integration of Stellar orbits,*Celest.Mech.*,**18**,311-318.
Huang,T-Y.and Innanen,K.A.,1983,The Accuracy Check in Numerical Integration of Dynamical Systems,*Astron.J.*,**88**,870-876.
Huang,T-Y.and Valtonen,M.J.,1988,Two Multiderivative Multistep (MDMS)Integrators,*Celest.Mech.*,in press.
Kalbouridis,T.,Papageorgiou,G.,and Simos,Th.,1988,On the Integration

of the Magnetic-Binary Problem by Explicit-Runge-Kutta Methods, *Dep.Mathematics,National Technical University of Athens*,TR-1/1988.

Kinoshita,H.,1976,Third-Order Solution of an Artificial-Satellite Theory, *in Dynamics of Planets and Satellites and Theories of Their Motions, ed.by Szebehely,D.Reidel Pub.Co.,Dordrecht-Holland*,241-257.

Kinoshita,H.,1968,Propagation of the Discretization Error in the Two-Body Problem.*Publ.Astron.soc.Japan*,**20**,1-23.

Kinoshita,H. and Nakai,H.,1984,Motions of the Perihelions of Neptune and Pluto,*Celest.Mech.*,**34**,203-217.

Krogh,F.T.,1970,Variable Order Integrators for the Numerical Solution of Ordinary Differential Equations,*JPL Technical Utilization Document*,No.CP-238.

Lundberg,J.B.,1981,Multistep Integration Formulas for the Numerical Integration of the Satellite Problem,*IASOM,TR* 81-1.

Menyuk,C.R.,1984,Some Properties of the Discrete Hamiltonian Methods, *Physica*,**11D**,109-129.

Mikkola,S.,1985,A Practical and Regular Formulation of the N-Body Equations,*Mon.Not.R.Astr.Soc.*,**215**,171-177.

Milani,A.and Nobili,A.M.,1988,Integration Error over Very Long Time Spans,*Celest.Mech.*,(in press).

Moran,P.E.,1973,Studies in the Application of Recurrence Relations to Special Perturbation Methods II,*Celest.Mech.*,**7**,122-135.

Moran,P.E.,Roy,A.E.,and Black,W.,1973,Studies in the Application of Recurrence Relations to Special Perturbation Methods II,*Celest.Mech.*,**8**,405-428.

Neri,F.,1987,Lie Algebras and Canonical Integration,*University of Arizona,Dept. of Physics Technical Report*,pp.25.

Parp,K.A.,Innane,K.A.,and Patrick,A.T.,1978,A Comparison of Five Algorithms for Numerical Orbit Computation in Galaxy Models, *Celest.Mech.*,**18**,277-286.

Parp,K.A.,Innane,K.A.,and Patrick,A.T.,1980,A Comparison of Five Algorithms for Numerical Orbit Computation in Galaxy Models, *Celest.Mech.*,**21**,277-286.

Rabe,E.,1961,Determination and Survey of Periodic Trojan Orbits in the Restricted Problem of Three Bodies,*Astron.J.*,**66**,500-513.

Ruth,R.,1983,A Canonical Integration Technique,*IEEE Trans.Nucl. Sci.*,**30**(4),2699-2671.

Schubart,J.and Stumpf,P.,1966,On an N-Body Program of High Accuracy for the Computation of Ephemerides of Minor Planets and Comets, *Verof.Astron.Rechen-Inst.,Heidelberg,Nr*.18.

Steffensen,J.F.,1957,On the Problem of Three Bodies in the Plane,*Math. Fys.Medd.Dansk.Vid.Selskap.*,**31**,No.3.

Stetter,H.J.,1970,Symmetric Two-Step Algorithms for Ordinary Differential Equations,*Computing*,**5**,267-280.

Stiefel,E.L.and Scheifele,G.,1971,*Linear and Regular Celestial Mechanics,Springer,Berlin*.

Sussman,G.J. and Wisdom,J.,1988,Numerical Evidence That the Motion of Pluto is Chaotic,*Science*,**241**.433-437.

Szebehely,V. and Bettis,D.G.,1972,Recent Developments of Integrating the Gravitational Problem of N-Bodies,*in Gravitational N-Body Problem, ed.by Lecar,D.Reidel,136-147*.

Wisdom,J.,1982,The Origin of the Kirkwood Gaps:A Mapping for Asteroidal

Motion Near the 3/1 Commensurability, Astron. J., **87**, 577–593.

(Textbooks)

Butcher, J.C., 1987, *The Numerical Analysis of Ordinary Differential Equations: Runge-Kutta and General Linear Methods*, John-Wiley & Sons. (C)

Dekker, K. and Verwer, J.G., 1984, *Stability of Runge-Kutta Methods for Stiff Nonlinear Differential Equations*, North-Holand. (C)

Gear, C.W., 1971, *Numerical Initial Value Problems in Ordinary Differentail Equations*, Prentice=hall, Englewood Cliffs. (A)

Gekeler, E., 1984, *Discretization Methods for Stable Initial Value Problems*, Springer-Verlag. (C)

Hairer, E., Norsett, S.P., and Wanner, G., 1984, *Solving Ordinary Differential Equations I: Nonstiff Problem*, Springer-Verlag. (B)

Henrici, P., 1961, *Discrete Variable Methods in Ordinary Differential Equations*, John Wiley and Sons, New York. (B)

Henrici, P., 1963, *Error Propagation for Difference Methods*, John Wiley and Sons, New York. (B)

Lambert, J.C., 1973, *Computational Methods in Ordinary Differential Equations*, Johm Wiley and Sons, London. (A)

Shampine, L.G. and Gordon, M.K., 1975, *Computer Solutions of Ordianry Differential Equations*, Freeman, San Francisco.

Stetter, H.J., 1973, *Analysis of Discretization Methods for Ordinary Differential Equations*, Springer-Verlag, Berlin. (C)

(A: elementary, B: intermidiate, C: not elementary.)

A Survey of Poisson Series Processors

Jacques Henrard
Facultés Universitaires de Namur
5000 Namur, Belgium

Abstract

The sheer magnitude of the work involved in the construction of perturbation theories (the "astronomical computations") made it inevitable that astronomers would very early become interested in the possibility of constructing them with the aid of computers. Since then general systems for symbolic manipulations have been developed and used widely. Nevertheless there still remain problems for which these general systems are not well adapted and many specialized systems of algebraic manipulation (mostly Poisson series processors) are in use. An attempt is made to review this field by sketching some of the ideas on which these Poisson series processors are built.

1 Poisson series processors

I would like to start this brief review by emphasizing why in my opinion, Poisson series processors are not, at least are not yet, part of the more general systems for "symbolic manipulations".

Those general systems (that many of you know and use, according to the survey conducted by H. Kinoshita) like MACSYMA, REDUCE or the still experimental SCRATCHPAD aim at embodying our knowledge of calculus (some say our knowledge of mathematics, but this I am afraid is a bit premature).

As such, in expert hands, they can be very efficient. The evaluation of indefinite integrals, for instance, can be considered as one of the most difficult tasks in calculus. Nowadays symbolic integrators are very easy to use and more reliable than tables. A "caveat" is necessary nevertheless. People disagree over what constitutes the simplest form of an expression; computers disagree even more. The simplification problem is still very much open and it is not clear that a purely algorithmic general solution to it exists.

In face of this (at least partial) success of the general systems for "symbolic manipulations", why do we need a specialized processor to manipulate a Poisson series of the form

$$P = \sum_{i_1,\cdots,i_n} \sum_{j_1,\cdots,j_n} (\text{numerical coefficient})\, x_1^{i_1} x_2^{i_2} \cdots x_n^{i_n} \frac{\cos}{\sin}(j_1\phi_1 + \cdots j_n\phi_n)\ ?$$

This type of series appears very often in celestial mechanics and more generally in perturbation theories for non-linear mechanics or non-linear differential equations.

Celestial Mechanics **45**: 245–253, 1989.

The "calculus" related to them, sum, multiplication, differentiation, integration does not present any difficulty. The simplification problems are (most of the time) absent as we have a nice canonical form. So most of the facilities of the general systems are simply not used when we deal with Poisson series.

But these facilities or at least the environment in which they thrive, are an impediment when it comes to manipulating the very large Poisson series generated in perturbation theories : series of 10,000 terms are not unusual. You do not want to handle them in an interactive environment, and you do not want to see them on your screen. In any case you will not have to because the general system will quit before even reading them.

So there is a real need for one (or several) specialized processor(s) for Poisson series (and maybe for other special algebras : Tensor analysis, non-commutative algebras of operators and so on). In my opinion they should be developed outside the frame of the general systems but of course the communications between general systems and specialized systems should also be developed. It is not rare that a problem will contain a part best suited for a general system and a part best suited for a specialized system. R.A. Howland will tell you of a recent problem he investigated in which he used Richardson's SAP specialized system to average an Hamiltonian and then the REDUCE system to integrate the resulting differential equations.

2 A little history

Of course celestial mechanics did not wait until this conference to feel the need for a Poisson series processor and to see the opportunity provided by computers to help them in their "astronomical computations".

In the sixties several investigators independently used computers to generate analytical expressions : Herget and Musen (1959), Iszak (1964), Contopoulos and Moutsoulas (1965), Henrard (1965), Roels (1965), Gustavson (1966), Chapront and Mangeney (1968), Seidelmann (1970) and so on. These were mainly ad-hoc programs designed to manipulate analytical expressions in a particular problem but they demonstrated the capabilities of computers in this domain : their power and their reliability.

Two contributions (Danby et al 1965, Barton 1966) merit a special mention as they laid the foundations of general purpose systems for manipulating Poisson series.

In the late sixties and early seventies, several of these general purpose systems were born : MAO (Mechanized Algebraic Operations, Rom 1970), TRIGMAN (Trigonometric manipulator), Jefferys 1970-1971-1972), NONAME (Broucke and Garthwaite - 1970), CAMAL (Cambridge Algebra System, Bourne and Harton 1971), SPASM (Smithsonian Package for Algebra and Symbolic Manipulation, Cherniack 1973) and were applied in earnest to problems in dynamical astronomy (Deprit et al. 1967, Broucke 1970-1971, Deprit and Rom 1970, Chapront 1970, Jefferys 1971, Gaposchkin et al 1971, Bourne 1972, Brumberg and Chapront 1972, and so on).

Let me outline a few of these applications to give the flavor of them.

The first one is concerned with the theory of the motion of the Moon. This problem has always been considered as an especially hard one : Newton declared that it gave him headaches. Delaunay invented his perturbation method (which I consider the father of all perturbation methods) to solve it and went on to spend 20 years of his live doing algebraic manipulations by hand to apply it to the problem. One of the first large scale application of a Poisson series processor was the reproduction and the prolongation of Delaunay's work by Deprit et al (1970-1971). The solution was pushed to order 25 (Delaunay's work went to order 7 with complements to order 8 and 9) by Henrard (1979), improved by iteration by Chapront-Touzé (1980) and completed by planetary and other perturbations (see a review by Chapront-Touzé 1982). More recently, Gutzwiller and Schmidt (1985) have completed a very accurate solution of the main problem. At present the most complete solution, ELP (Ephemeride Lunaire Parisien) contains some 50,000 periodic terms and reproduce the motion of the Moon within 20 meters. This is tantalizingly close (but still for away !) to the 20 cm accuracy one would like to achieve for geodetic applications. In any case a subset of ELP (see Chapront-Touzé and Chapront 1988) has proved itself very useful in the computation of old eclipses.

The recent history of planetary theories has been very similar. Although it has been very much concentrated geographically at the Bureau des Longitudes in Paris, it is also a story of trying different techniques, mixing the results, and starting again in another direction (a few checkpoints are : Bretagnon 1974, Duriez 1979, Bretagnon 1982, Laskar 1985).

Indeed in both of these problems we reach the limit, not so much of the Poisson series processors (although Laskar final results contains more than 100,000 terms), but of the perturbation techniques themselves (do we really need 30,000 periodic terms to gain one decimal place in the position of the Moon, and what about the 100,000 terms smaller by a factor of 10 that are not computed. Do they ruin the validity of the solution ?)

The second type of application I would like to outline is somewhat different. There are problems, not especially at the frontier of perturbation methods but where a rigorous approach, well choosen variables, and the facilities provided by even simple-minded Poisson series processors can lead rather quickly to very useful results. Let me mention as an example, the theory of the rotation of the rigid earth (Kinoshita 1977-1982). The resulting series contain a few hundred terms (not very taxing for a Poisson series processor but certainly uncomfortable or even impossible on a general system) but they improved markedly on the previously internationally adopted formulae and were soon adopted by the I.A.U. In a slightly different class of problems I would mention the automatized expansion of the disturbing function by Murray (1985). This is a very useful tool mixing some rather heavy numerical computations with some facilities of algebraic manipulation. Here it is the heavy numerical computations which would be difficult in the context of a general system while an ad-hoc elementary Poisson series processor in FORTRAN is all it takes to handle the algebraic manipulation part.

What is presently the situation in the field of those Poisson series processors which can be so useful ? Well, I am sorry to say that the hundred flowers which were sprouting in the early seventies have produced many more flowers (some of them very beautiful) but no trees. Many processors are in use but they do no speak to each other. Most of them are machine dependent to various degrees and are not very well documented. In the survey conducted by H. Kinoshita, sixteen investigators report using their own Poisson series processor and ten report using an imported one. There are also eleven users of REDUCE and eight of MACSYMA. Quite often investigators report using both a general system (REDUCE or MACSYMA or sometimes mu MATH) and a specialized Poisson series processor.

3 The overall design of a Poisson series processor

In view of this state of affairs, it is difficult to learn from each other and to decide what should be the general design of a good Poisson series processor. Nevertheless I shall try to discuss some of the alternatives and indicate what my choice is.

I do feel the need for better exchanges of ideas and software between people involved in algebraic manipulations and if this paper can start a general discussion and some exchanges, even if none of my ideas are implemented in tomorrow Poisson series processors, I will feel very happy.

First I shall advocate that the Poisson series processor I have in mind, be sharply divided at least in two stages.

The first stage, let us call it the Basic Poisson series processor, would contain only the basic facilities : addition, multiplication, multiplication by a scalar, copying and erasing Poisson series, coding and decoding a Poisson series term by term. It contains of course the data management facilities (garbage collector of some sort, traffic between files and virtual memory) although these kinds of facilities can in turn constitute a separate substage.

The second stage will be constituted by higher level functions : substitution, truncated series manipulation, two-body expansions , and so on, and can again be divided in substages (general purpose facilities and facilities specific to celestial mechanics for instance).

The reason for this sharp division is that it enables the higher stage to be less dependent upon a specific system and thus facilitate exchanges or transfer to a new system. The TRIGPROC general precompiler of Ricklefs et al (1983) is built with this in mind.

I shall go further and advocate the need for at least two versions of the first stage.

One version, the mini-basic-Poisson series processor does not need to be very sophisticated or very efficient but needs to be highly portable. This is the version that would go with higher level functions in case of exchange enabling the receiver to test the higher level functions and to familiarize himself with Poisson series processors. This is often the only version needed as quite a few applications (as we pointed out in the preceeding section) are not very demanding upon the Poisson series processor. This is also a very good teaching tool as the student (or the colleague) can really look inside the black box if he so desires.

The second version would of course be more sophisticated. This is the one that would be used for large scale projects, when a typical series has more than say a thousand terms or that the set of series to be manipulated do not fit in virtual memory. For this version the data management substage is of course particularly important. For instance, in the latest version of the Poisson series processor (the MS) we have developed in Namur, all series are considered as disk-files and brought automatically in virtual memory page by page when needed. This may seem a bit radical but it does help when the series is large.

4 How to represent a term and implications of doing so

The algorithms that can be implemented to perform the basic manipulations on a series and their efficiency are very much dependent upon the way a term of a series is coded.

Let us review some possibilities and indicate their good and bad points. For convenience of notation, I shall consider here only polynomials in several variables. The extension to Poisson series is trivial.

Let us consider a specific polynomial in two variables

$$P = X^2 + XY + X^2Y + XY^2 + Y^3 \tag{1}$$

A first scheme, let us call it the scheme with fully spelled keys, will make you code (a) the name of the variables, (b) the exponent and (c) the numerical coefficients. Each term is a variable length record and the series is a list of such records

$P \longrightarrow$

X	2	1.··· DO		
X	1	Y	1	1.··· DO
⋮				
Y	3	1.··· DO		

This is close to the representation used by the general systems, although general systems will also usually code operation signs : exponents, multiplications, and so on which are implicitly assumed here.

A second scheme, let us call it the scheme with coded keys, will make the assumption that this is a polynomial in X and Y and that the names of the variables do not have to be repeated. The assumption could be valid for the full application or just for this series (and then the names of the variables will appear somewhere in the heading of the series).

$P \longrightarrow$

the variables are X, Y

(X)	(Y)	(coeff)
2	0	1.··· DO
1	1	1.··· DO
⋮		
0	3	1.··· DO

A third scheme, let us call it the "tabular" scheme, will set up, in some way a one-to one correspondance between the possible terms in the series and a cell in an area of the virtual memory.

$P \longrightarrow$ double entry table indexed by the exponents of X and Y the maximum of which are 3

	(X^0)	(X^1)	(X^2)	(X^3)
(Y^0)	0.	0.	1.	0.
(Y^1)	0.	1.	1.	0.
(Y^2)	0.	0.	0.	0.
(Y^3)	1.	0.	0.	0.

Let us remark that from one scheme to the next we actually code less information and make more of it implicit. In the last scheme for instance, only the numerical coefficient is actually coded. The names of the variables and the powers at which they are raised are deduced implicitly from the address in virtual memory of the coefficient.

The implicit character of an information leads, in general, to more efficient algorithms but, of course, to much less flexible ones.

The last scheme for instance leads to very efficient algorithms because the location of the coefficient of a given term can be computed, while with the other schemes it has to be found by a search algorithm. But of course unless the polynomials are full (i.e. most of the possible terms have a non-zero coefficient) or can be made full by some special coding (but there goes the generality of the processor) this scheme can be very wasteful in virtual memory resources. Nevertheless such a representation enabled Laskar (1985) to manipulate the very large series of his general planetary theory.

Most of the Poisson series processors have adopted something close to the second scheme which, is in some sense, a compromise.

It is not wasteful in virtual memory resources unless the polynomials are very sparse not only in the exponents, but also in the variables that actually appear. Let us take for instance a polynomial in the 26 variables : $A, B, C, D, \cdots X, Y, Z$:

$$P = A + C \cdot D + Z^2$$

Its representation by the first scheme

A	1	1.··· DO		
C	1	D	1	1.··· DO
Z	2	1.··· DO		

is of course much shorter than its representation in the second scheme

(A)	(B)	(C)	(D)	$\cdots$	(X)	(Y)	(Z)	coefficient
1	0	0	0	$\cdots$	0	0	0	1.···DO
0	0	1	1	$\cdots$	0	0	0	1.···DO
0	0	0	0	$\cdots$	0	0	2	1.···DO

These representations with keys can be made reasonably efficient if the key (the coding of the exponents or of the variables and exponents) is used to define a canonical order between the terms in a series. Such an order can be used to arrange the terms in a binary tree structure and this leads to very efficient algorithms for locating a given term in the series. Many Poisson series processors use now such a structure which, to my knowledge, was first implemented in this context in Schmidt's POLYPAK Processor.

There is another way to look at polynomials in n -variables. It is to consider them as polynomials in $p(p \leq n)$ variables, the coefficients of which are polynomials in $(n - p)$ variables. This is a very flexible recursive representation and it is implemented in the new MAO developed by Deprit and Miller in LISP. In this context the polynomial (1) could be represented as

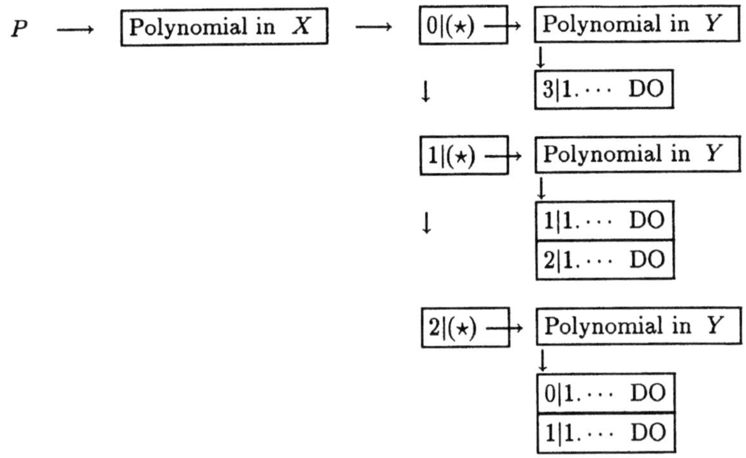

This idea of manipulating abstract algebras the domain of which can again be algebras is a very fruitful one, especially if at each level the implementation of the algebra can be choosen independently as any one of the scheme described previously.

5 Algorithms

I shall point out only to two basic algorithms which for large scale problems can make a big difference.

The first one is concerned with the product of truncated series. We often manipulate Poisson series which are not exact but truncated infinite series. The number of terms is often a fast growing function of the size of the coefficients. Let us consider as a typical example a Poisson series which contains 10^n terms, the coefficients of which are of the order of 10^{-n} , and is truncated at 10^{-4} . It does contain 1 term of order one, 10 terms of order 0.1 and so on up to 10,000 terms of order 10^{-4} . In order to multiply two such series we have to consider more than 10^8 elementary products.

But the resulting series will certainly not be accurate at the level of 10^{-8} , which is the size of most of the elementary products considered. We can avoid computing and even considering the small elementary products if the series are previously ordered according to the size of the coefficients. Even if we compute all the elementary products down to 10^{-5} to allow for some accumulation, we have only to consider $4\ 10^4$ of these elementary products.

The second algorithm is concerned with the search of a particular term in a series. This function is a crucial one as it is nested deeply in many other functions. For instance, in a product $A \cdot B \longrightarrow C$, every elementary product of a term of A by a term of B is followed (in most algorithms) by a search of C to check whether or not the resulting term is already present. In a sequential search the number of comparisons can be as high as the number of terms already present in C (say N) while in a dichotomic search the maximum number of comparisons is reduced to $log_2 N$. In case $N = 10,000$, this is a reduction by a factor of 1000. Such a dichotomic search can be implemented by the balanced tree algorithm (see Knuth 1973) using the lexicographic order on the keys of the terms.

The implementation of both these algorithms has consequences on the representation of the series themselves on the computer. Indeed they necessitate not only the coding of the terms themselves but also of some ordering (sometimes lexicographic, sometimes according to the size of the coefficient) of these terms.

Aknowledgement

I thank very much H. Kinoshita who sent me the answers to the survey he made about the use of algebraic manipulation in dynamical astronomy and the 35 persons who responded to his survey. I thank also J.C. Agnese, H. Caprasse, A. Deprit, R.H. Howland, C.D. Murray, D.L. Richardson, D. Schmidt, I. Tupikova who sent me long letters and/or documentation on this topic. These persons helped me significantly in writing up this review but of course I bear the sole responsibility of its shortcomings.

References

Barton, D. : 1966, 'Lunar disturbing function', *Astron. J.*, **71**, 438-439.

Bourne, S.R. : 1972, 'Literal expressions for the coordinates of the Moon,' *Celest. Mech.*, **6**, 167-186.

Bourne, S.R. and Harton, J.R. : 1971, 'The design of the Cambridge Algebra System.' Proc., *SYMSAM II.*

Bretagnon, P. : 1974, 'Termes à longues périodes dans le système solaire,' *Astron. Astrophys.*, **30**, 141-154.

Bretagnon, P. : 1982, 'VSOP 82,' *Astron. Astrophys.*, **114**, 278.

Broucke, R. : 1970, 'How to assemble a Keplerian Processor,' *Celest. Mech.*, **2**, 9-20.

Broucke, R. and Garthwaite, K. : 1969, 'A programming system for analytical series expansions on a computer,' *Celest. Mech.*, **1**, 271.

Broucke, R. and Smith, G. : 1971, 'Expansion of the planetary disturbing function,' *Celest. Mech.*, **4**, 490-499.

Brumberg, V. and Chapront, J. : 1972, 'Construction of a General Planetary Theory of the first order,' *Celest. Mech.*, **8**, 335-355.

Caviness, B.F. : 1986, 'Computer Algebra : past and future,' *J. Symb. Comp.*, **2**, 217-236.

Chapront, J. : 1970, 'Construction d'une théorie littérale planétaire jusqu'au second ordre des masses,' *Astron. Astrophys.*, **7**, 175-203

Chapront, J. and Mangeney, L. : 1968, 'Applications of literal series to the main problem of lunar theory,' *Astron. J.*, **73**, 214-216.

Chapront-Touzé, M. : 1980, 'La solution ELP du problème central de la Lune,' *Astron. Astrophys.*, **83**, 86.

Chapront-Touzé, M. : 1982, 'Progress in the analytical theories for the orbital motion of the Moon,' *Celest. Mech.*, **26**, 53-62.

Chapront-Touzé, M. and Chapront, J. : 1988, 'ELP 2000-85. A Semi Analytical Lunar Ephemeris adequate for historical times,' *Astron. Astrophys.* in print.

Cherniak, J.R. : 1973, 'A more general system for Poisson series manipulation,' *Celest. Mech.*, **7**, 107-123.

Contopoulos, G. and Moutsoulas, M. : 1965, 'Resonance case in Third Integral,' *Astron. J.*, **70**, 817.

Danby, J.M.A., Deprit, A. and Rom, A.R.M. : 1965, 'The symbolic manipulation of Poisson's series,' *BSRL note 423*.

Deprit, A., Henrard, J. and Rom, A. : 1967, 'Trojan orbits : Birkhoff's normalization,' *Icarus*, **6**, 381-406.

Deprit, A., Henrard, J. and Rom, A. : 1970, 'La théorie de la lune de Delaunay et son prolongement,' *C.R. Acad. Sc. Paris*, **271**, 519-520.

Deprit, A., Henrard, J. and Rom, A. : 1971, 'Analytical Lunar Ephemeris : Delaunay's Theory,' *Astron. J.*, **76**, 269-272.

Deprit, A. and Rom, A. : 1970, 'The Main Problem of Artificial Satellite Theory for small and moderate eccentricities,' *Celest. Mech.*, **2**, 166-206.

Duriez, L. : 1979, 'Approche d'une théorie générale planétaire en variables elliptiques héliocentriques,' Thesis, Lille University.

Gaposchkin, E.M., Cherniack, J.R., Briggs, R. and Benima, B. : 1971, 'Third order oblateness perturbation for an artificial satellite,' *Am. Geophys. Union. meeting.*.

Gustavson, F.G. : 1966, 'On constructing formal integrals of an Hamiltonian system near an equilibrium point,' *Astron. J.*, **71**, 670-686.

Gutzwiller, M.C. and Schmidt, D.S. : 1985, 'The motion of the Moon as computed by the method of Hill, Brown and Eckert,' *Astron. Papers. Am. Ephemerides.*

Henrard, J. : 1965, 'Solution générale au voisinage des équilibres colinéaires du problème restreint,' Thesis, Louvain University.

Henrard, J. : 1979, 'A new solution to the main problem of the Lunar Theory,' *Celest. Mech.*, **19**, 337.

Herget, P. and Musen, P. : 1959, 'The calculation of literal expansions,' *Astron. J.*, **64**, 11-20.

Iszak, I. : 1964, *Smithsonian Astrophys. Obs. Special Rep.*, **140**.

Jefferys, W.H. : 1970, 'A Fortran based list processor for Poisson's series,' *Celest. Mech.*, **2**, 474-480.

Jefferys, W.H. : 1971, 'Automated Algebraic Manipulation in Celestial Mechanics,' *Comm. A.C.M.*, **14**, 538-541.

Jefferys, W.H. : 1972, 'A precompiler for the formula manipulation system TRIGMAN,' *Celest. Mech.*, **6**, 117-124.

Kinoshita, H. : 1977, 'Theory of the rotation of the Rigid Earth,' *Celest. Mech.*, **15**, 277-326.

Kinoshita, H. : 1982, 'Note on Nutations in Ephemerides,' *Pub. Intern. Observ. Mizusawa*, **XII**, 71-108.

Knuth, D.E. : 1973, 'The Art of Computer Programming,' *Addison Wesley*.

Laskar, J. : 1985, 'Accurate methods in general planetary theory,' *Astron. Astrophys.*, **144**, 133-146.

Murray, C.D. : 1985, 'A note on Leverrier's expansion of the disturbing function,' *Celest. Mech.*, **36**, 163-164.

Ricklefs, R.L., Jefferys, W.H. and Broucke, R.A. : 1983, 'A general precompiler for algebraic manipulation,' *Celest. Mech.*, **29**, 179-190.

Roels, J. : 1965, 'Les phénomènes de résonances au voisinage des équilibres équilatéraux,' Thesis, Louvain University.

Rom, A. : 1970, 'Mechanized Algebraic Operations (MAO),' *Celest. Mech.*, **1**, 301-319.

Rom, A. : 1971, 'Echeloned Series Processor (E.S.P.),' *Celest. Mech*, **3**, 331-345.

Seidelmann, P.K. : 1970, 'An iterative method of general perturbations programmed for a computer,' *Celest. Mech.*, **2**, 134-146.

A FORTRAN-BASED POISSON SERIES PROCESSOR AND IT'S APPLICATIONS IN CELESTIAL MECHANICS

R. A. Broucke
The University of Texas
Austin, Texas

ABSTRACT The present article describes the design and applications of our Poisson series subroutine package developed and maintained since 1968. The programs are written in standard FORTRAN-77 and are almost independent of the wordlength of the particular computer. The system has no restriction on the number of polynomial and angular variables and the storage allocations are completely automatic, invisible to the user.

The nucleus of the system consists of about 20 basic "traffic subroutines" that handle the terms of the different series. Besides these subroutines, we have a number of I/O routines as well as arithmetic subroutines and a large number of Celestial Mechanics applications such as the classical expansions of the Kepler Problem and several expansions of Disturbing Functions. A preprocessor has also been built, allowing the user to write code in a high-level language, such as Jefferys' Trigman, and then translate it in our call-statements.

The system was developed on several different computers: first on the Univac 1108 at the Jet Propulsion Laboratory in 1968, then the IBM 360-91 at UCLA, Los Angeles, California and finally the CDC 6600 and CYBER 170/750 at the University of Texas, Austin, Texas. The latest version is entirely geared towards the IBM-PC and compatibles.

1. The Basic Series Manipulations System

We will first enumerate here the most important characteristics of our set of computer subroutines for the manipulation of Poisson series. All the subroutines are written in standard FORTRAN language, the latest version being in FORTRAN-77, which is as transportable and machine-independent as possible.

Celestial Mechanics **45**: 255–265, 1989.

The main goal of the system is to allow the user to manipulate Fourier and power series in a flexible way, especially without having to set up dimension statements in the main program. The internal storage allocations are dynamical, within a large *stack* of fixed dimension (which can be changed by the user).

The system automatically drops all terms whose coefficient is below a certain preassigned epsilon, in order to save space.

Although the user can completely ignore the internal mechanism of the storage allocations, we will mention that the storage technique for the series is based on the following principles:

(a) Each series has a name which is an integer FORTRAN vasriable. When the series is "DEFINED," a value is given to this variable. However the user never has to "DEFINE" a series: this is done internally.

(b) Each series is represented by two integer numbers which occupy one row of a double-dictionary DIC1 and DIC2. These two numbers indicate the position of the coefficients of the series in an array or stack and the number of terms.

(c) The terms of all the series are stored consecutively in two stacks: STACKC with the double-precision coefficients, and STACKA with the arguments or indices in a packed form.

(d) When a series is "ERASED" from the stacks, the value of the FORTRAN name and the corresponding row in the dictionary are set equal to zero, while the space occupied by this series in the stacks becomes available for new series.

(e) Several internal named COMMON-statements are used to allocate the different storage arrays. These arrays have only one subscript and no matrices are used, in order to make it easier for the user to change the maximum dimensions and the available storage according to each particular computer.

(f) The indices of the trigonometric arguments and the exponents of the polynomial variables are packed in INTEGER*2 Arrays, at the rate of one small integer per byte. Therefore these integers can vary only from -63 to +64. Internally, a bias of 63 is added to all these integers in order to have to pack only positive numbers.

2. The Arithmetic Capabilities of the System

Besides the standard arithmetic operations on series, such as addition, subtraction and multiplication, a large number of more specialized operations are available to the user. For instance there are subroutines giving the series for the sine, cosine, exponential, arc-tangent and arc-sine of another series.

Some other available operations are the substitution of a series for a polynomial variable in a series and the factorization of a series S into series A, B such that $S = A + Bp^n$ where p is a given polynomial variable and n it's given exponent. Several subroutines for the operations of Integration and Differentiation have also been made. There are first of all the routines for total differentiation and integration with respect to time t, under the assumption that all the trigonometric variables (angles) are linear functions of the time. Secondly there are routines for partial differentiation and partial integration, as well with respect to polynomial as trigonometric variables. The n-th partials can be taken in a single operation or call statement, both for differentiation and integration.

It is well known that it is impossible to divide Poisson series in a simple closed-form operation. Therefore we constructed several subroutines which are oriented towards the generation of an approximate reciprocal of a series. Three separate approaches have been followed. The first consists in the binomial expansion of a series having a large leading term, (Broucke, 1971). Note that this method also gives the square root or inverse square root of a series. The second approach consists in performing symbolic Newton iterations towards the inverse of a series. Finally the third approach uses the geometric series $(1-z)^{-1} = 1 + z + z^2 + \ldots$, simultaneously with Euler's method for the acceleration of the convergence.

Finally, the system implements the expansion of the most important mathematical functions, which are actually of crucial importance in Celestial Mechanics; Legendre Polynomials, Bessel functions and Laplace series, (Laplace coefficients), see Brouwer and Clemence (1961) or Broucke (1971). More details about the basic structure of the system have been published almost 20 years ago (Broucke and Garthwaite, 1969).

3. The Keplerian Processor

The first application of the Algebraic Processor has been the construction of a Keplerian processor: a package of FORTRAN-routines for the literal expansions of all the basic functions of the unperturbed two-body problem. Such a package is a necessary tool for the construction of general perturbation solutions in all Celestial Mechanics Problems.

The Keplerian processor is essentially based on the classical Bessel function expansions in the eccentricity. Bessel functions are used to generate the expansions of a/r, r/a, $\cos nE$, $\sin nE$, $\cos nv$ and $\sin nv$ in the mean anomaly M. Other expansions give $E - M$, essentially the solution of Kepler's equation. Besides the classical expansion, our package also contains a solution of Kepler's equation through symbolic Newton Iterations (Broucke, 1969). Besides the Bessel Expansions, we also implemented several Keplerian expansions, with the Lagrange's formula. This formula requires computing the successive m-th derivatives of a generating function, which is especially well-suited for a computerized algebraic processor.

In the next stage, we generate the expansions for the coordinates $\vec{r} = (x, y, z)$ of a satellite in a two-body orbit. This requires first the elementary expansion of the unit-vectors $\vec{P}$ and $\vec{Q}$ in the three Euler angles (i, ω, Ω). Note that these quantities are closed-form expressions in *cosi* and $\sin i$, but they are infinite series in $\sin i$ alone.

One of the first medium-scale applications of the Kepler package was the fashionable application of the late sixties (see Barton, 1966 and Chapront, 1968): the expansion of the Lunar disturbing function with the Legendre series. The basic quantity, the cosine of the angle S between the Moon and the Sun, is represented by the formula

$$\cos S = \left(\frac{a}{r}\right) \cdot \left(\frac{a'}{r'}\right) \cdot \left(\frac{x}{a} \cdot \frac{x'}{a'} + \frac{y}{a} \cdot \frac{y'}{a'} + \frac{z}{a} \cdot \frac{z'}{a'}\right),$$

which involves five series multiplications (of dimensionless normalized quantities).

4. Expansion with Legendre Polynomials

The classical expression that is known for the disturbing function for the perturbation of a satellite ($\vec{r}$) by a third body ($\vec{r}_3$) is as follows:

$$R = r_3 \left[\frac{1}{\Delta} - \frac{\vec{r} \cdot \vec{r}_3}{r_3^3}\right] = \frac{r_3}{r_3} \left[\frac{r_3}{\Delta} - \frac{r}{r_3} \cos \psi\right],$$

where r_3 is the constant $Gm_3 = k^2 m_3$ and

$$\Delta^2 = r^2 + r_3^2 - 2rr_3 \cos \psi = |\vec{r}_3 - \vec{r}|^2; \qquad \vec{r} \cdot \vec{r}_3 = rr_3 \cos \psi,$$

ψ being the angle between the two vectors $(r < r_3)$.

Applying the classical expansion in Legendre polynomials, the expression that is found for the disturbing function may be presented as

$$R = \frac{r_3}{r_3} \sum_{n=2}^{\infty} \left(\frac{r}{r_3}\right)^n P_n(\cos \psi).$$

In order to obtain a better expression for $\cos \psi$, we introduce the coordinates of the different bodies in terms of their radius vector, r, right ascension, α, and declination, δ, so that:

$$\cos\psi = \sin\delta\sin\delta_3 + \cos\delta\cos\delta_3\cos(\alpha - \alpha_3).$$

This allows us to apply the standard addition theorem of the theory of Legendre polynomials, so that we can write:

$$P_n(\cos\psi) = \sum_{m=0}^{n}(2-\delta_{om})\frac{(n-m)!}{(n+m)!}\,P_{nm}(\sin\delta)P_{nm}(\sin\delta_3)\cos m(\alpha-\alpha_3),$$

where δ_{om} is a Kronecker delta and P_{nm} are *associated* Legendre functions.

Finally, we make one more simplification for the purpose of grouping the factors involving the satellite orbit elements together (excluding all other factors which only involve the perturbing body). It will be seen that this is in fact the most important step of all the operations. The disturbing function can be efficiently expanded to very high orders in the elements if the factors involving each body are explicitly separated:

$$R = \frac{r_3}{a_3}\sum_{n=2}^{\infty}\sum_{m=0}^{n} W_{nm}\epsilon^n(X_{nm}X_{nm}^{(3)} + Y_{nm}Y_{nm}^{(3)}),$$

where $\epsilon = a/a_3$ and

$$W_{nm} = (2-\delta_{om})\frac{(n-m)!}{(n+m)!}.$$

The following elementary two-body functions, called surface harmonics, have been defined:

$$X_{nm} = \left(\frac{r}{a}\right)^n Q_{nm}(\sin\delta)C_m(\alpha,\delta),$$

$$X_{nm}^{(3)} = \left(\frac{a_3}{r_3}\right)^{n+1} Q_{nm}(\sin\delta_3)C_m(\alpha_3,\alpha_3).$$

The Y-factors are similar, except that they use the S_m-functions:

$$C_m(\alpha,\delta) = \cos^m\alpha\cos m\alpha = \left(\frac{\tilde r}{r}\right)^m \cos m\alpha,$$

$$S_m(\alpha,\delta) = \cos^m\delta\sin m\alpha = \left(\frac{\tilde r}{r}\right)^m \sin m\alpha,$$

which can be computed with the recurrence relations:

$$C_m = C_1C_{m-1} - S_1S_{m-1},$$

$$S_m = C_1S_{m-1} + S_1C_{m-1},$$

which are especially easy to implement with an automatic processor. The symbol Q_{nm} stands for the m-th derivative of the Legrendre polynomial P_n.

What we have just described above is the separated form of the disturbing function: all the terms have been decomposed in two factors, each factor depending on only *one* of the two bodies, either the satellite *or* the perturbing body. These factors can be expanded separately in Poisson series. These series are relatively short, considering that they contain only a small number of orbit elements and angles.

Our Poisson Series Packages contains nine subroutines required to perform this factorized expansion of the disturbing function. Each one of these subroutines contains one particular series construction, such as the Legendre Polynomial P_n, the derivatives Q_{nm}, the numerical factor W_{nm}, the C_m and S_m-functions (generated, of course, with the recurrence relations), and finally, the X_{nm} and Y_{nm}-functions for both bodies. With the use of these subroutines it takes less than half a page of FORTRAN coding to obtain a complete expansion, for whatever purpose, such as averaging or perturbation calculations, for example.

5. Application of the Binomial Theorem to Generate the Disturbing Function

Let us assume that X is a given formal series and that we desire to construct the series X^m, where $m = -1/\alpha$ and α a positive integer.

We also assume that the series X may be decomposed in a sum of two series: a principal series X_o (for which X_o^m is known) and a relatively smaller series DX. We will then be able to apply the binomial theorem to the sum of two series:

$$X = X_o + DX = X_o(1 + DX.X_o^{-1}).$$

This will give us the m-th power of X in the form

$$X^m = X_o^m \left[1 + C_m^1 D + C_m^2 D^2 + \ldots\right],$$

The expression

$$D = DX.X_o^{-1} = X.X_o^{-1} - 1 = X_0(X_o^m)^\alpha - 1$$

can be computed entirely with the three basic Arithmetic series operations $(+, -, *)$. The quantity D requires only α multiplications. In Celestial Mechanics $\alpha = 1$ and 2 are the most important applications; X_o and X_o^m are given by the unperturbed Keplerian series.

6. The Newton Iteration Method for the Construction of X^m

An iterative method for obtaining X^m may be obtained by retaining only the two first terms of the above series. In order to simplify the writings let us use the notations $X^m = Y_{i+1}$ and $X_o^m = Y_i$ together with $m = -1/\alpha$. We derive then from the two first terms the following iteration formula:

$$Y_{i+1} = Y_i[1 + mD_i] = Y_i\left[1 - \frac{D_i}{\alpha}\right] = \frac{Y_i}{\alpha}\left[(1+\alpha) - XY_i^\alpha\right],$$

where the subscript i and $i+1$ are iteration numbers and D_i is the expression D, computed at the i-th iteration.

The iterative formula is also easily seen to be the Newton iteration method for the solution of the equation in Y:

$$f(Y) \equiv Y^{-\alpha} - X = 0,$$

which has quadratic convergence. More details about the iteration methods for the symbolic inversion of a series can be found in Broucke (1971). Both the binomial expansion and the Newton Iterations have been implemented and are an integral part of the Series Manipulation System.

7. The Classical Expansion of the Planetary Disturbing Function

The classical expansion of the planetary disturbing function with the Laplace coefficients (Broucke and Smith, 1971) has been incorporated as a standard feature in our Poisson Series Package. This expansion essentially gives the inverse distance Δ^{-1} between the two planets, which occurs as the direct term in the disturbing function:

$$\frac{r'}{\Delta} = \left[1 - 2\left(\frac{r}{r'}\right)\cos S + \left(\frac{r}{r'}\right)^2\right]^{-1/2}$$

In the lunar and satellite case this is usually expanded with the Legendre series, but in the planetary case, we rather use the Laplace identity:

$$(1 - 2\gamma\cos S + \gamma^2)^{-s} = \frac{1}{2}\sum_{k=0}^{\infty} b^0_{sk}\gamma^k + \sum_{\substack{j=1\\k=j}}^{\infty} b^j_{sk}\lambda^k \cos jS \ ,$$

where $\gamma = r/r'$, proportional to the small parameter a/a'. The b-coefficients are the so-called Laplace coefficients. There exists well-known recurrence relations for these coefficients. We also used the following formula, which relates them to the binomial coefficients:

$$b^{(j)}_{s,2k+j} = 2(-1)^j c^k_{-s} C^{k+j}_{-s}; \quad j,k = 0,1,2\ldots$$

The quantity $\cos(S)$ has been expanded with the formula that was given above, involving 5 series multiplications. Next, the different $\cos jS$-series can be obtained by one series multiplication per j, using the trigonometric identity:

$$\cos jS = 2\cos\theta\cos(j-1)S - \cos(j-2)S.$$

The series $\gamma = r/r'$ is first obtained by a multiplication of two Keplerian series (r/a) and (a'/r'). Next, all the different powers k are obtained, resulting in series containing three polynomial variables α, e, e' and two angles M, M'. All these operations are implemented in about 10 subroutines in our system, one of which gives the complete Laplace expansion of $1/\Delta$ in a single call statement.

However, it is important to mention that essentially two different sets of routines were made: one for the literal expansions and one for the semi-analytical approach where numerical values are substituted for all or most of the polynomial variables. The reader should also be aware of the scaling options which are available in order to shorten the series (see Broucke and Smith, 1971, page 496). If the numerical or semi-analytical approach is followed, with Fourier series, with purely numerical coefficients, the convergence and the size of the coefficients is an important point to consider. This is made clear in the following section on error control in the series manipulations.

8. Error Propagation in Manipulating Poisson Series

The truncation errors and the precision of a Poisson series can be studied by introducing the concept of *range* of a series, which is a measure of the number of useful significant digits. Let us take a Poisson series $X(a,b)$ with largest coefficient a and with truncation error b (both factors being absolute values). By definition, its range is the non-negative, real number:

$$\text{Range}(X) = \log_{10}(a/b) > 0.$$

We will usually assume that the truncation error b of a series is the same as the smallest term (in absolute value). Let us now assume that we have to add two series: $X(a,b) + Y(a',b')$: The largest term of the sum will be $a'' =$ largest (a,a'), while the smallest correct term of this sum will be $b'' =$ largest (b,b'). In particular, if the two series X and Y have the same truncation error, $b = b'$, then the sum $X+Y$ will also have this truncation error. The range of the sum, Range $(X+Y) = \log_{10}\left(\frac{a''}{b''}\right)$, could be any one of the following four numbers: Range (X), Range (Y), $\log_{10}\left(\frac{a}{b'}\right)$, $\log_{10}\left(\frac{a'}{b}\right)$.

It is thus clear that in general the ranges are changed by the operations of addition or subtraction . However, when we add up two series with the same range, then the range of $(X+Y)$ will always be the range of X or the range of Y, whichever has the largest terms. But when we add up two series with the same range and the same order of magnitude ($a = a'$; $b = b'$), then the range is conserved by the addition operation.

Let us also consider the range and truncation error of the product: $Z(a'',b'',) = X(a,b)\ Y(a',b')$. The largest term, a'' of Z is equal to the product of the largest terms: $a'' = aa'$. The truncation error, b'' (or smallest correct term), is $b'' =$ largest $(b'a, a'b)$, while the range of the product is then obtained by Range $(Z) = \log_{10}(a''/b'')$.

In performing the product of the two series, it will be quite useless to keep any term that is less than b'' (in absolute value). The formula for the range of XY may be written Range $(XY) = \log_{10} \frac{aa'}{\text{largest } (ab', a'b)} = \text{smallest } \left[\log_{10}\left(\frac{a'}{b'}\right); \log_{10}\left(\frac{a}{b}\right)\right]$. Thus we arrive at the result that the range of the product is equal to the smallest of the ranges of the two original series X and Y:

$$\text{Range } (XY) = \text{smallest } (\text{range } (X), \text{range } (Y)).$$

In the particular case where the two ranges are the same, we see that the product will have the same range. In this case, multiplication conserves the range.

9. A Trigonometric Preprocessor

The last major milestone in the evolution of our symbolic manipulation activities was the creation of a new high-level language, by R. L. Ricklefs, in UT's Department of Astronomy, under the supervision of W. H. Jefferys, (Ricklefs, Jefferys, Broucke, 1983). The new language is a FORTRAN Superset in the sense that all FORTRAN statements and features are preserved and available to the user. Its main goal is to facilitate the task of the programmer. A preprocessor was made to translate the high-level TRIGMAN-type language in a Fortran program with all the appropriate calls to our series manipulation routines.

The high level language looks similar to FORTRAN, except that three new declarations have been added: TRIG, POLY and SERIES. The first two of these, TRIG and POLY define the names of all the Trigonometric and Polynomial variables that will be used. The SERIES-declaration defines all the variable names that will be used as symbolic expressions, which are functions of the POLY- and TRIG-variables. The TRIG- and POLY-variables as well as the SERIES can all be dimensioned with up to three subscripts.

The new language allows for taking the derivatives or the integration of series with respect to any variable, as well as substitution or factorization.

As an illustration, we display here a short program which illustrates the type of statements used in this language. On the left side we show the source program which goes in the preprocessor. On the right-side we show the output from the program. The intermediate FORTRAN program that was created by the precompiler is not shown.

	X 2 TERMS
TRIG T	+(+A)
POLY A, B	
SERIES $X, S1$	+(+B) * COS(T)
$X = A + B * \text{COS}(T)$	
OUTPUT (C) X	S1 4 TERMS
$S1 = X * X$	
OUTPUT (6) $S1$	+(+1/2 * B^2 + A^2)
STOP	
END	+(+2 * A * B) * COS(T)
	+(+1/2 * B^2) * COS(2 * T)

10. Conclusions

The reader should be aware of the fact that the most efficient expansion for hand calculations is usually not the most efficient expansion with a computerized algebraic processor. Simplicity of programming is another factor that needs to be considered on a computer. In general, the computer will prefer simple recurrence relations. The Poisson series processor will usually have a certain number of standard operations available: repeated partial differentiation, substitution of a series for a polynomial variable, the binomial theorem and the Bessel series of the Kepler problem are all standard one-line operations with our processor. It is clear that many of the classical Celestial Mechanics problems must be completely re-formulated in the light of these modern tools.

It is under the guidance from these principles that we implemented our Poisson Series and Celestial Mechanics Algebraic Processor. Several other applications have been implemented, such as planetary theories and Lie-Hori perturbation expansions, but space limitations do not allow us to describe them here.

References

Barton, D., 1967, "On Literal Developments of Lunar Theory With the Aid of a Computer", *Astron. Journal*, Vol. 72, pp. 1281-1287.

Broucke, R. and Garthwaite, K., 1969, "A Programming System for Analytical Series Expansions on a Computer", *Celestial Mechanics*, Vol. 1, pp. 271-284.

Broucke, R., 1970, "On the Matrizant of the Two-Body Problem", *Astronomy and Astrophysics*, pp. 173-182.

Broucke, R., 1970, "How to Assemble a Keplerian Processor ", *Celestial Mechanics*, Vol. 2, pp. 9-20.

Broucke, R., 1971, "Construction of Rational and Negative Powers of a Formal Series", Communication of the ACM, Vol. 14, No. 1, pp. 32-35.

Broucke, R. and Smith, G., 1971, "Expansion of the Planetary Disturbing Function", *Celestial Mechanics*, Vol. 3, pp. 408-426.

Broucke, R., 1981, "Expansion of the Third-Body Disturbing Function", *Journal of Guidance and Control*, Vol. 4, No. 3, pp. 346-348.

Broucke, R., 1982, "Expansion of the Disturbing Function by Factorization", in *High-Precision Earth Rotation and Earth-Moon Dynamics*, Editor O. Calame, pp. 337-348.

Brouwer D. and Clemence, G., 1961, *Methods of Celestial Mechanics*, Academic Press, N.Y.

Chapront, J. and Maugeney-Ghertzman, L., 1968, "Applications of Literal Series to the Main Program of the Lunar Theory", *Astronomical Journal*, Vol. 73, pp. 214-216.

Ricklefs, R., Jefferys, W. and Broucke, R., 1983, "A General Precompiler for Algebraic Manipulation", *Celestial Mechanics*, Vol. 29, pp. 179-190.

PARSEC: An Interactive Poisson Series Processor for Personal Computing Systems

D. L. Richardson
Aerospace Engineering
University of Cincinnati
Cincinnati, Ohio USA 45221

ABSTRACT. PARSEC is a PC-based, interactive algebraic manipulation package designed to manipulate series of the kind frequently found in Celestial Mechanics applications and perturbation procedures. The system is fundamentally an input/command interpreter which allows the user to enter algebraic expressions and procedures and to control the flow of series manipulations interactively. The system is designed to allow easy manipulations of polynomial-trigonometric series within the environment of an electronic scratchpad.

1. The PARSEC Algebraic Manipulation System

PARSEC is an interactive, PC-based symbolic manipulation program designed to perform standard operations of algebra and calculus on the polynomial-trigonometric series commonly encountered in orbital and celestial mechanics research. Such series are sometimes referred to as *Poisson series* and are written in the general form

$$\sum_{i,j,k,\ldots} c_{ijk\ldots} x_1^{n_{1i}} x_2^{n_{2j}} x_3^{n_{3k}} \ldots \left\{ \begin{smallmatrix} \sin \\ \cos \end{smallmatrix} \right\} (m_{1i}x_1 + m_{2j}x_2 + m_{3k}x_3 + \ldots).$$

PARSEC manipulates these expressions assuming a certain well-defined structure and a few restrictions:

- The coefficients $c_{ij\ldots}$ are real or complex whose real and imaginary parts are real, rational numbers.
- Variables $x_1, x_2, \ldots$ are real and are classified as *polynomial* and/or *trigonometric* depending on usage in algebraic assignment statements.
- Exponents of the polynomial variables are real and rational.
- Linear combinations of variables define the trigonometric arguments. The coefficients of these variables are integers.

Current versions are operational over the entire IBM PC family and all PC clones that are 100% IBM compatible. A version which runs on the 68020-based Definicon 780 hardware card is also available. Because the run-time code contains numeric coprocessor instructions, the host hardware must include a coprocessor. Implementations that execute in the central memory of the IBM PC and its compatibles operate under DOS and require 512K bytes of free memory. (DOS and additional memory-resident programs would boost this total). The "DOS" and Definicon versions both limit the maximum number of *active* polynomial and trigonometric variables and the

Celestial Mechanics **45**: 267–274, 1989.

total number of functions, symbolic rationals, and user-defined procedures. These limitations are

- polynomial variables: maximum = 20
- trigonometric variables: maximum = 4
- symbolic rationals: maximum = 500
- functions plus procedures: maximum = 600
- term count (all functions): maximum = 3000 (DOS version)
  = 32000 (Definicon and DOS 80386)

PARSEC was never designed to serve as a general-purpose symbolic manipulation package with the features of MACSYMA or REDUCE. It does provide an easy to use, "friendly" PC environment for the manipulation of algebraic expressions that can be classified as Poisson series. There are a number of features which can be applied to facilitate perturbation analyses and to help with the development and use of perturbation procedures. The user can fully exploit these aspects by building a library of specialized procedures. The major features of the system with its strengths and weaknesses are addressed below.

1.1. SYSTEM COMPONENTS

There are four main components of the PARSEC system:

1. *The Communications Manager* is a series of modules which handle user input, video display, and hardcopy output. The display drivers produce a limited set of internally generated, bit-mapped characters to an EGA screen set in high resolution (640×350) graphics mode. The hardcopy output drivers generate code appropriate for Hewlett-Packard LaserJet printers or for printers that can emulate the HP LaserJet. Dot-matrix printers are not (yet) directly supported. When the system is placed in its input mode, the input stream (barring user-generated interrupts) is saved and reformed into an intermediate code suitable for the Command Interpreter or the Assignment-Statement Parser. After the input is processed, it can be recalled for a re-submission or for limited editing.

2. *The Command Interpreter* receives input streams that begin with the $ character, determines if a valid command is requested, checks the validity of the operands, and initiates the command execution. Errors encountered during command execution are overseen by the Error Manager.

3. *The Assignment-Statement Parser* also receives its input from the Communications Manager. The input stream may signal the execution of a procedure (the first character is a backslash \), or define an algebraic assignment statement having the general form

<function or symbolic rational> = <algebraic expression>.

The Parser's main task is the decomposition of the algebraic expression into a series of binary operations suitable for use by the algebraic code generator. The code generator performs the actual symbolic manipulations and returns its output to the Communications Manager. The Parser uses a recursive-descent algorithm based on a "current operator-next operator" (CONO) decision table. Such procedures have been used extensively in the development of simple compilers and interpreters. The ideas are old, but in this case, they provide rapid compilation, easy error diagnostics, and ease of extension as it becomes necessary.

4. *The Error Manager* can be invoked by any of the main components. Its primary function is to issue warnings and syntax error messages at the direction of the Command Interpreter or the

Parser. In addition, the Error Manager's *error recovery subsystem* performs "garbage collection procedures" whenever necessary to return PARSEC to its pre-error state so as to avoid possible system shutdown. "Integer overflow" and "term count overflow" are the most likely error conditions which could arise during an expression evaluation. These situations require that the current expression, any of its algebraic code, and all internal transient productions be flushed. The command $Q is used to display the after-error storage system state and general usage statistics. The error recovery system is not bulletproof, but it does provide the user protection from the most common and potentially catastrophic error sources.

2. PARSEC Syntax and Operations

PARSEC performs manipulations on symbolic expressions consisting of the following symbolic types:

- integer constants
- symbolic rationals
- polynomial and trigonometric variables
- functions
- intrinsic functions

Algebraic assignment statements are input to PARSEC from the keyboard in standard infix notation and are always parsed left to right. The equations

$$n = 8/3$$
$$f = (x + y/2 - 3z)^{3n} \sin(6nx - t)$$

would be entered at the keyboard (using upper case letters) as

```
N = 8/3;
F = (X+Y/2-3*Z)^(3*N)*\sin(6*N*X-T);
```

Here, we would identify 2, 3, 6, and 8 as integer constants, N as a symbolic rational, F as a function (or series name), Y and Z as polynomial variables, T as a trigonometric variable, and \SIN as an intrinsic function. The variable X is used as both a polynomial and a trigonometric variable.

The PARSEC character set consists of upper case letters, digits, and certain special characters. Valid input characters are

alphabetic:	A B C ... Y Z
digits:	0 1 2 ... 9
special characters:	space ! $ ( ) * + , - . / ; = ? @ [ ] \ ^ %

User-defined functions, symbolic rationals, and variables are character strings consisting of alphanumerics having no more than 14 characters. All types must begin with an upper case alphabetic character. Symbolic types beginning with the @ special character are PARSEC-defined *reserved* functions or symbolic rationals.

Algebraic expressions are constructed by combining various symbol types with standard unary and binary operators:

+	addition	−	subtraction or unary negation
/	division	*	multiplication
^	exponentiation	!	unary factorial

The hierarchy of elementary operations is the same as that used in the FORTRAN programming language: The ! operator has the highest precedence, followed by ^, * and /. Of lowest precedence are + and −. Productions are assembled from the parser as it moves along the input statement from left to right. Intrinsic functions are treated as parenthesized expressions. Potential ambiguities are avoided by the use of parentheses. Parentheses can be nested no greater than a depth of 10.

Input statements must conclude with either a semicolon (;) or a period (.):

```
APPROXEXPX=\SUM(K=0,10)[X^K/(K)!];
```

The result of the manipulation of an algebraic statement is echoed to the screen if a semicolon has terminated the input. The results are not echoed if a period is used as the termination character. All operations are terminated and the user is returned to the DOS prompt whenever the escape key (ASCII 27_{10}) is pressed. Ongoing manipulations can be stopped and "flushed" by pressing the sequence "CNTRL Q". Input statements can be modified inline or previous statements recalled and edited using various character-control sequences similar to those available in the line editors of earlier days.

2.1. SYSTEM CONSTRAINTS AND LIMITATIONS

A valid PARSEC expression is a sequence of unary or binary operations applied to functions, variables, integers, and/or symbolic rationals. The construction of valid expressions is contingent upon the following major restrictions:

1. Exponentiation within an exponent is not allowed.
2. Exponents must be internally reducible by PARSEC to integers or rational numbers.
3. Fractional exponents which produce rational numbers are valid constructions:

   ```
       N = 27/8;
   ROOT1 = N^(1/3);
   ```

4. Any expression enclosed within parentheses is tagged as a "transient" function, and PARSEC will not manipulate functions raised to a fractional power unless the function has only one non-trigonometric term and its coefficient raised to the fractional power produces a rational number:

   ```
   OK=(25/16*X^3/Z)^(1/2);
   NOTOK=(25/16*X^3/Z+Z)^(1/2);
   ```

   The second equation would produce an error.
5. Division by a function is not allowed except if the function has only one term which does not contain a trigonometric function.
6. The factorial operation can only be applied to parenthesized expressions which immediately precede the ! symbol. These expressions must be reducible by PARSEC to an integer.

PARSEC manipulates symbolic expressions having coefficients which are rational numbers. Normal internal operations perform integer arithmetic only on integers of 32 bits or less. Floating-point arithmetic operations are not supported. The only exception is the intrinsic function \FLOAT(EXPRESSION) which is used to display a rational integer expression as a floating-point number. Extended-precision manipulations using integers of up to 128 bits can be performed on rational integer expressions using the \BIGNUM(EXPRESSION) intrinsic function.

3. Commands, Intrinsic Functions, and Procedures

PARSEC command statements begin with the $ sign at the prompt and continue with the appropriate character string. Commands can have several operands which are always delimited by the space character. The command line input is terminated with a semicolon. For example, to replace the polynomial variable X everywhere in function F2 with the function F1 would be accomplished by the $R command through the input

$R X WITH F1 IN F2;

The current command list numbers over 25 and includes commands for procedure definition and control, "DOWHILE" iteration, hardcopy and FORTRAN output, system restart and resume, variable and series substitution, and logic for variable and series truncation.

The intrinsic functions in PARSEC begin with the backslash character (\) and can be used as algebraic elements in any expression. The only exception is FLOAT which is implemented as a procedure, and therefore cannot be used within algebraic expressions. The following is a small sample from the current list:

\DD(VARIABLE,COUNT)(EXPRESSION)	differentiation
\EXP(ORDER)(EXPRESSION)	expand as an exponential series
\INT(VARIABLE,A,B)(EXPRESSION)	definite or indefinite integration
\SUM(I0=I1,I2,I3)(ARGUMENT)	DO-LOOP like summation
\XPAND(ORDER)(EXPONENT)(EXPRESSION)	binomial expansion

Expansion by binomial series is quite often used in the construction of truncated multivariate power-series expansions. The XPAND intrinsic function has been developed for this purpose. XPAND constructs a binomial series for

$$(\mathrm{EXPRESSION})^{\mathrm{EXPONENT}}$$

where EXPRESSION is assumed to have the decomposition

$$\mathrm{EXPRESSION} = 1 + \mathrm{THEREST}.$$

The development is truncated at order ORDER in THEREST. THEREST can be any valid PARSEC expression (except those coming from BIGNUM or FLOAT). The operands are enclosed in parentheses as indicated above and are defined below:

ORDER = an integer or symbolic rational giving the order at which the series is to be truncated (must be ≥ 0).

EXPONENT = any algebraic expression involving integers or symbolic rationals which can be reduced by PARSEC to a symbolic rational.

EXPRESSION = any algebraic expression which has the form (1 + expression).

3.2.1 *Example.* The complete elliptic integral of the first kind is defined by the definite integral

$$K(k) = \int_0^{\pi/2} \frac{dx}{\sqrt{1 - k^2 \sin^2 x}}.$$

A power series in k^2 through degree 10 in k is generated by the following PARSEC input:

K1 = \INT(X,0,@PIHALF)[\XPAND(5)(-1/2)(1-K^2*\SIN(X)^2)];

(The variable @PIHALF is a reserved variable (symbolic for $\pi/2$) intended to be used as either an upper or lower limit in definite integrals.) PARSEC returns its construction showing each term has a factor of $\pi/2$. Removing this factor by the statement

K1BYHALFPI = K1/@PIHALF;

gives

$$\text{K1BYHALFPI} = 1 + \frac{1}{4}k^2 + \frac{9}{16}k^4 + \frac{25}{256}k^6 + \frac{1225}{16384}k^8 + \frac{3969}{65536}k^{10}.$$

3.1. PROCEDURES

The user can build specialized procedures of varying complexity using the $PROC command. Procedures can be saved for a later restart or for use in another PARSEC session. Procedures function as stand-alone reentrant algebraic code modules. A procedure is a named sequence of assignment and command statements defined by the following syntax:

```
$PROC(PROCNAME);
   ⋮
   assignment and command statements with comments
   ⋮
$ENDPROC;
```

PROCNAME is the procedure name and is any valid PARSEC name up to 14 characters in length. A PROCNAME cannot be used within algebraic assignment statements. A procedure definition begins with the $PROC command and ends with $ENDPROC. Statement comments begin with the % character and can be placed inline after the statement. The procedure is invoked at the input prompt by the statement

\PROCNAME;

Procedures can invoke other procedures as well as themselves.

Once procedures are developed they can be saved for future use by storing them in a disk file using the $SP command. Filed procedures are retrieved using the $LP command. Procedures are saved as ASCII files which allows them to be created or modified using a standard ASCII text editor.

3.3.1 *Example.* The solution to Kepler's equation

$$M = E - e\sin(E)$$

(assuming appropriately bounded e) is given by the infinite series

$$E = M + 2\sum_{n\geq 1} \frac{1}{n} J_n(ne)\sin(nM),$$

where J_n is the Bessel function of the first kind of order n. Writing J_n in its integral representation

$$J_n(ne) = \frac{1}{\pi}\int_0^{\pi} \cos(nx - ne\sin x)\,dx,$$

gives us a method for constructing the solution using two short procedures. The first is called BESSEL which returns the function JN (= $J_n(ne)$):

```
$PROC(BESSEL);
COS = \COSINE(ORDER)(N*E*\SIN(X));      %SET N, ORDER
SIN = \SINE(ORDER)(N*E*\SIN(X));
INTEGRAND = \COS(N*X)*COS+\SIN(N*X)*SIN;
JN = (\INT(X,0,@PI)(INTEGRAND))/@PI;
$ENDPROC;
```

The procedure is based on the decomposition

$$\begin{aligned}\cos(nx - ne\sin x) &= \cos(nx)\cos(ne\sin x) + \sin(nx)\sin(ne\sin x),\\ &= \cos(nx)\sum_{k=0,2,4,\ldots}^{\text{ORDER}} \frac{1}{k!}(ne\sin x)^k + \sin(nx)\sum_{k=1,3,5,\ldots}^{\text{ORDER}} \frac{1}{k!}(ne\sin x)^k.\end{aligned}$$

Inputs are n and ORDER which is the expansion order for the intrinsics SINE and COSINE. BESSEL is called from procedure KEPLER:

```
$PROC(KEPLER);
N = N+1;                    %ENTER WITH N=0, SUM=0, SET MAX
\BESSEL;                    %RETURNS JN
SUM = SUM+JN*\SIN(N*M)/N;
$EXITIF N EQ MAX;           %SUM UP TO MAX
\KEPLER;                    %REPEAT
$ENDPROC;
```

The command $EXITIF is a procedure flow-control command and is used for procedure exit if the logical operation defined in its 3 operands is true. To execute the procedures, display the results, and save on disk as the FORTRAN subroutine KEPLER.FOR would require the input:

```
N = 0;
SUM = 0;
ORDER = 7;
MAX = 7;
\KEPLER;
ANSWER = M+2*SUM;
$PF ANSWER KEPLER.FOR;
```

Setting MAX = 7 gives the solution valid through degree 7 in e. The actual FORTRAN code is listed below. (The subroutine name SAP215 is generated internally):

```
SUBROUTINE SAP215(RESULT)
IMPLICIT REAL*8(A-H,O-Z)
COMMON/SAPBLK/V1,V2,V3,V4,T1,T2
```

```
C         V1=X, V2=@PI, V3=E, V4=M, T1=X, T2=M
      DATA
     $ D1,D2,D3,D4,D5,D6,D7,D8,D9,D10,D11,D12,D13,D14,D15/
     $3.647352430555555D-01,3.375000000000000D-01,3.390842013888889D-01,
     $3.255208333333333D-01,2.666666666666666D-01,3.333333333333333D-01,
     $4.746093750000000D-02,2.109375000000000D-01,3.750000000000000D-01,
     $2.083333333333333D-02,1.666666666666666D-01,5.000000000000000D-01,
     $1.085069444444444D-04,5.208333333333333D-03,1.250000000000000D-01/
      V3P2=V3**2
      V3P3=V3**3
      V3P4=V3**4
      V3P5=V3**5
      V3P6=V3**6
      V3P7=V3**7
      T2P2=2.D0*T2
      T2P3=3.D0*T2
      T2P4=4.D0*T2
      T2P5=5.D0*T2
      T2P6=6.D0*T2
      T2P7=7.D0*T2
      W1=D1*V3P7*DSIN(T2P7)+D2*V3P6*DSIN(T2P6)+(-D3*V3P7+D4*V3P5)*DSIN(T
     $ 2P5)+(-D5*V3P6+D6*V3P4)*DSIN(T2P4)+(D7*V3P7-D8*V3P5+D9*V3P3)*DSIN
     $ (T2P3)+(D10*V3P6-D11*V3P4+D12*V3P2)*DSIN(T2P2)+(-D13*V3P7+D14*V3P
     $ 5-D15*V3P3+V3)*DSIN(T2)+V4
      RESULT=W1
      RETURN
      END
```

The comment statements indicate the correct associations between the COMMON block variables and the user-defined polynomial and trigonometric variables. A call to SAP215 computes ANSWER (= W1) which is returned to the calling module through the argument RESULT. (V3, V4 and T2 would be preset as indicated).

4. Current Use

PARSEC has been used to simplify expressions in the closed-form solution for the two-body motion in the post-Newtonian approximation [1]. It is being used currently by astrodynamics students at the University of Cincinnati who are studying nonlinear dynamics and perturbation procedures. PARSEC is proving useful in the planetary theory for the investigation of various simplifications, intermediaries, and normal forms. Its interactive nature allows the user relative freedom to search for and test possible simplifications as opportunities or needs arise. Such freedom has never been easily available using non-interactive (or batch oriented) algebraic processors. Having the capability to pursue a "tangent" or a whim in an expeditious manner becomes an essential ingredient in the development of efficient higher-order perturbation theories. PARSEC provides us a tool for the practical and efficient exploration of possible methodologies.

5. References

[1] Richardson, D.L. and Kelly, T., "Two-Body Motion in the Post-Newtonian Approximation," *Celes. Mech.*, in press.

The problem of the Eulerian oscillations : a weakness of numerical versus analytical methods

Odile CALAME
CERGA
Avenue Copernic
06130 Grasse
FRANCE

J. Derral MULHOLLAND
CERGA and
Space Astronomy Laboratory
Gainesville, Florida 32609
USA

ABSTRACT . The application of computer technology has permitted more and more problems of dynamical astronomy to be solved more easily, quickly, and accurately. In this area, numerical integration is often very efficient, and sometimes essential. There is often, however, a temptation to choose numerical integration simply because it is the easiest way to attack the problem. Sometimes this works to the detriment of a satisfactory understanding of the physics of the problem under study. It is particularly the case for the "free", or Eulerian, oscillations. The forces that create such a motion are not of gravitational origin and are not even conservative. The theory can only specify the frequencies of oscillation, not their amplitudes nor phases. The case is complicated when the free oscillations interact with gravitationally-forced oscillations, a situation that is almost inevitable, since nothing is isolated in the Universe. The first author has particularly studied this problem in the case of the rotation of the Moon, and published the first credible determinations of the lunar free libration. In this kind of problem, the observations have to be used and care must to be taken to create no spurious free librations in the results by using numerical integrations to describe the other related motions. A differential correction of the starting conditions to fit the observations does not necessarily give any valid information on the real free oscillation contained in the data. An analytical model is necessary, if the goal of the research is to understand the origins and characteristics of an Eulerian oscillation in such a system.

I — INTRODUCTION

Computer technology has permitted more and more problems of dynamical astronomy to be solved more easily, quickly and accurately. Numerical integration is a good example. It is a temptation to choose the numerical integration rather than to treat the problem by solving the equations of motions by analytical methods. It is often very efficient and sometimes essential, but we must understand that it is not an universal panacea. Sometimes, numerical integration works to

Celestial Mechanics **45**: 275–280, 1989.

the detriment of a satisfactory understanding of the physics of the problem under study. In particular, there is a category of problems for which it does not work at all : the "free" (or Eulerian) oscillations in the rotational motion of a body. The first author has made a particular study of the lunar rotation, including the first determination with modern observations of the free libration [e.g. Calame 1975, 1976, 1977], and the following discussion will be based on that experience.

II — DESCRIPTION OF EULERIAN OSCILLATIONS

The phenomenon of free oscillation exists in the rotational motion of a body around its center of mass. Mathematically, such oscillations are the solutions to the homogeneous parts of the equations of motion. The equations give, however, only the permitted frequencies, not the amplitudes nor the phases. The origin of the forces that provoke these motions may be more or less complex, but not conservative and thus not gravitational. They may be due, for example, to meteorite impacts on the surface, or to tectonic events in the interior, which jostle the behavior produced by the gravitational forces. They are usually episodic. The body reacts to these attacks in a manner that is controlled by its internal structure, its material composition, and other factors. In the case of impact, the velocity, mass and incidence angle of the impactor is of key importance. The plastic rheology of the body plays an essential role in the temporal evolution of the response. Consequently, there exists a viscous damping phenomenon that progressively decreases the effect of the trigger event with the passage of time. Therefore, observations made at an ulterior epoch can permit a determination only of the remanent effects at that date, not of the initial condition. Non-conservative systems leave no unique memory in the observations.

Added to the problem that one can only determine an instantaneous effect, with essentially no information about its cause, is the fact that nothing is isolated in the Universe. With the development of ever higher precision in observing techniques, this has sometimes become a major problem. It was once possible — indeed imperative — to treat a free oscillation independently of the forced oscillations, and both independently of the orbital motion. This is no longer true, and there are sometimes strong correlations in the frequencies of the various components of the total motion. Such a situation is encountered especially in the Earth-Moon system.

III — DETERMINATION OF THE FREE LIBRATIONS OF THE MOON

As with any application of dynamics to a real physical problem, the determination of the free oscillations requires a combination of accurate observations and as fine a modelling as possible, taking into account all of the other effects that influence the quantity measured. For the Moon, such effects include the orbital motion and the forced gravitational libration. An adequate theory will describe them completely, unlike the Eulerian motion, for which the theory gives only the periods (or frequencies). Because we cannot know the details of the exciting phenomena, the free amplitudes and phases are obtained only by analysis of otherwise inexplicable residuals in the observations. In the past, the three motions have been treated by superposition of solutions, without accounting for the fact that their interactions are quite significant. Because of advances in observing technique, this is no longer possible; a "correction" to one of them implies observable changes in the others. The root of our problem lies there.

The rotational motion is represented by a system of second-order differential equations expressing the three Euler angles as functions of various parameters describing the non-sphericity of the body, its orbit, and the gravitational influences of Sun and planets.

The associated homogeneous system may be written, after linearization, as follows:

$$\ddot{\tau} + \gamma\epsilon_2 n^2 \tau = 0$$

$$\ddot{p}_1 - (1-\beta)n\dot{p}_2 + (1+\epsilon_1)\beta n^2 p_1 = 0$$

$$\ddot{p}_2 + (1-\alpha)n\dot{p}_1 + \alpha n^2 p_2 = 0$$

where τ is the libration in longitude and p_1, p_2 the librations in the direction cosines of the ecliptic pole with respect to the first two principal axes of inertia, corresponding to the principal moments A and B; n is the sidereal mean motion of the Moon; ϵ_1, ϵ_2 are two constants depending on the orbital motion; and α, β, γ are related to the three principal moments of inertia.

The first equation is independent, while the two others are coupled. The frequencies of these three modes are given by the roots of the characteristic equations and, for the Moon, correspond to periods of 2.9 years in longitude, 27.3 days and 75 years in latitude. As indicated earlier, the amplitudes and phases must be determined from observation, by comparisons with a mathematical model that tries to account for all of the other effects. The residuals thus computed obviously contain the signatures of the free libration phenomena, but are also contaminated by observational errors and deficiencies in the mathematical model.

The two major aspects of that model are the lunar orbit and the description of the forced rotation. Although the choice of lunar orbital ephemeris is of extreme importance in many applications, it has very little influence on the free libration. Numerous solutions have been made for their amplitudes and phases with a wide variety of both analytic and numerical ephemerides of the lunar orbit, without much effect on the resulting values. The same is far from true for the rotation.

IV — ANALYTICAL VS. NUMERICAL: A PROBLEM OF CHOICE

There are, as in all dynamical problems, two major approaches to describe the gravitationally-forced perturbations in the rotational motion of the Moon: analytical theory or numerical integration. Until about 1974, only analytical (or semi-analytical) methods were used [e.g. Eckhardt 1965, 1982], apparently because of the limitations on computer speed and capacity. Such theories are expressed as series whose coefficients are expressed as functions of the parameters of the lunar orbit and of the lunar potential field. These same parameters directly influence the frequencies of the free oscillations. Therefore, in analyzing the observational residuals, it is necessary to determine some of these parameters simultaneously with the solution for the free librations. An analytic (or semi-analytic) theory permits this readily, since the partial derivatives are also (semi-) analytic series, derived directly from the theory.

The problem becomes quite different when one uses a numerical integration of the lunar forced rotation, as has become most common today. While such an ephemeris permits a net global improvement in the representation of the total motion compared to current analytical models, it is not possible to adjust the coefficients of the lunar gravity field and other parameters pertinent to the free librations without also adjusting the starting conditions. The partial derivatives are computed by numerical integration of the variational equations, which include automatically the effects of the free librations. Thus, the adjustment of the starting conditions for a numerically-generated rotational model does not permit a reliable separation of the Eulerian part from the gravitational part [King 1982].

The problem is all the more critical in the lunar case, as some of the free frequencies are close to some of the forced libration terms due to the orbital motion.

V — COMPUTER ARTIFACTS IN NUMERICAL SOLUTIONS

Finally, there is another aspect of the problem that does not have its origin in either the physics or the mathematics of the dynamical system. It is the purely electronic problem of numerical artifacts produced by the architecture of the computer. The physical quantities used to define the computational model, such as the gravitational harmonic coefficients, are inevitably given to a finite number of digits. The starting conditions, the accelerations, and the partial derivatives will suffer inconsistencies due to inexact decimal-binary conversions in the next higher digit. For the forced terms, this only produces accumulation of round-off error, because the forced frequencies are gravitationally imposed in the accelerations. Since the Eulerian frequencies are partial solutions to the equations of motion, the computer will try to absorb the inconsistencies by creating spurious free librations. Whether such numerical artifacts influence the apparent amplitudes and phases at a detectible level, contaminating the determination of the "true" values, depends on the structure of the program, the system, and the machine.

Whether it be from this cause or from the nature of the variational equations discussed above, solutions for the free amplitudes and phases made with different numerically-integrated libration models (from JPL and MIT) have produced practically random results. By comparison, many tests have been made with semi-analytic rotation theories, including simultaneous solutions for nearby forced terms, and the results have been quite stable [Calame 1977, Mulholland 1980]. This experience, based on lunar laser ranging data, suggests strongly that numerical integration is simply not well adapted to this sort of study.

VI — CONCLUSION

It is necessary to bear in mind that studies of a dynamical system such as Earth-Moon assume several forms, depending on the goals of the specific research. The methods best suited to one goal may be totally inappropriate to another. The description of the motions in the Earth-Moon system provide a good example. At present, and possibly forever in the future, the best possible numerical modelling of the geocentric motion of a point on the lunar surface requires a simultaneous numerical integration of the lunar orbit and rotation, taking into account all known influences and adding *post hoc* empiric terms for the free libration. If, however, the goal is to have a real knowledge of the origins and characteristics of such phenomena as the Eulerian oscillation, it is necessary to develop accurate analytical theories for orbit and forced libration. The two approaches must continue to be developed in parallel.

VII — ACKNOWLEDGEMENTS

JDM is presently Astronome Associé at the Observatoire de la Côte d'Azur/CERGA, on leave from the University of Florida, under the auspices of the French Ministère de l'Éducation Nationale.

REFERENCES

Calame, O. 1975: "Étude des mouvements libratoires lunaires et localisation des stations terrestres à partir des mesures laser de distance", thèse d'état, Université de Paris VI.

Calame, O. 1976: "Free librations of the Moon determined by an analysis of laser range observations", Moon **15**, 343.

Calame, O. 1977: "Free librations of the Moon from lunar laser ranging", in SCIENTIFIC APPLICATIONS OF LUNAR LASER RANGING (J. D. Mulholland, ed.), Reidel, Dordrecht, p. 53.

Eckhardt, D. H. 1965: "Computer solutions of the forced physical librations of the Moon", Astron. J. **70**, 466.

Eckhardt, D. H. 1982: "Planetary and Earth figure perturbations in the librations of the Moon", in HIGH-PRECISION EARTH ROTATION AND EARTH-

MOON DYNAMICS (O. Calame, ed.), Reidel, Dordrecht, p. 193.

King, R. W. 1982: in discussion of Eckhardt [1982], *op. cit.*, p. 198.

Mulholland, J. D. 1980: "Scientific Achievements from Ten Years of Lunar Laser Ranging", Rev. Geophys. and Space Phys. **18**, 549.

A SEMI-ANALYTICAL METHOD TO STUDY PERTURBED ROTATIONAL MOTION

R. VILHENA DE MORAES
Instituto Tecnológico de Aeronáutica
CTA-ITA-IEA-IEAB
12225-São José dos Campos-SP, Brasil

ABSTRACT. A semi-analytical method is presented to study the system of differential equations governing the rotational motion of an artificial satellite. Gravity gradient and non gravitational torques are considered. Operations with trigonometric series were performed using an algebraic manipulator. Andoyer's variables are used to describe the rotational motion. The osculating elements are transformed analytically into a mean set of elements. As the differential equations in the mean elements are free of fast frequency terms, their numerical integration can be performed using a large step size.

1. INTRODUCTION

The use of analytical methods to analyse the rotational motion of an artificial satellite has been of prime importance in connection with many satellite missions. However an analytical solution becomes very cumbersome when we use realistic models for non gravitational perturbative torques.

We present here a semi-analytical method to propagate the elements that describe the attitude of an artificial satellite. The method is based on the theories of Liu and Alford (1980) and Hoots (1983) for the translational motion of a low Earth orbiting satellite. The solution is obtained by first using an averaging method to eliminate short period terms. This mean set of differential equations is then integrated numerically or, analytically, when it is possible.

2. EQUATIONS OF MOTION

Using Andoyer's variables L_i and ℓ_i $(i = 1,2,3)$, the study of the rotational motion of an artificial satellite under the influence of the gravity gradient torque coupled with a non conservative torque $\vec{N}$, leads us to the analysis of the following system of differential equations in the extended canonical form:

Celestial Mechanics **45**: 281–284, 1989.

$$\frac{d\ell_i}{dt} = \frac{\partial H}{\partial L_i} + P_i$$
$$(i = 1,2,3). \qquad (1)$$
$$\frac{dL_i}{dt} = -\frac{\partial H}{\partial \ell_i} + Q_i$$

Here $H = H_0 + H_1$, H_0 is the Hamiltonian of the torque free motion, and

$$P_i = \sum_{j=1}^{3} N_j p_j$$
$$(i = 1,2,3) \qquad (2)$$
$$Q_i = \sum_{j=1}^{3} N_j q_j$$

where p_j and q_j are functions of the Andoyer's variables. The N_i are the components of $\vec{N}$ in a system with its origin at the mass center of the satellite with its axes parallel to the principal axes of inertia of the satellite. For our purposes we assume that P_i and Q_i are of the same order (or less) than H_1.

3. AVERAGED EQUATIONS

The solution of system (1) can be written in the following form (Vilhena de Moraes, 1981):

$$\ell_i = \ell_i^* + \delta\ell_i$$
$$(i = 1,2,3) \qquad (3)$$
$$L_i = L_i^* + \delta L_i$$

where $\delta\ell_i$ and δL_i come from the solution of the same system when $P_i = Q_i \equiv 0$ and ℓ_i^*, L_i^* are solutions of the system

$$\dot{\ell}_i^* = \langle \frac{\partial \bar{H}}{\partial L_i^*} \rangle + \bar{P}_i - \sum_{k=1}^{3} \{\bar{P}_k \frac{\partial(\delta\ell_i)}{\partial \ell_k^*} + \bar{Q}_k \frac{\partial(\delta\ell_i)}{\partial L_k^*}\}$$
$$(4)$$
$$\dot{L}_i^* = \bar{Q}_i - \sum_{k=1}^{3} \{\bar{P}_k \frac{\partial(\delta L_i)}{\partial \ell_k^*} + \bar{Q}_k \frac{\partial(\delta L_i)}{\partial L_k^*}\}$$

The bar over a function means that this function must be evaluated at L_i^* and ℓ_i^* and the symbol $< \partial\bar{H}/\partial L_i^* >$ means the non periodic part of $\partial\bar{H}/\partial L_i^*$.

The solution of system (1), when $P_i = Q_i = 0$, can be obtained by Hori's method (Hori, 1966). After a lengthy calculation (Algebraic Manipulator TRIGPROC (Ricklefs et al., 1983) has been used) we get (Kinoshita, 1972):

$$\begin{aligned} L_i &= L_i^* + \delta L_i \\ \ell_i &= \ell_i^* + \delta\ell_i \end{aligned} \qquad (i = 1,2,3) \qquad (5)$$

where L_i^* are constants; $\ell_i^* = \ell_{io} + n_i t$; ℓ_{io} are constants, δL_i and $\delta\ell_i$ are periodic functions and $n_i = n_i(L_j^*)$ $(i,j) = 1,2,3$.

System (4) gives: a) secular perturbations due to the conservative and non conservative torques, b) periodic perturbations due to P_i and Q_i and, c) the coupling of the conservative and non conservative torques. A transformation must be introduced to avoid spurious terms (Ferraz-Mello, 1981).

The secular perturbations can be obtained solving the system:

$$\begin{aligned} \dot{\ell}_i^* &= < \frac{\partial\bar{H}}{dL_i^*} > + < \bar{P}_i > \\ \dot{L}_i^* &= < \bar{Q}_i > \end{aligned} \qquad (6)$$

The averaged system (6) can be integrated numerically with large step size (Crenshaw and Fitzpatrick, 1968). The integration can be performed even if P_i and Q_i are known only by their averaged values. For sophisticated models a numerical integration may be the most feasible approach to estimated the means values $< \bar{P}_i >$ and $< \bar{Q}_i >$.

If $< \bar{P}_i >$ and $< \bar{Q}_i >$ are replaced by numerical values the system can be integrated analytically.

Neglecting terms of higher order the constants of integration of system (6) are given by

$$L_i^*(o) = L_i(o) - \delta L_i(L_i(o), \ell_i(o)); \ell_i^*(o) = \ell_i(o) - \delta\ell_i(L_i(o), \ell_i(o))$$

4. CONCLUSIONS

A semi-analytical solution was obtained to study the rotational motion of an artificial Earth satellite. The method has a more general characteristic and can be applied to several problems where the perturbations have their first part well-known and the remaining part has only the averaged contribution known. Explicit functions of the time were obtained for the elements describing the perturbed rotational motion. The elements can be propagated if the numerical values of the mean contributions of the remaining part of the perturbative torques are known.

The analytical expressions as given may be conveniently handled with algebraic manipulators.

5. ACKNOWLEDGEMENTS

This paper was sponsored by FAPESP under contract nº 88/2027-0.

6. REFERENCES

Crenshaw , J. W. and Fitzpatrick, P. M. (1968)'Gravity effects on the rotational motion of a unaxial artificial satellite', AIAA Journal, 6, 2140.

Ferraz Mello, S. (1981)'Elimination of secular terms generated by the coupling of perturbations', Cel. Mech., 25, 293.

Hoots, F. R. (1981)'Theory of the motion of an artificial Earth satellite", Cel. Mech. 23, 307.

HORI, G. (1966)'Theory of general perturbations with unspecified canonical variables', Publ. Astron. Soc. Japan, Vol. 18, 4, 287.

Kinoshita, H. (1972)'First-Order perturbations of the two finite body problem', Publ. Astron. Soc. Japan, Vol. 24, 4, 423.

Liu, J. J. F. and Alford, R. L. (1980)'Semianalytic theory for a close-Earth artificial satellite'', J. Guidance and Control, Vol. 3, 4, 304.

Ricklef, R. L., Jefferys, W. H. and Broucke, R. A. (1983), A general precompiler for algebraic manipulation', Cel. Mech. 29, 179.

Vilhena de Moraes, R. (1981)'Combined solar radiation pressure and drag effects on the orbits of artificial satellites', Cel. Mech., 25, 281.

COMPUTATIONAL STUDIES OF CLOUDY GASEOUS GALACTIC DISKS

WILLIAM W. ROBERTS, JR. AND DAVID S. ADLER
Departments of Applied Mathematics and Astronomy
University of Virginia
Charlottesville, Virginia 22903

ABSTRACT. Computational studies are carried out to address questions centered on the clumpy cloudy interstellar medium, giant molecular clouds, and star formation in galactic disks. In application to galactic spirals, gaseous self-gravity is found to act on the "large scale" to enhance the overall collective gravitational field driving the gaseous response and thus help maintain the global spiral structure. On "local scales," gaseous self-gravity is found to aid the formation and assembling of massive aggregations of clouds into giant cloud complexes, spurs, and feather-like features. Striking is the local raggedness and patchiness of the computed distribution of gas and young stellar associations. Local spurs, feathers, and secondary features continually break apart and reform as the loosely-associated aggregations and giant complexes of clouds continually disassemble and reassemble over time. Such transient features give rise to local disorder within the global spiral structure and blur the global coherence. Of paramount importance are the nonlinear effects and the dissipative character of the cold cloudy galactic gas component, which largely distinguish it from the stellar component. Without the presence of a cold and dissipative gaseous component, galactic disks would be hard pressed to produce and exhibit sharp, clear-cut spiral structures on global scales.

This work is divided into three stages: 1) 'formulation' of the physical problem in mathematical terms, 2) 'solution' of the mathematical model thus formulated, and 3) 'interpretation' of the results. In the 'formulation' stage the focus is on dominant physical mechanisms and dynamical processes. These are outlined in Table 1, with particular emphasis on the self-gravitational effects, dissipative effects, and collisional dynamics of cloudy gaseous galactic disks. In the 'solution' stage, an N-body, cloud-particle computational code is developed for the purpose of isolating the role of gaseous self gravity from the roles of other dominant physical mechanisms and dynamical processes, e.g. the interstellar medium's collisional dynamics and dissipative processes (Roberts and Hausman, 1984; Hausman and Roberts, 1984; Roberts, Adler, and Stewart, 1988). Self gravitational effects of the interstellar medium's gas clouds are included in this work by means of Fourier Transform techniques, adapted from those developed by Miller (1976), Miller and Smith (1979a, b), and Smith and Miller (1986).

Celestial Mechanics **45**: 285–292, 1989.

TABLE I.
FORMULATION STAGE: PHYSICAL MECHANISMS AND DYNAMICAL PROCESSES

- Galactic Gravity. Gaseous Self-Gravitational Effects.
- Orbital Dynamics of Interstellar Gas Clouds. Finite Cloud Cross Sections; Representative Cloud Collisional Mean Free Paths.
- Inelastic, Energy Dissipating Cloud-Cloud Collisions.
- Supernova Explosions; Gas-Star Interactions. Replenishment of Random Kinetic Energy.
- Birth of Protostars. Stellar Associations.
- Time Delay before "Active Period" of Star Formation.
- "Refractory Period" for Gas Clouds Delaying their Participation in Subsequent Star Formation.

Figure 1 shows the results of one representative simulation in which the gas mass to stellar disk mass is 10%. Displayed in a photographic intensity map at one sample time epoch [800 Myr] during the computations are the computed global distributions of the system of gas clouds, represented by patches, and the system of young to middle-aged stellar associations active (with supernova events) during the past 60 Myr, represented by white dots. Gaseous self gravity is present, along with a prescribed spiral perturbation gravitational force field of magnitude adopted as 5% - 10% that of the central axisymmetric force field. The results show that both gas clouds and the stellar associations triggered from the clouds exhibit aggregations of giant complexes along the global, spiral-wave-arm structures. Stellar associations are strongly correlated with the gas, with few associations which are not adjacent to clouds. The important multifold role played by the gas is quite evident. On the "large scale," gaseous self-gravity acts to enhance the overall collective gravitational field driving the gaseous response and thus help maintain the global spiral structure. On "local scales," gaseous self-gravity aids the formation and assembling of massive aggregations of clouds into giant cloud complexes, spurs, and feather-like features. Such transient features give rise to local disorder within the global spiral structure and blur the global coherence (c.f., Roberts and Adler, 1988).

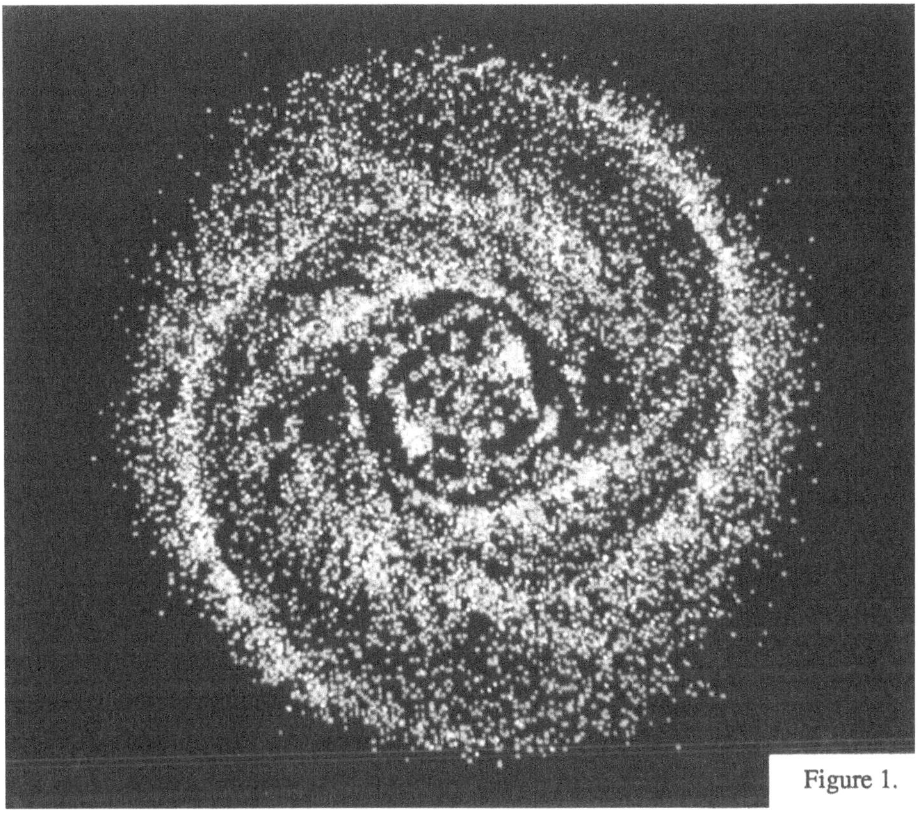

Figure 1.

Figure 2 illustrates in further detail this multifold role of the gas. Four cases with different gas-to-stellar disk mass fractions are displayed, all at the same sample time epoch [800 Myr]. The gas mass fraction increases along the sequence of cases from upper left, to upper right, to lower left, to lower right. The case at upper left is one for which gaseous self gravity is fully excluded. At upper right is the case displayed in Figure 1 with 10% gas mass fraction. The cases at lower left and lower right have 15% and 20% gas mass fractions respectively. It is evident that the effects of gaseous self gravity become increasingly important as gas mass fraction increases. The global coherence of the computed spiral structure in the "outer" disk is particularly strong in the self gravity cases with 15% and 20% gas mass fractions. These cases exhibit gas cloud density distributions with peak-to-mean values across the outer-disk spiral arms typically in the range of 2.8 to 2.9. In contrast in the case with self gravity excluded [upper left], corresponding peak-to-mean values across the outer-disk spiral arms are typically only 2.3. Likewise, within the "inner" disk the global gas cloud distribution is locally quite ragged under the influence of gaseous self gravitational effects, with massive aggregations and large complexes of clouds appearing throughout the global spiral structure particularly in the cases with higher gas mass fraction.

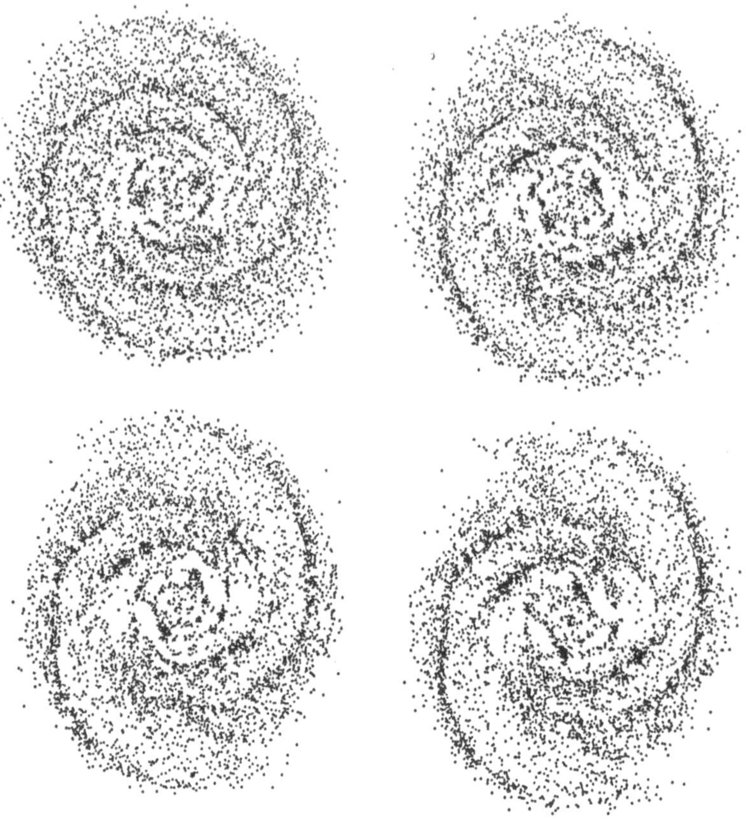

Figure 2.

Figure 3 displays the corresponding distributions of stellar associations computed at the same sample time epoch [800 Myr] in these four cases (upper left - self gravity excluded; upper right, to lower left, to lower right - 10%, 15%, and 20% gas mass fractions respectively). Note that the spiral structure is much more sharply defined by the system of young stellar associations (Figure 3) than by the system of gas clouds (Figure 2). This is presumably due to the collisional origin of many stellar associations (collision rate goes as the square of cloud number density) and the relatively short average delay time between star formation and supernova explosions (leading to strong correlations in the positions of successive generations of associations). Quite prominent in the "outer" disk is the global spiral arm structure exhibited by the distributions of stellar associations, particularly in those self gravitational cases for which the gas represents higher fraction of total galactic mass. Likewise, more striking in

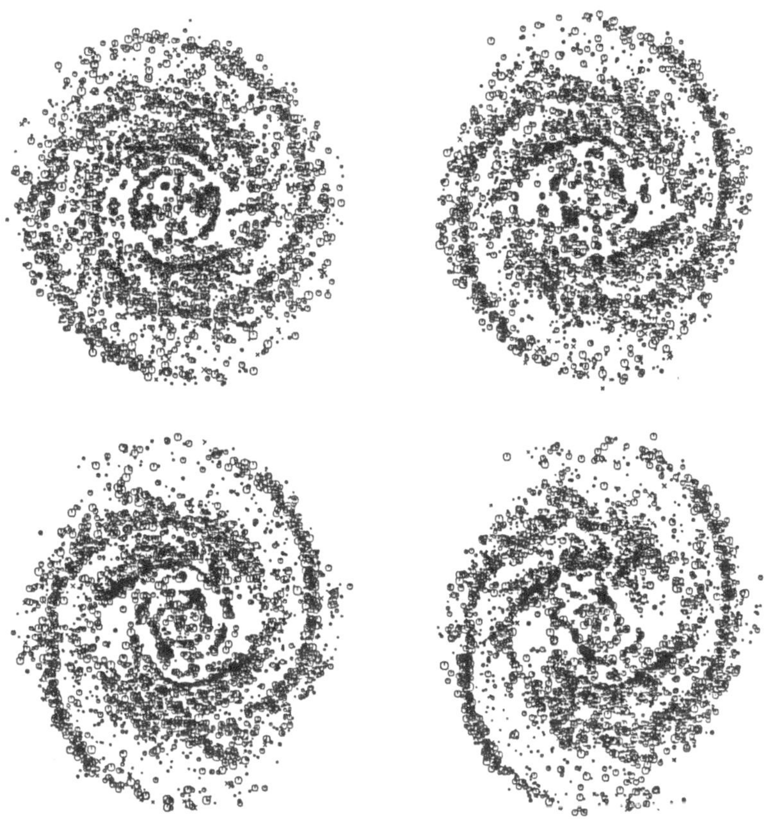

Figure 3.

the cases with higher gas mass fraction is the local raggedness and patchiness of the computed distribution of young stellar associations in the "inner" disk under the influence of the more pronounced gaseous self gravitational effects. Local spurs, feathers, and secondary features continually break apart and reform as the loosely-associated aggregations and giant complexes of stellar-association forming clouds continually disassemble and reassemble over time. It is evident that these continually-evolving, transient manifestations effectively perturb the global spiral structure on local scales in the "inner" disk.

It is critical to investigate the nonlinear dynamical nature of the gas and the extent to which nonlinear effects of the gaseous component may dominate galactic disks. Figure 4 provides a perspective of the degree to which nonlinear gaseous effects may come into play in real galactic disks. Displayed are detailed characteristics of selected physical quantities at three sample epochs computed in the computational studies for the representative self gravitational case with 10% gas mass fraction (displayed in Figure 1 and the upper right panels of Figures 2 and 3). Plotted with respect to spiral phase around a representative annulus in the model galactic disk are: cloud number density, components of velocity perpendicular and parallel to spiral equipotential loci, computed velocity dispersion among gas clouds, and computed distribution of young stellar associations currently active in supernovae explosions (SN rate). The density distribution of the self-gravitating cloud system is strongly-peaked with peak-to-mean values in the range of 2.4. to 2.8 and arm-to-interarm contrasts typically 6:1, with arm thicknesses on the order of a kpc. The sharp deceleration, just preceding 180^0 spiral phase, reflected in the $u_\perp$ velocity component from supersonic to subsonic is a striking manifestation, in these self-gravitating computations, with much-more-gradual-characteristic-rise downstream. This characteristic skewness in the $u_\perp$ velocity component as well as the characteristic asymmetry in the parallel velocity component together delineate the galactic shock structure formed. Such skewness is less apparent in the density distribution, with the density rise occurring over the broad shock width of a number of collisional mean free paths (e.g., consistent with quasi-analytical calculations of star-gas density waves over an annulus [Lubow, Balbus, and Cowie, 1986]).

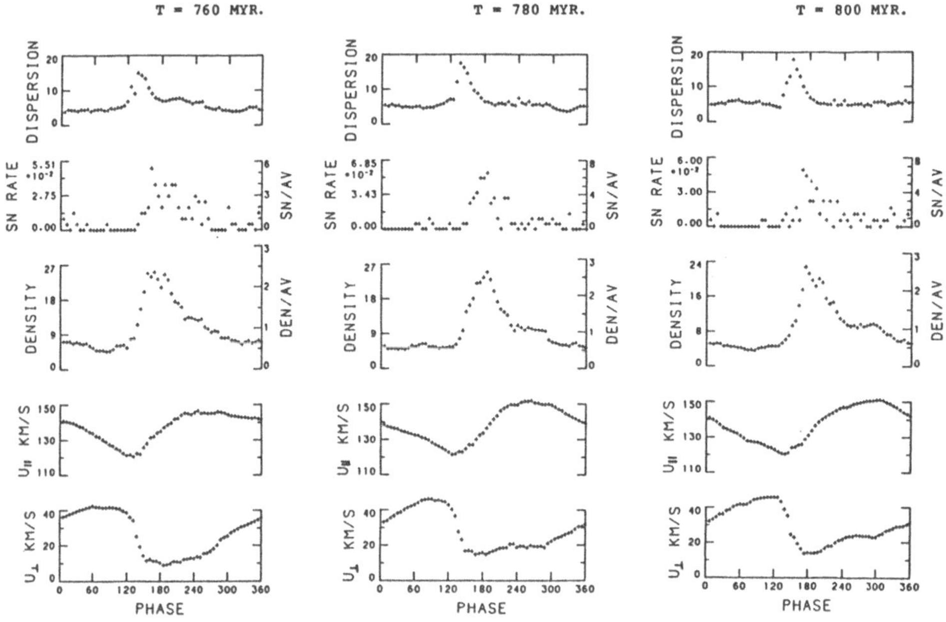

Figure 4.

Of paramount importance in these studies is the dissipative character of the cold cloudy component of the gaseous interstellar medium that largely distinguishes it from the stellar component (also see Kalnajs, 1972; Roberts and Shu, 1972; Lubow, 1986). First of all, it is the presence of this cold, cloudy component that makes possible an overall sufficiently-low velocity dispersion system to promote regimes of unstable, growing global spiral modes (Bertin, Lin, Lowe, and Thurstans, 1988). Secondly, in real galaxies it is likely to be the balance achieved between such moderately to rapidly growing global modes on the one hand and the dissipative nonlinear gaseous response on the other that allows the emergence of finely tuned, coherent grand designs of global spiral structures. Without the presence of a cold and dissipative gaseous component, real galactic disks may be hard pressed to produce and exhibit any such sharp, clear cut structures on global scales.

Roberts and Stewart (1987) have shown that the results of these computational simulations can be largely understood in terms of individual cloud-particle orbits. In particular, the upper panel in Figure 5 shows the computed radial position as a function of time for one sample cloud-particle in a stripped-down collisionless test case. The cloud-particle undergoes nonlinear epicyclic motion, characterized by repeatable sequences of radial oscillations. The spiral forcing modulates the epicyclic oscillations in a regular nonlinear manner, leading to "periodic trapping" of the cloud-

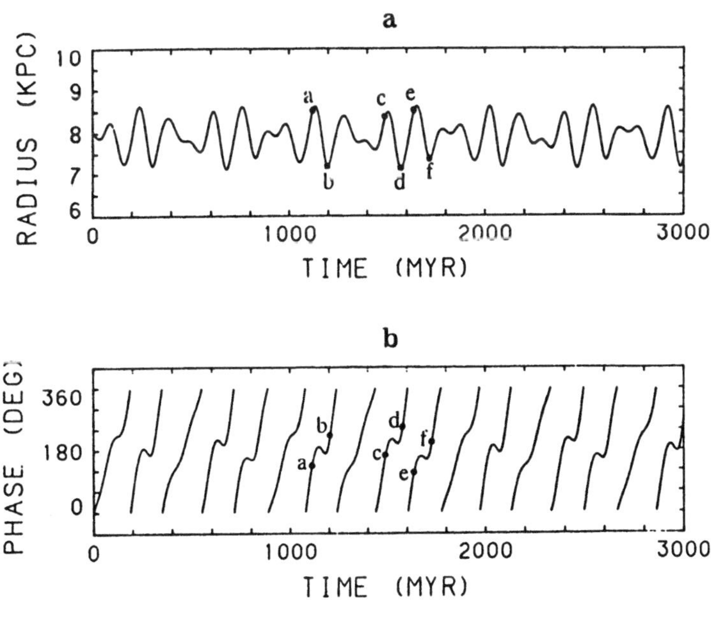

Figure 5.

particle in the arms. The lower panel shows the spiral phase of the cloud-particle versus time, where 180^0 marks the location of the spiral potential minimum. Each continuous segment represents the motion of the cloud-particle around half the disk (through a full 360^0 in spiral phase), from one interarm region to the next. The relative amount of time that the cloud-particle spends in any given interval of spiral phase is inversely proportional to the slope along each segment. Many of the segments exhibit "bumps" of retrograde motion where the cloud-particle temporarily becomes trapped and for a short time (about 50 Myr) moves backwards across the arm, before continuing in its forward motion. Such nonlinear trapping events cause the cloud-particle to spend a large fraction of its total orbital time within a spiral arm. It is evident that the strong global spiral response of the self-gravitating collective cloud system in the computational simulations can be largely attributed to this tendency of individual cloud-particles to become trapped and spend a large fraction of their time in the region of the spiral potential minima. On this basis, we speculate that the self gravitating, dissipative (cold), "nonlinear" gaseous component in galaxies may play a critically important role in the nonlinear locking of two or more density wave modes together over significant epochs.

Effective synergism of the computational studies with corresponding high resolution observational studies should help determine to what degree these characteristics apply or do not apply in real galaxies and to what degree our current understanding requires further refinement. Such synergism is expected to have strong impact on our deeper understanding of the clumpy, cloudy interstellar medium, star formation, and the underlying dynamical processes.

This work was supported in part by NSF under Grants AST-82-04256 and AST-87-12084 and NASA under Grant NAGW-929. The computations were carried out on the CDC 855 at the University of Virginia and the Cray X-MP at the Pittsburgh Supercomputing Center under Grant AST880019P.

References

Bertin, G., Lin, C.C., Lowe, S.A., and Thurstans, R.P. (1988), **Ap. J.**, (in press).
Hausman, M.A., and Roberts, W.W. (1984), **Ap. J., 282,** 106.
Kalnajs, A. (1972), **Astrophysical Letters, 11,** 41.
Lubow, S.H. (1986), **Ap. J. Letters, 307,** L39.
Lubow, S.H., Balbus, S.A., and Cowie, L.L. (1986), **Ap. J., 309,** 496.
Miller, R.H. (1976), **J. Comput. Phys., 21,** 400.
Miller, R.H. and Smith, B.F. (1979a), **Ap. J., 227,** 407.
Miller, R.H. and Smith, B.F. (1979b), **Ap. J., 227,** 785.
Roberts, W.W., and Adler, D.S. (1988), **Proc. Symp. Molecular Clouds in the Milky Way and External Galaxies,** Springer Verlag, Berlin, in press.
Roberts, W.W., Adler, D.S, and Stewart, G.R. (1988), **Proc. of the MIT Symposium honoring C.C. Lin,** World Science Pub. Co., Singapore, p. 373.
Roberts, W.W., and Hausman, M.A. (1984), **Ap. J., 277,** 744.
Roberts, W.W. and Shu, F.H. (1972), **Astrophysical Letters, 12,** 49.
Roberts, W.W., and Stewart, G.R. (1987), **Ap. J., 314,** 10.
Smith, B.F., and Miller, R.H. (1986), **Ap. J., 309,** 522.

SYNTHETIC SECULAR THEORIES OF THE PLANETARY ORBITS: REGULAR AND CHAOTIC BEHAVIOR

ANNA M NOBILI
Department of Mathematics, University of Pisa
Via Buonarroti 2, I-56100 Pisa, Italy
and
Department of Astronomy, Cornell University
Ithaca, NY 14853, USA

Modern computer technology allows dynamical astronomers to investigate the long term stability of real systems as thoroughly as ever. However, the process is not straightforward and new problems need to be solved. This work deals with only one such problem: the construction -from the numerical integration- of a secular perturbation theory that is able to describe the dynamical behavior of the system. The discussion refers to the outer planets and is based on the knowledge acquired by the author during her participation in project LONGSTOP. A digital filter is used in order to reduce the output and eliminate short periodic terms. Filtering uncovers long term variations in the semimajor axes. From the filtered output a secular perturbation theory is constructed in the assumption that the solution is regular, as secular perturbation theories can only be constructed for regular solutions. If we succeed, this means that the solution is indeed regular for the computed span of time; if not -and this can be established in a rigorous way- it has to be concluded a posteriori that the solution is not regular. The LONGSTOP 1A and 1B integrations show well that as the timespan of the integration increases it is possible to detect the non-regular behavior of the solution. This happens in the eccentricity of Saturn at the 10^{-4} level.

1. The need for filtering

Numerical simulations of dynamical models in the solar system have become a major topic in modern celestial mechanics. Powerful, fast and reliable electronic computers allow us to tackle problems that are far too hard to be tackled with purely analytical methods. However, as our goals also become more and more ambitious, new problems need to be faced. We are not concerned here with the choice of a realistic physical model, the development of a fast enough integration algorithm, or the control of integration error accumulation. Let us assume that these problems have already been solved and focus on the fact that the *potential* numerical output is far too big and needs to be reduced. In a computation of the orbits of the outer planets (from Jupiter to Pluto) for $10^8\,yr$, storing position and velocity at every step, for each body, amounts to about $200\,Gbytes$ ($1\,Gbyte = 10^9\,bytes$). Such an output is practically impossible to store and is far too big to handle. We are interested in the long term dynamics of the system, and this requires filtering out the short periodic variations, because their larger amplitude would otherwise mask

Celestial Mechanics **45**: 293–304, 1989.

long term features. Last but not least, a huge number of data points makes it impossible to represent the results in graphical form. A plot should not contain more than about 3,000 points, since more points would make it less meaningful.

Although there is recent evidence of short timescale dynamical phenomena in the solar system (e.g. in planetary rings), most of the *N–body* systems that we are interested in are characterized by small mutual perturbations. They are typically arranged in a *planetary type hierarchy* (increasing semimajor axes and no orbit crossing), as a result of which each 2–*body* subsystem is only slightly perturbed by the other bodies (Milani and Nobili 1983) This makes the system hierarchically stable (i.e. the hierarchical arrangement of the orbits is preserved) for very long timespans. The long term dynamics of such *N–body* systems is described by *secular perturbation theories,* which can be either analytic (and semianalityc) or synthetic, namely derived from the output of a numerical integration of the orbits. *N–body* systems with $N > 2$ are known to be non-integrable (Poincaré 1893). It is also known that in the phase space of a non-integrable dynamical system regular and chaotic regions of motion are mixed together in a very intricate way, as in the famous drawing by Arnold (1963). The KAM theorem has proven that *far enough* from small divisors and for *small enough* perturbations, the regular solution of the unperturbed problem remains regular (see Arnold 1976, Appendix 8). Although the hypotheses under which the KAM theorem is proven do not apply to real dynamical problems in the solar system, the fact that a hierarchical system has not suffered dramatic changes over a very long timespan suggests that, as long as mutual perturbations are small, the solution is regular. This is assumed as a working hypothesis which allows us to construct a secular perturbation theory that represents the solution as a convergent multivariate Fourier series with arguments that are linear functions of time; the spectrum of a regular solution is discrete. However, as the timespan of the computation increases, it may be possible to detect that the solution belongs to some chaotic region of the phase space. When this happens, the construction of a secular perturbation theory valid over that timespan is no longer meaningful.

Secular perturbation theories describe the long term changes of the orbits, namely their dimensions (through the semimajor axes a) and orientation in space (through the eccentricities e and inclinations I). Eccentricities and inclinations are usually replaced by the non singular variables h, k and p, q:

$$\begin{aligned} h &= e \sin \varpi & k &= e \cos \varpi \\ p &= \sin(I/2) \sin \Omega & q &= \sin(I/2) \cos \Omega \quad , \end{aligned} \tag{1}$$

with ϖ and Ω longitude of the pericenter and the node respectively. A secular perturbation theory consists of a few tables of numbers giving the fundamental frequencies of the system (with their phases), and the amplitudes of the spectral lines (combinations of the fundamental frequencies) which are present in the variables h, k, p, q and a. Altogether they amount to several tens of numbers only. In the case of *synthetic* secular perturbation theories these numbers must be extracted from a potentially huge output. The numerical ouput must be reduced without either losing valuable information or introducing artificial features. A detailed discussion on the use of digital filters in planetary dynamics can be found in Carpino et al. (1987), where reference is made to the literature available on filters.

2. Sampling and aliasing

The numerical output (position and velocity for each body, or the corresponding Keplerian elements) is a discrete time signal. The amount of output can be reduced by sampling it with a certain sampling period ΔT_s:

$$x_i = x(i\Delta T_s) \quad , \quad i \in Z \tag{2}$$

where x is one of the ouput variables. However, if two signals are sampled at intervals ΔT_s, they cannot be distinguished from one another if their frequencies differ by integer multiples of the sampling frequency $\omega_s = 2\pi/\Delta T_s$. In the frequency domain, sampling produces an infinite number of images of the spectrum of the original signal centered at integer multiples of ω_s. This is the well known *aliasing* phenomenon (Figure 1). The discrete Fourier transform of the signal (2) is:

$$\hat{x}(\omega) = \sum_{-\infty}^{+\infty} x_i e^{-j\omega i \Delta T_s} \qquad (j^2 = -1) \ . \tag{3}$$

The ordinary Fourier transform of the original continuous signal $x(t)$ —which is not available in our case— is:

$$\hat{x}_F(\omega) = \int_{-\infty}^{+\infty} x(t) e^{-j\omega t} dt \tag{4}$$

and it is related to the discrete Fourier transform (3) through the relation:

$$\hat{x}(\omega) = \frac{1}{\Delta T_s} \sum_{k=-\infty}^{+\infty} \hat{x}_F(\omega + k\omega_s) \quad , \tag{5}$$

which shows well the arising of the aliasing phenomenon. Every spectral line of the original signal with frequency $|\omega| > \omega_s/2$ will appear as a low frequency *alias* in the sampled signal. It is apparent that trying to reduce the amount of output by sampling is not the correct procedure to investigate the long term dynamics of the system, unless ways are found to avoid aliasing. Since we are interested in constructing a secular perturbation theory, there is a *maximum frequency of interest* ω_{int} for the system which is being investigated, corresponding to the shortest period that has to be retained in the secular perturbation theory.

If the spectrum of the original signal is band limited by a maximum frequency ω_{max}, i.e. if

$$\hat{x}(\omega) = 0 \qquad for \qquad |\omega| > \omega_{max} \ , \tag{6}$$

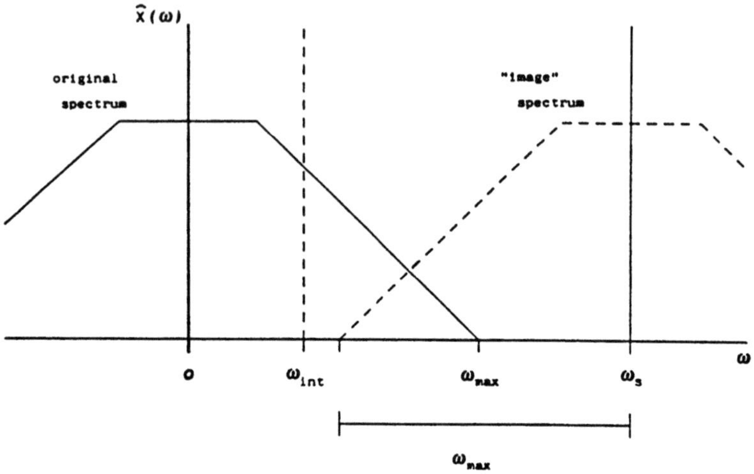

Figure 1: Sampling produces an infinite number of images of the spectrum $\hat{x}(\omega)$. Images are centered at the integer multiples of the sampling frequency ω_s of the signal. This phenomenon is known as aliasing. If the spectrum is band limited by a maximum frequency ω_{max}, the frequencies $|\omega| \leq \omega_{int}$ are free from aliasing provided the aliasing-free condition (7) is fulfilled.

then the band $|\omega| \leq \omega_{int}$ of the sampled signal is free from aliasing provided that:

$$\omega_s \geq \omega_{max} + \omega_{int} \quad , \tag{7}$$

(see Figure 1). If the spectrum of the original signal *is not* band limited, then the aliasing-free condition (7) requires all the frequencies smaller than ω_{max} to be filtered out.

The most natural sampling time of the ouput, before any further reduction is made, is the stepsize of the numerical integration algorithm. Since the stepsize is only about 1/100 of the shortest orbital period in the system, whereas an interesting integration timespan can be 10^7–10^8 times such period, it is apparent that a further reduction of the output is needed. Furthermore, it should be realized that the output of the numerical integration consists of Cartesian coordinates, while the filter must be applied to quantities whose spectrum is clearly divided into a short and a long-periodic part. This requires a coordinate transformation to be performed for each data point which is input to the filter, and therefore a preliminary decimation of the output would be advantageous. We have just shown that sampling generates the aliasing phenomenon, which must be avoided by fulfilling the aliasing–free condition (7), which in turns require the signal to be filtered. However, it is worth noticing that it may be possible to make a preliminary sampling of the signal —*before* applying the filter— without introducing aliases. In the case of the outer planets, where the stepsize is 40 d, if the signal is sampled every 50 data points —with a new sampling period of $\simeq 5.47\, yr$— all frequencies corresponding to periods shorter than $\simeq 10.95\, yr$ could appear as long period aliases. The procedure is correct provided one makes sure that these periods lie in the region of the spectrum which will be filtered out. For the outer planets we have checked that this is actually the case and the preliminary sampling can be performed. However, any further sampling would introduce truly dangerous aliases, unless condition (7) is fulfilled, which in turn requires that a filter is applied.

3. Digital filtering

The frequency response of an ideal low-frequency pass digital filter should be *one* in the pass band and *zero* in the dark band. A real filter will have a *ripple* around the pass band. It will also have an *attenuation factor* in the dark band that we can constraint to be of the order of the largest mutual perturbation in the system, in order to make sure that all short periodic effects are filtered out. Furthermore, for continuity reasons, the frequency response of the filter cannot be a step function. Therefore, there is a region in between the pass and the dark band where the frequency response of the filter assumes any value between $\simeq 1$ and $\simeq 0$; therefore the filtered output cannot be relied upon as far as frequencies in this region are concerned. However, if the original spectrum of the system has *a gap* in between the low and the high frequency regions, the filter can be designed in such a way that the region between the pass and the dark band —where the response of the filter is unreliable— coincides with the region of the spectrum which is devoid of lines. Since the filter has to be applied on line with the numerical integration in order to reduce the amount of output to be stored, this requires some knowledge of the spectrum of the system beforehand.

As discussed in §1, secular perturbation theories are constructed for *N–body* systems where mutual perturbations are small. If mutual perturbations are small, the 2–*body* sub-systems are only slightly perturbed by the remaining bodies and the frequencies of the pericenters and the nodes of each 2 *body* orbit (that are zero in absence of external perturbations) are very low; more so their combinations. They are certainly much lower than, and well separated from, the mean motions. However, as 2–*body* systems are coupled together, combinations of the mean motions appear as well which might be much lower than each of the mean motions. In the outer solar system the lowest mean motion is that of Pluto (with a period of $\simeq 250\,yr$), but smaller frequencies appear because of quasi-resonances in mean motion between the planets, such as the 5:2 quasi-resonance between Jupiter and Saturn (with a period of about $900\,yr$), and the 2:1 quasi-resonance between Uranus and Neptune (with a period of about $4{,}300\,yr$). In contrast, the highest frequencies of the long–periodic part of the spectrum are some combinations of the frequencies of the pericenters and the nodes with periods around $15{,}000\,yr$–$20{,}000\,yr$, so that the short and long–periodic part of the spectrum are well separated by a gap about $10{,}000\,yr$ wide. This separation of the spectrum between mean motion and mean motion resonances or quasi-resonances on one side, and secular frequencies on the other side, has only one exception, namely the 3:2 resonance in mean motion between Neptune and Pluto which has a period of about $19{,}900\,yr$ and therefore happens to be in the long–periodic part of the spectrum. It is worth noticing that Pluto is the only planet in the solar system which does not fulfill the planetary hierarchy structure (its orbit crosses the orbit of Neptune), and this makes mutual perturbations stronger. In any case, the spectra of N–*body* systems, for which secular perturbation theories can be constructed, tend to be made of a short and a long-periodic part, and are separated by a gap. For these systems it is also possible to design a low–frequency pass digital filter.

For a gap between the dark and the pass band of the filter of width $\Delta\omega$, the aliasing–free condition (7) becomes

$$\omega_s \geq 2\omega_{int} + \Delta\omega \quad . \tag{8}$$

The maximum frequency of interest ω_{int} is given by the physics of the problem, while the sampling frequency ω_s must be chosen according to (8). One would like to have ω_s as small as possible (so as to have the sampling period as large as possible), but also the gap width should be as large as possible as this would result in a filter of shorter length. However, we see from (8) that there is little advantage —in terms of getting a smaller value for ω_s— in choosing a value of $\Delta\omega$ much smaller than ω_{int}. Therefore, if the system has a gap between the short and the long-periodic part of the spectrum at least as wide as ω_{int}, it is a reasonable compromise to choose $\Delta\omega \simeq \omega_{int}$.

The output signal y_m of a linear filter is obtained by the discrete convolution of the input signal x_m with a suitable finite succession d_m:

$$y_m = (d * x)_m = \sum_{m'=-M}^{+M} d_{m'} \cdot x_{m-m'} \qquad m, m' \in Z \tag{9}$$

($L = 2M + 1$ is the length of the filter). The frequency response of the filter is the discrete Fourier transform $\hat{d}(\omega)$ of the impulse response d_m, and fulfills the relation:

$$\hat{y}(\omega) = \hat{d}(\omega) \cdot \hat{x}(\omega) \quad . \tag{10}$$

The filter must satisfy 2 conditions. *i*) It must not modify the phases of the Fourier components of the signal; this is obtained by requiring that $d_m \in R$ and $d_m = d_{-m}$. *ii*) For a given cutoff frequency ω_{max}, the frequency response of the filter must be $\hat{d}(\omega) = 0$ for $\omega > \omega_{max}$, and $\hat{d}(\omega) = 1$ for $\omega \leq \omega_{max}$; this requirement can be met only approximately, in the sense that the frequency response of the real filter will have an *attenuation factor* larger than zero in the dark band and a *ripple* around the unity response of the pass band. It can be shown that there is a unique choice of the coefficients d_m which minimizes the maximum error of the frequency response of the filter with respect to the ideal response. The time duration of the filter $(L-1)\Delta T_{input}$, where ΔT_{input} is the input sampling period, must be much shorter than the timespan of the numerical integration. Designing a filter of length $L = 2M + 1$ requires solving an interpolation problem for a polynomial of degree M; if M is very large, it is in fact better to design a 2–*stage* filtering–decimation procedure which accomplishes the same results.

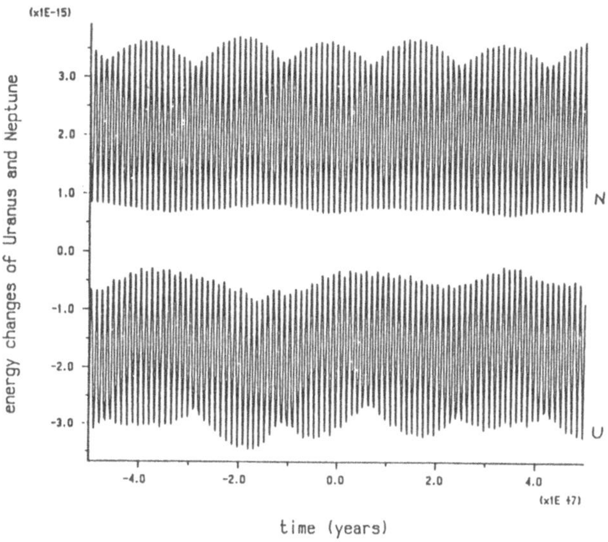

Figure 2: Energy changes of Uranus and Neptune over $100\,Myr$ as obtained from the filtered output of the LONGSTOP 1B numerical integration. The average is -3.364×10^{-10} for Uranus and -2.535×10^{-10} for Neptune. The units are $M_{\odot} AU^2/day^2$ ($M_{\odot}$ is the mass of the Sun and AU the astronomical unit). The main oscillation —in antiphase— has a period of $1{,}106{,}994.6\, yr$, corresponding to the frequency combination $g_5 - g_7$ (the fundamental frequencies related to the pericenters of Jupiter and Uranus respectively). The reason why this term is so important is because it contains the small divisor, to power 3, of the 2:1 quasi-resonance in mean motion between Uranus and Neptune.

4. Filtering uncovers long term variations in the semimajor axes of the planets

Filtering has proven essential in investigating one of the oldest problems of celestial mechanics, namely the variation of the semimajor axes of the planets over very long timespans. Tisserand (1889) improved the famous Lagrange theorem showing that, to order 2 in the smallness parameters μ (the planet-to-Sun mass ratios), the differential equations for the semimajor axes contain only short periodic terms, i.e. terms in which the frequencies of the pericenters and the nodes always appear together with non-zero combinations of the mean motions. He concluded that:

$$\dot{a}_{long\,term} = O(\mu^3) \quad . \tag{11}$$

Nevertheless, a perturbation of order μ^3 acting with a slow frequency —hence accumulating over a timespan of the order of a typical orbital period divided by μ— would in fact give

$$\Delta a_{long\,term} = O(\mu^2) \quad . \tag{12}$$

It has been found that indeed the semimajor axes of the planets undergo variations to order μ^2 (Message 1976; 1982), but nobody succeeded in computing these terms, until they appeared in the *filtered* output of the LONGSTOP numerical integrations of the orbits of the outer planets (Milani et al. 1986; 1987; Nobili et al. 1988). The semimajor axes of Uranus and Neptune show a dominant term, in antiphase, with a period of 1.1 Myr and a relative amplitude of about 10^{-6} (Figure 2). In contrast, the unfiltered energy —shown in Figure 3 for a shorter timespan— is dominated by a short-periodic effect about 100 times larger. The same 1.1 Myr period appears in e^2, suggesting that the semimajor axes variation must be of order 2 in the smallness parameters and of degree 2 in the eccentricities. However, this does not explain a relative amplitude of 10^{-6}, unless some *small divisor* appears and amplifies it. It can be shown analytically that the 2:1 quasi-resonance in mean motion between Uranus and Neptune is responsible for amplifying the semimajor axes effect to the observed level (Milani et al. 1987). Figure 2 suggests that longer term effects are also present. They become apparent by filtering out periods shorter than, or equal to, 1.1 Myr. Figure 4 shows the residuals in the energy of Uranus, in which spectral lines with periods of more than 10 Myr are present. (Nobili et al. 1988, Table 6). Similar long-periodic effects are present in the energies of the other planets.

5. Fourier analysis and synthetic secular perturbation theories: regular and chaotic solutions

A synthetic secular perturbation theory for the system is constructed by performing a Fourier analysis of the filtered output. However, because of the finite span of data, each spectral line is surrounded by a diffraction pattern which limits the accuracy with which its frequency and phase can be detected; a low amplitude line close to a higher one can be seriously distorted and even impossible to detect. A procedure of successive detection-and-elimination of the spectral lines, by which each line –starting from the one with the largest amplitude– is measured and eliminated so that the next line can be measured and eliminated from the residuals, is bound to propagate any measurement error.

To make line measurement more reliable, particularly in the case of low-amplitude lines, a *numerology table* is computed listing all the combinations of fundamental frequencies which are theoretically permitted according to the conservation laws and symmerty properties of the system (the *D'Alembert rules*). The maximum number of fundamental frequencies in each combination must increase with the timespan of the numerical integration, as higher and higher order terms have to be taken into account if the theory has to be valid for a longer and longer interval of time. A least squares fit to the data of a general trigonometric model taking into account D'Alembert's rules can be devised. Starting with approximate values of the fundamental secular frequencies of the system, the fitting program supplies improved values for the fundamental frequencies with their amplitudes and phases, plus the time series of the residuals. The residuals are Fourier analysed and the most relevant lines are checked against the permitted lines of the *numerology table* for identification. A line is identified *with certainty* with a theoretically permitted one when a line with similar frequency and phase exists in the *numerology table*

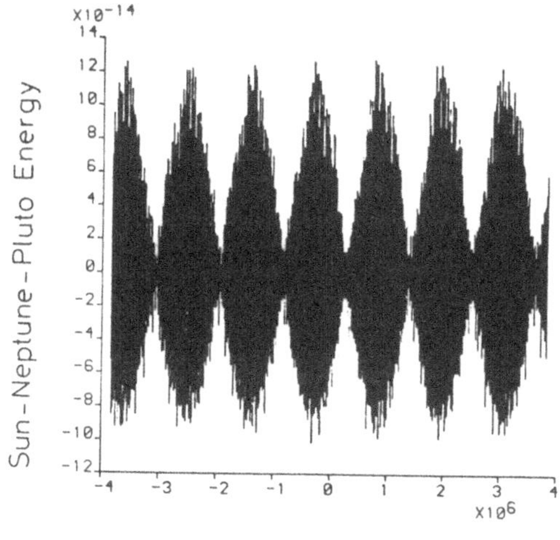

Figure 3: This is how the energy variation of Neptune looks if the filter is not applied. The timespan is about $8\,Myr$. The dominant effect is due to 2:1 quasi-resonace in mean motion between Uranus and Neptune, with a period of $\simeq 4{,}300\,yr$. Its amplitude is about 100 times larger than the amplitude of the oscillation shown in Figure 2, which therefore is completetly hidden here. A modulation with that period is present, but nothing more can be said from this plot. The units are the same as in Figure 2. Pluto was included in order to eliminate the period of the 3:2 resonance in mean motion between Neptune and Pluto.

while all others are farther away. The lines which have been successfully identified are added to the trigonometric model and the fitting program is run again. The procedure ends when the main spectral line in the residuals can no longer be identified *with certainty*. This method has been successfully applied to the LONGSTOP 1A numerical computation of the orbits of the outer planets spanning $9.3\,Myr$. The resulting secular perturbation theory is given by three tables (for the spectral lines in the h, k , p, q and a), plus the fundamental frequencies of the system (Carpino et al. 1987; Milani et al. 1987). Twenty five lines are enough to describe the dynamic behavior of the planetary orbits over such a timespan, and the main limitation to the recovery of smaller amplitude lines is the length of the integration interval. It is concluded that the solution is regular for the computed timespan.

If the timespan of the numerical integration is increased, then it becomes possible to detect longer period lines. If the solution is regular, then the number of lines at a fixed amplitude level must not increase with the interval of time for which the solution is computed. In a regular solution, longer period lines come from higher order effects and should therefore have smaller amplitudes. The filtering and Fourier analysis techniques discussed above have been applied to the LONGSTOP 1B numerical integration of the outer planets spanning $100\,Myr$. In order to better account for the perturbations due to the inner planets and to model the general relativistic advance of pericenters, the physical model is no longer a pure *N–body* model. Other perturbations have been investigated to make sure that

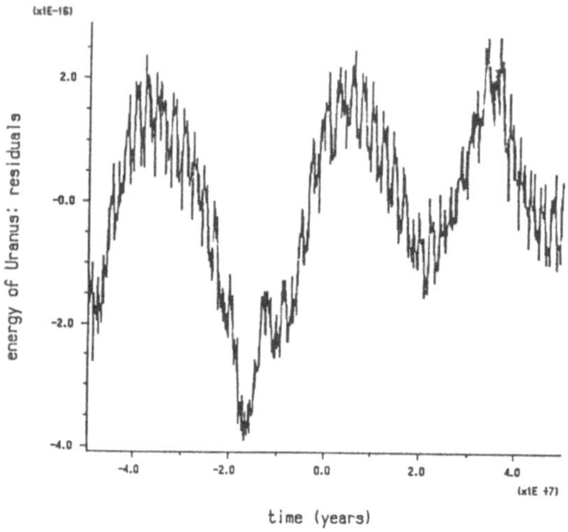

Figure 4: Residuals in the energy of Uranus after all periods up to $1.1\,Myr$ have been filtered out. The effects which are left have periods of $10\,Myr$ or more. Units as in Fig. 2.

this is a realistic model for the outer solar system as far as a timespan of $100\,Myr$ is concerned (Nobili et al. 1988, §2).

The LONGSTOP 1B secular perturbation theory shows that there are regions where many lines of comparable amplitude accumulate. These regions have a typical *multiplet* structure with lines symmetrically arranged on either side of the location of an identified line (see Figure 5). Such multiplets are known to be formed by the combination of many long-periodic lines with a short-periodic one; each combination generates two lines with periods close to the short one and with opposite phases. The higher the number of long-periodic lines, the higher the number of lines in the multiplets and therefore they should be observed more easily in the output of longer numerical integrations. However, it is surprising that multiplet lines have amplitudes of comparable size; it is also surprising that they have no counterpart in the *numerology table* which lists all the permitted combinations of up to 12 fundamental frequencies, in spite of the fact that their amplitude is comparable to that of lines which have been successfully identified with combinations of a small number of fundamental frequencies. Multiplets of lines are found in the h, k variables around g_6 and g_7 (the fundamental frequencies related to the pericenters of Saturn and Uranus), in the p, q variables around s_8 (the fundamental frequency related to the node of Neptune) and in the semimajor axes around $g_6 - g_5$ (the difference between the fundamental frequencies related to the pericenters of Saturn and Jupiter) and its higher harmonics. The best observed multiplet of lines is the one around g_6. In the eccentricity of Saturn, at the 10^{-4} level, there is a significant increase in the number of lines compared to the LONGSTOP 1A synthetic theory. They have no theoretical counterparts up to combinations of 12 secular frequencies, and therefore the rigorous procedure for the construction of a synthetic secular perturbation

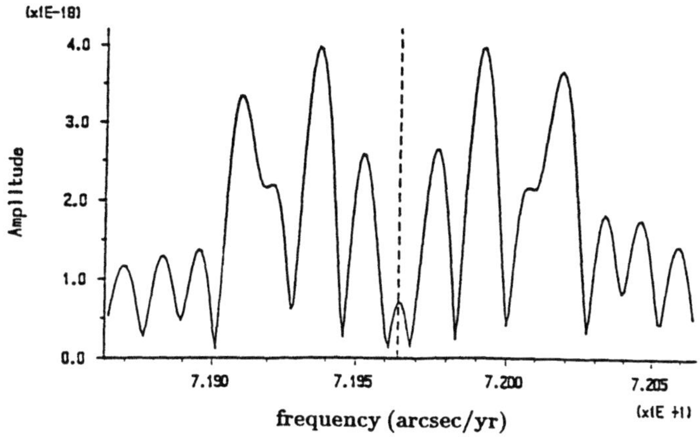

Figure 5: The spectrum of Saturn's energy around the combination $3(g_6 - g_5)$, which has been filtered out (its location is shown by the dotted line). g_6 and g_5 are the fundamental frequencies related to the pericenters of Saturn and Jupiter. The multiplet structure of the lines is apparent. The same structure appears also around $2(g_6 - g_5)$ and $g_6 - g_5$, with larger amplitudes. This spectrum was obtained from the residuals after terminating the procedure of construction of a secular perturbation theory for the semimajor axes.

theory is no longer applicable. We must conclude that the assumption under which the secular perturbation theory is constructed, namely that the solution is regular, is no longer valid (Nobili et al. 1988).

Recent numerical integrations of two massless Plutos for 845 Myr carried out on the Digital Orrery have shown that the motion of Pluto is chaotic (Sussman and Wisdom 1988). The distance between the two Plutos grows exponentiallly up to a saturation value of $\simeq 45\,AU$ and the Lyapounov exponent is positive and large ($1/20\,Myr$). Yet the authors see no sign of instability; however they notice that the spectrum is unusually dense of lines, very much unlike the discrete spectrum which is typical of regular solutions. Milani et al. (1988) have analysed the orbit of Pluto computed in the LONGSTOP 1B numerical integration. By looking at all the main resonances that Pluto is involved in (including two newly discovered super-resonances), and by comparison with the results published by Sussman and Wisdom, they show that indeed there is no instability. The saturation distance of $45\,AU$ —and the related positive Lyapounov exponent— simply amounts to the fact that the two Plutos have opposite phases in the 3/2 mean motion resonance with Neptune, while still maintaining the same libration amplitude of about 80°. In any case, the strenght of Pluto's perturbation on Saturn is $\simeq 10^{-10}$ only, and therefore it cannot be responsible for the indications of chaotic motion in the orbits of the major planets which have been found in LONGSTOP 1B.

Although the subject seems destined to bring many surprises, as for now we conclude as follows (1) a spectrum in which lines accumulate is most probaby related to non-regular motion; (2) if the orbits are computed for a long enough time, non-regular regions of the phase space are likely to be detected; (3) once chaos is found, then the problem arises of understanding its dynamical origin and relevance. Even for Pluto the chaotic region turns out to be quite narrow.

References

Arnold, V. 1976. *Le Méthodes Mathématiques de la Mécanique Classique,* Editions Mir, Moscou

Arnold, V. 1963. Small denominators and problems of stability of motion in classical and celestial mechanics, *Russian Mathematical Surveys* **18**, 85-191

Carpino, M., Milani, A. and Nobili, A.M. 1987. Long-term numerical integration and synthetic theories for the motion of the outer planets, *Astron. Astrophys.* **181**, 182-194

Message, P.J. 1976. Formal expressions for the motion of N planets in the plane, with the secular variations included, and an extension to Poisson's theorem, in *Long-Time Predictions in Dynamics, Szebehely and Tapley eds., Reidel Pu. Co., Dordrecht, Holland*, 279-293

Message, P.J. 1982. Asymptotic series for planetary motion in periodic terms in three dimensions, *Celestial Mechanics* **26**, 25-39

Milani, A. and A.M. Nobili. 1983. On the stability of hierarchical 4–*body* systems, *Celest. Mech.***35**, 241-291

Milani, A., Nobili, A.M., Fox, K. and Carpino, M. 1986. Long term changes in the semimjor axes of the outer planets, *Nature* **319**, 386-388

Milani, A., Nobili, A.M. and Carpino, M. 1987. Secular variations of the semimajor axes: theory and experiments, *Astronomy and Astrophysics* **172**, 265-279

Milani, A., Nobili, A.M. and Carpino, M. 1988. Dynamics of Pluto, submitted to *Icarus*

Nobili, A.M., Milani, A., and Carpino, M. 1988. Fundamental frequencies and small divisors in the orbits of the outer planets. *Astron. Astrophys.*, in press

Poincaré, H. 1893. *Méthodes Nouvelles de la Mécanique Célest, Vol. 1*, Gauthiers–Villars, Paris

Sussman, G.J. and Wisdom, J. 1988. Numerical evidence that the motion of Pluto is chaotic, *Science* **241**, 433-437

Tisserand, F.1889. *Traité de Mécanique Céleste,* Gauthier-Villars, Paris

NUMERICAL REPRESENTATION OF PLANETARY EPHEMERIDES

X X NEWHALL
Jet Propulsion Laboratory
California Institute of Technology
4800 Oak Grove Drive
Pasadena, CA 91109-8099

ABSTRACT. The Jet Propulsion Laboratory provides high-precision numerically integrated planetary and lunar ephemerides in support of spacecraft navigation and other activities relating to solar system bodies. Hundreds of users around the world have requested copies of the ephemerides. In the interests of compactness and utility, techniques have been developed for (1) the generation of the coefficients of an interpolating polynomial based on output from the integrator, and (2) transformation of the contents of an ephemeris file to a standard form usable on virtually any computer.

1. Representation by Chebyshev Coefficients

1.1. THE NEED FOR INTERPOLATION

The numerical integration program carries the instantaneous state of the solar system as polynomials in the form of position, velocity, acceleration, and up to 14 modified divided backward differences of acceleration for each cartesian component of the nine planets and the Moon. Saving the difference arrays at every integration step would result in prohibitively large files. Most applications of the ephemeris require only the positions of designated bodies. Considerable economy of file size is achieved by obtaining polynomial representations of the positions valid over a certain time span.

1.2. CHEBYSHEV POLYNOMIALS

Chebyshev polynomials are the functions of choice for ephemeris representation: they are stable during evaluation, they give a near-minimax representation, and they provide a readily apparent estimate of neglected terms on interpolation error. For an extensive discussion of these polynomials, see Rivlin, 1974.

The nth Chebyshev polynomial $T_n(t)$ is defined by the recursion formula

$$T_n(t) = 2\,t\,T_{n-1}(t) - T_{n-2}(t), \qquad n = 2, 3, \ldots \tag{1}$$

with $T_0(t) = 1$ and $T_1(t) = t$. The applicable range of t for interpolation is $-1 \leq t \leq 1$.

Celestial Mechanics **45**: 305–310, 1989.

Any given function $f(t)$ has an approximate Nth-degree expansion in Chebyshev polynomials:

$$f(t) \doteq \sum_{n=0}^{N} a_n T_n(t), \tag{2}$$

and, when differentiated,

$$\dot{f}(t) \doteq \sum_{n=1}^{N} a_n \dot{T}_n(t), \tag{3}$$

where the a_n are chosen in a manner appropriate for $f(t)$ and $\dot{f}(t)$. In the present case, where $f(t_j)$ and $\dot{f}(t_j)$ denote a coordinate and its derivative computed at discrete times t_j by the integrating program, the a_n serve to *define* the function $f(t)$ as a polynomial. The task becomes the determination of a set of a_n that provide interpolated values suitably approximating those available from the original backward-difference representations carried by the integrator. The following section details the generation of the a_n for ephemeris body coordinates. (It should be noted that the use of Chebyshev polynomials to repesent ephemerides is not new. The Jet Propulsion Laboratory has distributed Chebyshev-constructed files since 1974; the *Connaissance des Temps* has been available as Chebyshev polynomials since 1978, in both printed and machine-readable form.)

1.3. CHEBYSHEV COEFFICIENT GENERATION

The subroutine PVCH was developed to provide efficient generation of the Chebyshev polynomial coefficients representing the cartesian coordinates of the ephemeris bodies. The full span of, say, sixty years of an ephemeris file is segmented into contiguous intervals, or granules, of fixed length. (The actual length of a granule depends on the body; see Table 1 for details.) For each coordinate of an ephemeris body the Chebyshev coefficients a_n that define the interpolating polynomial valid over a given granule must be produced. There are as many sets of coefficients representing each coordinate as there are granules covering the ephemeris span.

PVCH accepts a pair of position and velocity values from the integrator for a given granule at each of the nine (equally spaced) normalized times: $1, 3/4, 1/2, 1/4, \ldots, -3/4, -1$. The output is the set of a_n for the polynomial that is an exact fit to the position values at the end points ($t = 1$, $t = -1$) and a least-squares fit to the interior positions, and whose differentiated polynomial is an exact fit to the velocities at the end points and a least-squares fit to the interior velocities. This approach has the advantage that interpolated position and velocity are continuous at the common end point of adjacent granules; it also minimizes the effects of noise that would otherwise degrade the interpolated velocity obtained from differentiation of a polynomial based on position values alone.

PVCH uses a set of multipliers c_k to obtain the a_n as a linear combination of the input positions and velocities:

$$\begin{aligned} a_n(N) = {} & c_1(N)\,p(1) + c_2(N)\,v(1) + c_3(N)\,p(3/4) + c_4(N)\,v(3/4) \\ & + \cdots + c_{17}(N)\,p(-1) + c_{18}(N)\,v(-1) \end{aligned}$$

The $c_k(N)$ are unique for each a_n and for each polynomial degree N. In the application of the following steps the $c_k(N)$ were formed and encoded as DATA statements intended for PVCH. That subroutine is subsequently used to form the entire set of Chebyshev coefficients constituting an ephemeris file. The $c_k(N)$ were obtained as follows: from Eq. (2) there are 18 conditions on the a_n:

$$\begin{aligned}
p(1) &= \sum_{n=0}^{N} a_n T_n(1) & v(1) &= \sum_{n=0}^{N} a_n \dot{T}_n(1) \\
p(3/4) &= \sum_{n=0}^{N} a_n T_n(3/4) & v(3/4) &= \sum_{n=0}^{N} a_n \dot{T}_n(3/4) \\
&\vdots & &\vdots \\
p(-1) &= \sum_{n=0}^{N} a_n T_n(-1) & v(-1) &= \sum_{n=0}^{N} a_n \dot{T}_n(-1)
\end{aligned} \tag{4}$$

This system can be expressed in matrix and vector notation as $\mathsf{Ta} = \mathbf{f}$, or

$$\begin{bmatrix} T_0(1) & T_1(1) & T_2(1) & \dots & T_N(1) \\ \dot{T}_0(1) & \dot{T}_1(1) & \dot{T}_2(1) & \dots & \dot{T}_N(1) \\ T_0(3/4) & T_1(3/4) & T_2(3/4) & \dots & T_N(3/4) \\ \dot{T}_0(3/4) & \dot{T}_1(3/4) & \dot{T}_2(3/4) & \dots & \dot{T}_N(3/4) \\ \vdots & \vdots & \vdots & \ddots & \vdots \\ T_0(-1) & T_1(-1) & T_2(-1) & \dots & T_N(-1) \\ \dot{T}_0(-1) & \dot{T}_1(-1) & \dot{T}_2(-1) & \dots & \dot{T}_N(-1) \end{bmatrix} \begin{bmatrix} a_0 \\ a_1 \\ a_2 \\ \vdots \\ a_N \end{bmatrix} = \begin{bmatrix} p(1) \\ v(1) \\ p(3/4) \\ v(3/4) \\ \vdots \\ p(-1) \\ v(-1) \end{bmatrix} \tag{5}$$

The matrix T is $18 \times (N+1)$, implying that the system is overdetermined for $N < 17$ and must be solved by a least-squares method. In addition the requirement that the end-point positions and velocities be reproduced exactly imposes four constraints on the system.

1.4. SOLUTION WITH LAGRANGE MULTIPLIERS

We want to minimize the Euclidean norm $\| \mathsf{Ta} - \mathbf{f} \|^2$, subject to the four constraints

$$\begin{aligned}
g_1(\mathrm{a}) &= \sum_{n=0}^{N} a_n T_n(1) - p(1) = 0, & g_3(\mathrm{a}) &= \sum_{n=0}^{N} a_n T_n(-1) - p(-1) = 0, \\
g_2(\mathrm{a}) &= \sum_{n=0}^{N} a_n \dot{T}_n(1) - v(1) = 0, & g_4(\mathrm{a}) &= \sum_{n=0}^{N} a_n \dot{T}_n(-1) - v(-1) = 0.
\end{aligned} \tag{6}$$

Following the standard use of Lagrange multipliers, for each a_n we have

$$\frac{\partial}{\partial a_n}\Big(\| \mathsf{Ta} - \mathbf{f} \|^2 + \sum_{i=1}^{4} \lambda_i g_i(\mathrm{a}) \Big) = 0. \tag{7}$$

Differentiating with respect to each of the a_n separately yields $N+1$ equations in $N+5$ unknowns (the λ_i must be included). Appending the four constraint equations (6) produces the $(N+5)\times(N+5)$ system:

$$\left[\begin{array}{cccc|cccc} \multicolumn{4}{c|}{} & T_0(1) & \dot{T}_0(1) & T_0(-1) & \dot{T}_0(-1) \\ \multicolumn{4}{c|}{\mathsf{T^*WT}} & T_1(1) & \dot{T}_1(1) & T_1(-1) & \dot{T}_1(-1) \\ \multicolumn{4}{c|}{} & \vdots & \vdots & \vdots & \vdots \\ \multicolumn{4}{c|}{} & T_N(1) & \dot{T}_N(1) & T_N(-1) & \dot{T}_N(-1) \\ \hline T_0(1) & T_1(1) & \dots & T_N(1) & 0 & 0 & 0 & 0 \\ \dot{T}_0(1) & \dot{T}_1(1) & \dots & \dot{T}_N(1) & 0 & 0 & 0 & 0 \\ T_0(-1) & T_1(-1) & \dots & T_N(-1) & 0 & 0 & 0 & 0 \\ \dot{T}_0(-1) & \dot{T}_1(-1) & \dots & \dot{T}_N(-1) & 0 & 0 & 0 & 0 \end{array}\right] \begin{bmatrix} a_0 \\ a_1 \\ \vdots \\ a_N \\ \lambda_1 \\ \lambda_2 \\ \lambda_3 \\ \lambda_4 \end{bmatrix}$$

$$= \left[\begin{array}{ccccccc} \multicolumn{7}{c}{\mathsf{T^*W}} \\ \hline 1 & 0 & 0 & \cdots & 0 & 0 & 0 \\ 0 & 1 & 0 & \cdots & 0 & 0 & 0 \\ 0 & 0 & 0 & \cdots & 0 & 1 & 0 \\ 0 & 0 & 0 & \cdots & 0 & 0 & 1 \end{array}\right] \begin{bmatrix} p(1) \\ v(1) \\ p(3/4) \\ v(3/4) \\ \vdots \\ p(-1) \\ v(-1) \end{bmatrix} \tag{8}$$

The notation T^* denotes the transpose of T. The matrix W is a diagonal weighting matrix, included to allow control of the relative effects of position and velocity values. It was found experimentally that the best results are obtained with velocity weighted at .4 relative to position, giving W the form diag(1.0, 0.16, 1.0, 0.16, ...).

The above matrix equation (8) can be written as $\mathsf{C}_1\mathbf{a}_\lambda = \mathsf{C}_2\mathbf{f}$, where $\mathbf{a}_\lambda$ is the vector $\mathbf{a}$ augmented by the λ_i. The solution is $\mathbf{a}_\lambda = (\mathsf{C}_1^{-1}\mathsf{C}_2)\mathbf{f}$. Row $n+1$ of the product matrix $(\mathsf{C}_1^{-1}\mathsf{C}_2)$ contains the multipliers $c_k(N)$ for each $a_n(N)$. It is these $c_k(N)$ that are formatted for the DATA statements in PVCH.

The solutions for the λ_i are also part of the result. However, they have no useful interpretation in this context and are ignored.

1.5. INTERPOLATION ERROR AND POLYNOMIAL DEGREE

It is essential to have a quantitative estimate of the maximum error expected from the interpolation process when the polynomials described above are used to extract coordinate values at arbitrary times. (It should be noted that the term "error" here refers to the difference between interpolated and integrator-supplied values; it does *not* indicate the degree of accuracy to which the original integrated ephemeris represents the dynamical state of the solar system.)

The Chebyshev polynomials provide a convenient and reliable estimate of interpolation error. An arbitrary function has the exact representation as an infinite Chebyshev expansion

$$f(t) = \sum_{n=0}^{\infty} a_n T_n(t). \tag{9}$$

The maximum value of $T_n(t)$ is unity on the interval $[-1, 1]$, the domain of validity for interpolation. Therefore, when a function is approximated by an Nth-degree polynomial, as in Eqn. (2), the maximum error ϵ arising from the omitted remainder of the series has the upper bound

$$\epsilon = \left| \sum_{n=N+1}^{\infty} a_n T_n(t) \right| \leq \sum_{n=N+1}^{\infty} |a_n| |T_n(t)| \leq \sum_{n=N+1}^{\infty} |a_n|. \tag{10}$$

Investigation has shown that the granule length and the polynomial degree N can be chosen so that $|a_{n+1}/a_n| \approx 0.1$ or less for $n \geq N$, implying that the maximum expected interpolation error is about one tenth the magnitude of the highest retained coefficient a_N.

The accuracy criterion for standard JPL ephemerides is that the interpolation error for all coordinate values must be less than 0.5 millimeters. (The DE102 ephemeris covers 4400 years; in the interests of providing a significantly compressed file the interpolation-error criterion was relaxed by defining polynomials of a given degree that span granules of twice the length of those on other JPL ephemerides [Newhall *et al.*, 1983].) The minimum degree N of the interpolating polynomial is 3, as the requirement that the end-point position and velocity values be matched exactly yields four constraints; the 18 combined position and velocity values permit a maximum degree of 17. Table 1 lists the granule length and polynomial degree for each body on the JPL ephemeris files.

Table 1. Granule Length and Polynomial Degree for the 11 Ephemeris Bodies

Body	Granule Length (days)	Polynomial Degree N	Body	Granule Length (days)	Polynomial Degree N
Mercury	8	13	Saturn	32	6
Venus	16	9	Uranus	32	5
Earth-Moon Barycenter	16	12	Neptune	32	5
			Pluto	32	5
Mars	32	10	Moon	4	12
Jupiter	32	7	Sun	16	10

2. Exporting the Ephemeris

2.1. PORTABILITY

Because of the accuracy requirments of space navigation the JPL ephemerides are suffient to satisfy the most stringent demands of any application. Hundreds of users from all over the world, representing virtually every type of computer available, have requested a copy of the ephemerides. It is essential that a thoroughly portable, machine-independent version of the files and software be produced.

2.2. DATA REPRESENTATION

The JPL ephemeris files consist of character, integer, and double-precision floating point data types. The files are originally produced on a Univac 1100/91 computer. When transformed for an export tape the file contents are written on a tape in what amounts to a formatted dump of the file. Character data are written as `CHARACTER*6` variables; integers are written as integers.

Double-precision data present a problem. Fortran floating point printed numbers are in general not exact representations of the binary quantities. On the Univac, floating point numbers are represented as a sign bit, an 11-bit biased exponent of 2, and a 60-bit mantissa, with the binary point assumed to be at the left of the mantissa. (The exponent bias is 1024_{10}, or 2000_8.) In the export format each double-precision number is written as three integers: N_1 = the exponent k with the bias removed, followed by the mantissa expressed as two 30-bit integers N_2 and N_3. When reconstructed on the receiving machine, the formula is

$$\text{Double-precision number} = N_2 \times 2^{N_1-30} + N_3 \times 2^{N_1-60}$$

As an illustrative example, on DE200 the conversion factor between AU's and kilometers is 149597870.66. It has the octal floating-point representation

2034 4352535272 5075341217.

When converted to export format the three-integer representation becomes

$$\begin{aligned} N_1 &= 28 && (= 34_8) \\ N_2 &= 598391482 && (= 4352535272_8) \\ N_3 &= 687194767 && (= 5075341217_8) \end{aligned}$$

2.3. THE EXPORT SOFTWARE

The software package on the export tape sent to users contains a program for decoding the tape and creating a direct-access file on mass storage. It also includes subroutines for subsequent reading and interpolating the ephemeris file.

3. Acknowledgment

The work described in this paper was carried out by the Jet Propulsion Laboratory, California Institute of Technology, under contract with the National Aeronautics and Space Administration.

4. References

Rivlin, T. J., *The Chebyshev Polynomials*, John Wiley & Sons, 1974.

Newhall, X X, Standish, E. M., and Williams, J. G.: "DE 102: a numerically integrated ephemeris of the Moon and planets spanning forty-four centuries," *Astron. Astrophys.* **125**, 150–167, 1983.

COMPRESSED PLANETARY AND LUNAR EPHEMERIDES

PETER KAMMEYER
United States Naval Observatory
Washington, D.C.
United States of America

ABSTRACT. A package of FORTRAN software has been developed which provides planetary and lunar positions, with respect to the solar system barycenter, for all times in the interval 1801-2049; positions agree to 1 milliarcsecond with those generated by Jet Propulsion Laboratory Development Ephemeris 200 (DE200). The system consists of approximately 800 kilobytes of ephemeris files and 40 kilobytes of programs, totalling 5% of the storage required by DE200. After removal of reference orbits, segments of DE200 positions were fitted by finite Chebyshev series of degree 40. The Chebyshev coefficients were rounded to integer multiples of a suitable unit and packed to form the ephemeris files.

1. Introduction

The compressed ephemeris system is a software package which can be used in microcomputers or mainframe computers to provide high accuracy positions and velocities for the planets and the Moon. It covers the interval from the year 1801 to the year 2049. Planetary positions are given in harmonic coordinates with origin at the solar system barycenter and differ from the corresponding DE200 positions by no more than $.7 \times 10^{-8}$ times the radius of the orbit. Such position differences correspond approximately to an angle of 1 milliarcsecond and are much smaller than the error reported by Standish (1986) and Seidelmann et al. (1986) for DE200 positions. The compressed ephemeris system gives values for the difference between the barycentric coordinates of the Earth and the Moon which again differ from the DE200 values by at most 1 milliarcsecond.

The compressed ephemeris system stores planetary and lunar positions as finite Chebyshev series. Lanczos (1952,1956) advocated the use of Chebyshev series for the representation of functions. Clenshaw and Norton (1963) studied the application of Chebyshev series to the iterative solution of nonlinear ordinary differential equations. Carpenter (1966) produced iterative solutions of the equations of celestial mechanics in Chebyshev series, and Broucke (1969) generalized this work. Corio (1973) used Chebyshev series to represent numerically integrated communications satellite orbits, and a number of other applications in the early 1970's are mentioned in Deprit et al. (1979)

By the middle of the 1970s, the use of Chebyshev series was well established in celestial mechanics. In 1975, the Jet Propulsion Laboratory produced an ephemeris, DE96, which used Chebyshev series to represent the positions and velocities of the Sun, the Moon, and the planets in barycentric harmonic coordinates (Standish et al. 1976). The United States Naval Observatory

Celestial Mechanics **45**: 311–316, 1989.

published in 1976 the Almanac for Computers (Kaplan et al. 1976), providing Chebyshev series for the geocentric apparent positions of the Sun, the Moon, and the planets.

The United States Naval Observatory first produced in 1986 its Floppy Almanac, a package of software which displays on a microcomputer screen tables of most of the quantities tabulated in the printed Astronomical Almanac. One 360 kilobyte floppy disc contains the Floppy Almanac software and data for a period of one year. The successor of the Floppy Almanac, the Interactive Computer Ephemeris, will use four 360 kilobyte discs to store the compressed ephemeris and data from the end of 1800 to the beginning of June 2049.

2. Structure of the Compressed Ephemeris System

The compressed ephemeris system consists of 830 kilobytes of ephemeris data, a subroutine SOLSYS which generates from the ephemeris data the position and velocity of a specified object at a specified time, and 12 kilobytes of parameters required by SOLSYS. The compiled version of SOLSYS requires 40 kilobytes of storage.

The data for the compressed ephemeris system is divided into 24 blocks, each of which contains all ephemeris data for the Sun, the Moon, and the planets from the block starting time to the block ending time. When data from a block is required, the block is read entirely into computer memory. The two most recently used blocks are stored in memory; after the compressed ephemeris system crosses the boundary between two adjoining blocks, both are available in memory. Two blocks of ephemeris data, together with the software required to produce positions and velocities, occupy less than 200 kilobytes of microcomputer memory.

Each block is stored on disc as a separate file, requiring less than 37 kilobytes of disc storage. The beginning of the first block is Julian Date 2378850.0, and the end of the last block is Julian Date 2469600.0; all other block boundaries occur at Julian Date values which are even multiples of 4000.0 days.

Ephemeris data for any particular object is a single, continuous part of a block. This data is divided into a succession of ephemeris segments, each containing all ephemeris data from the segment starting time to the segment ending time. For Mercury, Venus, the center of mass of the Earth-Moon system, Mars, the outer planets, and the Sun, the ephemeris data in a segment consists of the coefficients of degree 40 Chebyshev series approximations to three functions of barycentric dynamical time, from which the DE200 barycentric harmonic coordinates may be determined. The expansion interval extends before the starting time of the segment and after the ending time of the segment by 5 % of the segment length.

For the Moon the ephemeris data in a segment consists of degree 40 Chebyshev series approximations to three functions of barycentric dynamical time, from which the differences between the DE200 coordinates of the Moon and the Earth may be determined. The expansion interval is identical to the segment.

Segments of the compressed ephemeris are much longer than the expansion intervals used in DE200, which are at most 32 days and are four days for the Moon. For the outer planets, all blocks contain one segment. The first segment of the Sun's ephemeris in a block starts at the beginning of the block; following segments begin at intervals of 200 days. Except for the beginning of the first segment and the end of the last segment in a block, which coincide with block boundaries, segment boundaries for the Moon and the inner planets occur at pericenter passages of a mean orbit.

3. Removing Common Parts

The compressed ephemeris, if it consisted of series approximations to the harmonic coordinates and the coefficients were represented by 8-byte reals, would require about one third of the storage required by DE200. Each of the blocks except the first and the last contains between 254 and 257 segments. There are three coordinates, and there are 41 coefficients in each coordinate series. The storage required for a block would be approximately

$$256 \times 3 \times 41 \times 8 \text{ bytes} = 251904 \text{ bytes},$$

a total of about six megabytes for the 24 blocks.

The compressed ephemeris actually stores only enough bits of each Chebyshev coefficient to represent DE200 within 1 milliarcsecond. Two strategies are used to reduce the magnitudes of the Chebyshev coefficients stored, and thus the number of bits required for the coefficients. For the Moon and for each planet, new coordinate axes XYZ are introduced so that motion occurs near the XY plane. The axes used for the Moon and for the inner planets vary from segment to segment, while those used for the Sun and the outer planets have constant directions. From the Chebyshev series expansions of the X and Y position coordinates of the Moon or an inner planet are subtracted reference orbit Chebyshev series, different for each segment. The reference orbit series for each segment are obtained by rotating two Chebyshev series, depending only on the object and representing an orbit of fixed semi-major axis and eccentricity, to obtain an orbit with pericenter location matching that of the mean orbit at the middle of the segment.

4. Packing Coefficients

The Chebyshev series $P(s)$ ($s = 1$ at the beginning of the segment and $s = -1$ at the end) which represents a function $F(s)$ in a segment is of the form (Fox and Parker, 1968; Broucke, 1973)

$$P(s) = \sum_{K=0}^{40}{}' C_K \; T_K(s),$$

where

$$C_K = \frac{1}{20}\sum_{J=0}^{40}{}'' F\left(\cos\left(\frac{J\pi}{40}\right)\right) \cos\frac{JK\pi}{40}$$

and T_K is the Chebyshev polynomial of degree K. The single prime means that the $K = 0$ term is to be divided by two, and the double prime means that both the $J = 0$ and the $J = 40$ terms are to be divided by two. The coefficients C_K are bounded above in absolute value by twice the radial distance.

The first step in reducing each coefficient to only the necessary number of bits is to divide it by a unit of length and then round it to produce an integer coefficient. For each object the unit is 10^{-9} times an approximate least upper bound on the radial distance. The integer coefficients are thus between -2×10^9 and 2×10^9 and are represented in four bytes.

Only some of the bits in the four bytes of each integer coefficient contain useful information. For example, a coefficient equal to the integer 1 may be represented using a sign bit and one additional bit. The significant bits of each integer coefficient are extracted and packed tightly into a sequence of four-byte integers, the stored ephemeris.

5. Evaluation of Position and Velocity

To evaluate the position and velocity of an object at a specified time, the compressed ephemeris system unpacks the required Chebyshev series, adds reference orbit series if necessary, and performs rotations to generate Chebyshev series for the DE200 coordinates in an interval about the specified time. These series are stored in the computer's memory. The series for the DE200 coordinates are then evaluated, using subroutines developed by Broucke (1973), to obtain the position and velocity desired.

6. Error in Representing DE200

As stated above, the compressed ephemeris represents DE200 positions with an error of 1 milliarcsecond. The error in representing outer planet velocities is at most 10^{-9} astronomical unit per day. The error in representing the velocities of the inner planets is larger, reaching 3×10^{-9} astronomical unit per day for Mars. Velocity representation errors for the geocentric motion of the Moon are at most 2×10^{-10} astronomical unit per day.

Figure 1 shows the distance in kilometers between the positions of Saturn calculated by the compressed ephemeris system and by DE200 (note that 1 astronomical unit is 1.5×10^{8} kilometers). Position errors appear random.

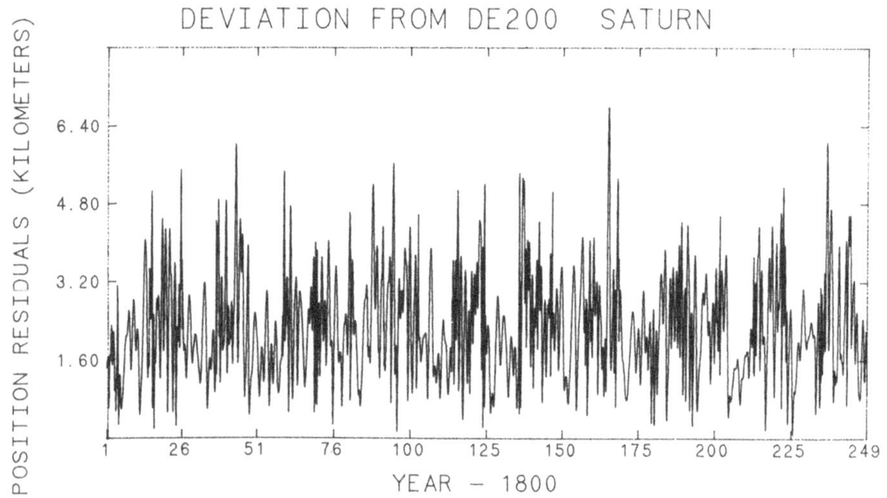

Figure 1. Differences between DE200 and Compressed Ephemeris Positions for Saturn

The magnitude of the velocity difference, shown in Figure 2, is dominated by evenly spaced spikes. Rounding the Chebyshev coefficients introduces Gibbs phenomenon oscillations at the boundaries of segments. To reduce the velocity errors to the level shown, the expansion intervals for Chebyshev series approximations are made longer than the segments on which the series are to be used.

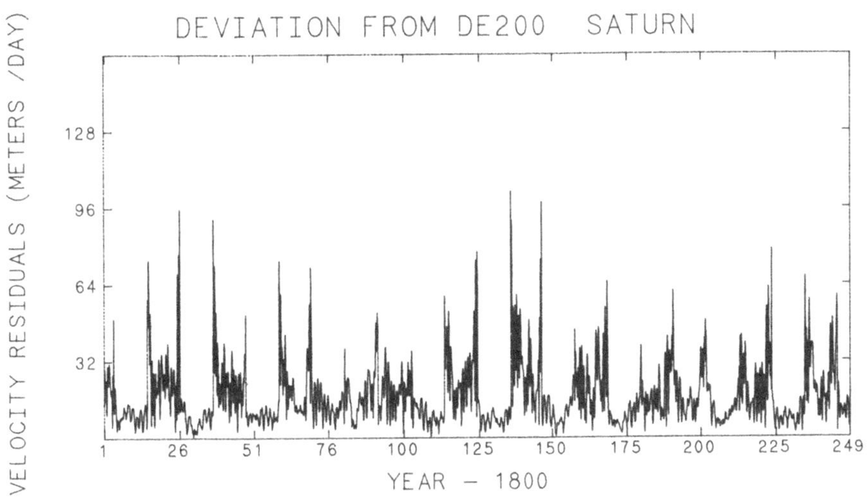

Figure 2. Differences between DE200 and Compressed Ephemeris Velocities for Saturn

Aknowledgments

The author would like to thank Mr. John A. Bangert of the Nautical Almanac Office, U.S. Naval Observatory, for explaining the structure of the DE200 ephemeris. He would like also to thank Dr. Andre Deprit of the National Bureau of Standards for his help in improving both the oral and written presentations of this paper.

References

Broucke, R. (1969) 'Perturbations in Rectangular Coordinates by Iteration', Celestial Mech., 1, 110-126.

Broucke, R. (1973) 'Ten Subroutines for the Manipulation of Chebyshev Series', Comm. of the ACM, 16, 254-256.

Carpenter, L. (1966) Planetary Perturbations in Chebyshev Series, NASA TN D-3168, National Aeronautics and Space Administration, Washington.

Clenshaw, C. W. and Norton, H. J. (1963) 'The solution of nonlinear ordinary differential

equations in Chebyshev series', Computer J., 6, 88-92.
Corio, A. J. (1973) 'The use of Chebyshev polynomials for satellite ephemerides', COMSAT Tech. Rev. 3, 411-418.
Deprit, A., Pickard, H., and Poplarchek, W. (1979) 'Compression of Ephemerides by Discrete Chebyshev Approximations', NAVIGATION: Journal of the Institute of Navigation, 26, 1-11.
Fox, L. and Parker, I. B. (1968) Chebyshev Polynomials in Numerical Analysis, Oxford Univ. Press, London.
Kaplan, G. H., Doggett, L. E., Seidelmann, P. K. (1976) Almanac for Computers, 1977 , U.S. Naval Observatory Circular No. 155, U.S. Naval Observatory, Washington.
Lanczos, C. (1952) 'Tables of Chebyshev Polynomials' Applied Math. Series, U.S. Bureau of Standards, 9, Government Printing Office, Washington.
Lanczos, C. (1956) Applied Analysis, Prentice Hall, Inc., Englewood Cliffs, N. J.
Seidelmann, P. K., Santoro, E. J., and Pulkkinen, K. F. (1986) 'Systematic differences between planetary observations and ephemerides', in J. Kovalevsky and V. A. Brumberg (eds.), Relativity in Celestial Mechanics and Astrometry, Proceedings of the 114th Symposium of the International Astronomical Union, D. Reidel Publishing Co., Dordrecht, pp. 99-103.
Standish, E. M. (1986) 'Numerical Planetary and Lunar Ephemerides: Present Status, Precision, and Accuracies', in J. Kovalevsky and V. A. Brumberg (eds.), Relativity in Celestial Mechanics and Astrometry, Proceedings of the 114th Symposium of the International Astronomical Union, D. Reidel Publishing Co., Dordrecht, pp. 71-83.
Standish, E. M., Jr., Keesey, M. S. W., and Newhall, XX (1976) Jet Propulsion Laboratory Development Ephemeris Number 96, NASA Tech. Rep. 32-1603, Jet Propulsion Laboratory, Pasadena.

AN ANALYTICAL REPRESENTATION OF EPHEMERIS DATA

B. JOVANOVIĆ
Astronomical Observatory
Volgina 7
Yu-11050, Belgrade
Yugoslavia

ABSTRACT. An algebraic method for the resolution of tabulated function into a class of elementary functions is examined on the high precision ephemeris data from the JPL DE 200.

1. Introduction

Approximating a large sample of numerical data by a real function is the common way to compress them with minimal loss of accuracy and/or to facilitate their use in specific applications. The various functions can be introduced for such purposes, but when one deals with the planetary ephemeris, efficient Chebyshev expansions are most frequently employed.

The aim of the present paper is to demonstrate that it is possible to achieve the high quality approximation of the precise planetary ephemeris by means of another procedure. The method enables the representation of tabulated equidistant data, in general case, as sum of linear function, harmonics and exponential functions. It can be classified into the algebraic methods of the type being used in the beginning of the century mostly for harmonic analysis (Kühnen (1909), Bruns (1911), Dale (1914)). However, lack of the adequate computing devices and appropriate numerical algorthims limited their application, and consequently they have been gradually abandoned.

2. Method

The real functions L, H and E, defined as

$$L(x) = C_1 x + C_2, \quad H(x) = C_1 \sin \omega x + C_2 \cos \omega x, \quad E(x) = C_1 \lambda^x + C_2 \lambda^{-x} \tag{1}$$

can be represented through their discrete values, with argument step ε, by recurrence relation

$$f(x) = kf(x - \varepsilon) - f(x - 2\varepsilon), \tag{2}$$

where k is the parameter that determines the analytical type of function $f(x)$

$$H : k \in (-2, 2), \; L : k = 2, \; E : k \in (2, \infty).$$

Celestial Mechanics **45**: 317–320, 1989.

Here, the parameters ω, λ are defined by

$$\omega = \frac{1}{\varepsilon} \arccos \frac{k}{2}, \quad \lambda = \left(\frac{k + \sqrt{k^2 - 4}}{2} \right)^{1/\varepsilon}. \tag{3}$$

and C_1, C_2 are arbitrary constants.

If one assumes that a N–element data set $\{F(x_i)\}$, $i = 1 \ldots N$, is given, with constant argument step $\varepsilon = x_i - x_{i-1}$, and where unknown function $F(x)$ is defined as sum of n ($\leq N/3$) functions (2), i.e.

$$F(x) = \sum_{i=1}^{n} f_i(x), \tag{4}$$

then it is possible to express $F(x)$ in the form of a linear difference equation, whose solution, in general case, is determined as

$$F(x) = L(x) + \sum_i H_i(x) + \sum_j E_j(x). \tag{5}$$

The procedure for determination of $F(x)$ can be realized as follows:

1) for a given n, the matrix A is created by

$$A_{ij} = \sum_{l=0}^{j-1} \binom{j-1}{l} F(x_{n+i+j-1-2l}), \qquad i = 1 \ldots N - 2n, \quad j = 1 \ldots n+1$$

and linear system is solved

$$-\sum_{j=1}^{n} A_{i,n-j+1} P_j = A_{i,n+1}, \qquad i = 1 \ldots N - 2n; \tag{6}$$

2) thence one solves the algebraic equation

$$\sum_{i=1}^{n+1} B_i k^{n-i+1} = 0, \tag{7}$$

where $B_1 = 1$ and $B_i = P_{i-1}$ for $i = 2 \ldots n+1$;

3) each real zero k (> -2) defines one particular function (2) in (4); if m such zeros are found, taking into account (1) and (3), $2m$ constants (C_1, C_2) are to be calculated from the linear system

$$\sum_{j=1}^{m} f_j(x_i) = F(x_i), \qquad i = 1 \ldots N \tag{8}$$

by which the expansion (5) is fully determined.

The main advantage of the method is the resolution of the input data into three types of component functions, since the other methods usually involve just one function type. In turn, this provides the more flexible approximation and the more accurate interpolation. Especially important are H components, which enable precise identifying of the hidden periodicities, while L, E components are more intended for the description of trends.

The periodicity T $(= 2\pi/\omega)$ of an extracted H term belongs to the broad interval $(2\varepsilon, \infty)$, but also it may belong to the series of diminishing intervals $(\frac{2\varepsilon}{i+1}, \frac{2\varepsilon}{i}), i = 1, 2, \ldots$, i.e. it can have any of the values T_i :

$$T_1 = \frac{T\varepsilon}{T-\varepsilon}, \quad T_2 = \frac{T\varepsilon}{T+\varepsilon}, \quad \ldots \quad, T_{2i-1} = \frac{T\varepsilon}{iT-\varepsilon}, \quad T_{2i} = \frac{T\varepsilon}{iT+\varepsilon}, \quad \ldots \tag{9}$$

Providing that one is aware of existence of any small periodicity T_i $(< 2\varepsilon)$, it can be recognized in its corresponding value T and taken into account accurately. This fact emphasizes the importance of the method for interpolation purpose.

The drawbacks of the method are connected mainly with the well-known difficulties that arise in solving the large linear systems, as well as with finding the close roots of high degree polynomials. Therefore, application of the method is limited to relatively small data sets. In case of the large data sets, however, these problems can be avoided by computing with long enough real variables and by using the most appropriate algorithms.

3. Results

The above procedure has been applied to derive the approximate functions for representation of planetary rectangular equatorial heliocentric coordinates JPL DE 200. Input data for each particular planet were picked with the step of ε days $(\varepsilon > 1)$, in order to increase the time span and to decrease the quantity of input data. Since the source data are tabulated with one day step, this gives the opportunity to check the interpolation capability of the resulting functions.

In the Table 1. the results are shown for an experiment pertaining to coordinates X, Y, Z of a number of planets in 1988.

TABLE 1. Some features of the obtained expansions for the span of 366 days.

ε – time step of input data in days;
nH, nE – total number of extracted H and E terms;
Δ_1 – maximum error of the approximate positions in 1×10^{-10} AU for $\varepsilon = 1$;
C – compression percentage $= 100(1 - (\sum nH + nE)/366)$

	X				Y				Z				
planet	ε	nH	nE	Δ_1	ε	nH	nE	Δ_1	ε	nH	nE	Δ_1	C
VENUS	13	10	0	14	12	11	0	22	13	9	1	24	92
EARTH+MOON	8	14	1	8	10	10	2	28	14	10	0	49	90
MARS	10	11	1	5	10	10	2	7	9	12	1	2	90
JUPITER	20	7	0	23	20	7	0	19	20	7	0	11	95
SATURN	20	7	0	32	20	5	1	33	20	7	0	11	95
URANUS	20	7	0	33	20	5	1	34	20	7	0	11	95

As one can see, no expansion has a linear term. This could have been expected due to the nature of the data and to the fact that linear term is boundary case.

No specific investigations concerning the optimum of the terms number, the step size ε , and the approximation accuracy have been done. Only two limitations were considered: in order to avoid numerical instability of computing procedure, the number of input data was chosen to be relatively small, and the maximum error of the obtained functions was not to be greater then 1×10^{-8}, that enables the precise calculation of planetary apparent places.

Although there are some inhomogeneities, the obtained results are illustrative enough for the possibilities of the method. Especially, the good interpolation is noticeable that can hardly be achieved by some other method. In this experiment, none of relations (9) was used.

All calculations have been carried out in FORTRAN double precision format with input data being rounded off to 14th decimal place (10^{-14} AU). The linear systems (6),(8) have been solved by means of the least square method, and the algebraic equation (7) by means of the Jenkins-Traub algorithm (IMSL routines : LLSQF and ZRPOLY).

4. Conclusions

The obtained preliminary results prove that it is possible to represent accurately enough the planetary motion by the presented method.

The approximate functions can be used to condense the ephemeris data for a specified time span and to interpolate them, as is the practice with other methods in astronomical and nautical almanacs. There, it is not neccesary to make any normalization of the argument, and their use is reduced to simple call of a functional statement with one input variable.

The method may prove useful if applied to represent Keplerian elements of planetary orbits, as their secular and periodic terms can be clearly separated in an efficient way.

Still better expansions of the approximate functions can be expected if one makes use of the programming languages permitting extra-long real variables.

References

Bruns, H. (1911) 'Über die Analyse periodischer Vorgänge', *Astronomische Nachrichten*, **188**, 55–60.

Dale, J. B. (1914) 'The resolution of a compound function into simple periodic functions', *Monthly Notices Roy. Astron. Soc.*, **74**, 628–648.

IMSL Library: (1981) 'FORTRAN Subroutines for Mathematics and Statistics' (VAX), IMSL Inc., Houston

Jovanović , B. (1987) 'An Approximation of Tabulated Function', *Publ. Inst. Math. Belgrade*, **41** (55), 143–148.

JPL DE200/LE200 Ephemeris: (1984) USNO Tapes NA0109 & NA0110.

Kühnen, F. (1909) 'Methode zur Aufsuchung periodischer Erscheinungen in Reihen äquidistanter Beobachtungen', *Astronomische Nachrichten*, **182**, 1–12.

Numerical Simulations of Narrow Planetary Rings : An Animation

R. A. Kolvoord and J. A. Burns
Department of Theoretical and Applied Mechanics
Cornell University
Ithaca, N.Y. 14853

The availability of bigger and more powerful computers has been a boon to the study of dynamical simulations. Models can now be extended both in size and time with a concommitant increase in output. This plethora of data has inspired the development of new methods for analysis and display of results. Animation has proven to be a very useful tool in this context, but it has traditionally been extremely difficult and tedious to produce films. In this paper, we present a simple and easy-to-use method for representing the output of a complex discrete dynamical simulation by animation on a personal computer. Using a Macintosh and the commercially available software MacMovie, we will demonstrate the usefulness of a simple image animation process for analyzing some results of complicated computer models.

Narrow shepherded planetary rings, such as Saturn's F ring or the Encke gap ringlet, show surprisingly complicated structures that include such features as kinks, clumps, and braids. In an attempt to understand this strange array of behavior, we have numerically modelled a discrete particle system including the perturbations of shepherding satellites on eccentric orbits. Lagrange's planetary equations were used to solve for the perturbations to the orbits of three strands of non-interacting ring particles, each containing 800 particles (cf. Showalter and Burns, 1982 and Kolvoord, 1987). Collisions and self-gravity are ignored. This method is computationally very efficient as it uses an interpolation scheme to compute the perturbations rather than direct integration for each interaction. This pre-calculation is possible due to the use of a geometrical transformation that links any and all configurations of ring particle and satellite to a reference state that has already been analyzed. This scheme allows us to study more particles for a longer time than would ordinarily be possible.

Our calculations reconfirm the presence of clumps and wave-like features with wavelength, $\lambda = 3\pi\Delta a$ where Δa is the separation in semi-major axis of the perturber and the ring particle, as shown by Showalter and Burns. We also note that these features are washed out by Kepler shear in a relatively short number of orbits. Braiding is not observed in our model even with the inclusion of embedded moonlets. Our results continue to show substantial azimuthal and temporal variability. They also depend strongly on the initial configuration and the relative positions of the perturbers.

The results of the simulations can be particularly difficult to interpret in a static display. However, using software available for the Macintosh, an animation can be created that renders the spatially and temporally varying features of the model much more readily apparent. The movie is created by capturing individual plots of the results at discrete time intervals and then compressing them to enable quicker recall. The procedure is straightforward and easy to use. The animation itself demonstrates the importance of the particular viewing geometry in interpreting spacecraft results. Groups of particles that appear to

Celestial Mechanics **45**: 321–322, 1989.

be clumped at a particular longitude have no coherence at another longitude. This again points to the importance of the exact geometries of the particles. The clumping and unclumping can be understood by looking at the relative alignment of pericenters and the radial epicyclic motion of individual particles in the ring.

We are extending our geometries to include small inclination orbits for each of the participants. The full three-dimensional case presents some added complications; what is the most efficient viewing geometry in which to animate the results, how does varying the longitude of nodes affect the ouput? We are also trying to compare our models with actual Voyager images of the F ring, which we are examining in detail. We further need to examine how much our simplifying assumptions (no collisions or self-gravity) affect the final results. However, the use of animation has proven to be both useful and straight-forward in aiding our analysis of a complicated dynamical simulation.

Kolvoord, R.A. (1987). The effect of embedded and shepherding satellites on a narrow ring: a numerical simulation. **BAAS** 18, 891.

Showalter, M.R. and J.A. Burns (1982). A numerical study of Saturn's F ring. **Icarus** 52, 526-544.

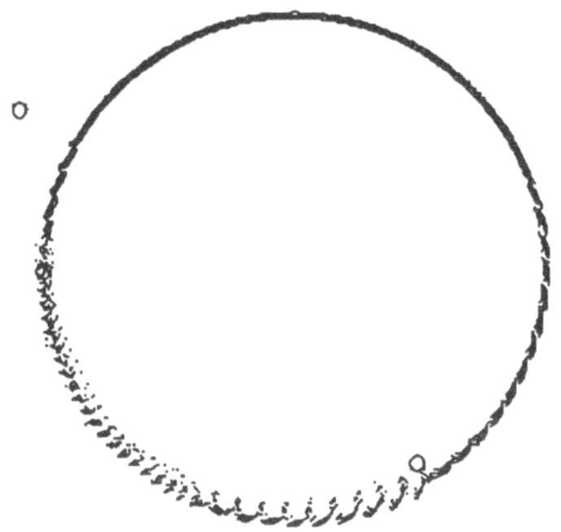

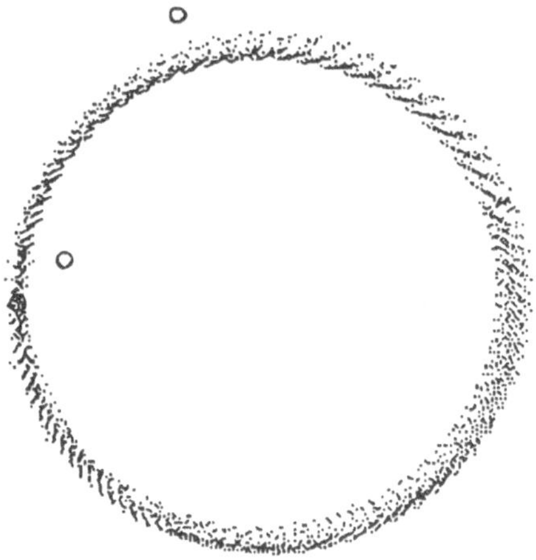

1a. 1b.

Figs. 1a and 1b. Sample output showing the system after 50 and 110 orbital periods of the ring particles. Kepler shear has washed out the initial clumping (seen at 7 o'clock in Fig. 1a) in this time interval. The simulations are viewed in a frame that corotates with the center strand of the ring, whose semi-major axis is normalized to 1. The positions of both the shepherding satellites ($m/M \sim 10^{-9}$, a = .994 and 1.01, e = $2.5 \cdot 10^{-3}$ and $4.3 \cdot 10^{-3}$) and the embedded satellite ($m/M \sim 10^{-13}$, a = .999925) are indicated by the small circles on the diagram.

ELECTRONIC ALMANACS — MATING THE MESSAGE AND THE MEDIUM

L. E. DOGGETT, T. S. CARROLL, J. A. DEYOUNG, J. R. ROHDE, J. A. BANGERT, W. T. HARRIS, S. P. PANOSSIAN, W. J. TANGREN, and P. C. KAMMEYER
Nautical Almanac Office
U.S. Naval Observatory
Washington, DC 20392
USA

ABSTRACT. Before computers, astronomical data were transmitted orally, then in manuscripts and finally in printed almanacs. The structure of the data was determined by the needs of users and by the nature of the medium of transmission. As computers have evolved in the past two decades we have learned to utilize them in the transmission process. A new series of electronic almanacs produced by the Nautical Almanac Office provides data to high precision and exploits computers as the medium of transmission.

1. Almanac for Computers

In the mid-1970s the Nautical Almanac Office began to investigate means of presenting astronomical data for efficient use with small computers and pocket calculators. This work was inspired by the efforts of Carpenter (1966), Broucke (1969) and others in developing planetary theories in the form of Chebyshev series. As a result Kaplan et al. (1976) published the first Almanac for Computers. Parallel efforts were also going on in the Bureau des Longitudes (1977) and H. M. Nautical Almanac Office (Yallop, 1981).

In the Almanac for Computers, short Chebyshev series are given for the apparent, geocentric coordinates of the Sun, Moon and planets. These series are constructed, using a least squares algorithm (Fox and Parker, 1968), to approximate tabular values in the Astronomical Almanac. The precision of the approximations is about one tenth of an arc second. Succinct formulas, to a precision of better than one tenth of an arc minute, are given for the apparent places of a list of 176 stars. These data are prepared annually and made available in printed form, on magnetic tape, and now on floppy disk. The printed Almanac for Computers also provides a collection of algorithms for basic calculations of spherical astronomy, including rising and setting of the Sun and Moon, and navigational sight reduction.

The Almanac for Computers has been prepared annually since 1977 with only minor changes. Despite recent developments in computer technology, it seems destined to serve a continuing need as direct input to programs that manipulate astronomical data.

Celestial Mechanics **45**: 323–326, 1989.

2. Floppy Almanac

During the early 1980s the proliferation of IBM PCs and clones running under MS–DOS established a standard calculating environment. For this environment, Kaplan et al. (1986) created an integrated package of software and astronomical data on a single floppy disk, the Floppy Almanac. Like the Almanac for Computers, the Floppy Almanac was prepared annually. However, the Floppy Almanac was not based on approximations to the Astronomical Almanac. Instead, Chebyshev series represented the barycentric rectangular coordinates and velocities of the Sun, Moon and planets. These were based on least squares approximations of short segments of JPL's DE200/LE200 numerical integration. Mean places and proper motions of stars were taken from the FK4 Catalog. Apparent places for all objects were calculated rigorously to the precision of the Astronomical Almanac. For any specified geographic location, the program calculated times of rise, set and transit, and altitudes and azimuths for specific times, for all objects. Finally, the program calculated physical ephemerides of the Moon and planets.

The Floppy Almanac was menu-driven, with key word entries invoking the algorithms. Although the Floppy Almanac was designed for use with microcomputers, the code was ported to run on IBM mainframes running under VM, VAX computers under VMS, and HP1000 minicomputers. The annual Floppy Almanac is being discontinued to make way for the Interactive Computer Ephemeris.

3. Interactive Computer Ephemeris

Recent software and hardware developments are making possible a new generation of almanacs for microcomputers. On the software front, Kammeyer (1989) has produced a new approximation and compression of JPL's DE200/LE200 numerical integration. This makes it possible to supply nearly the complete span of the integration, in this case 1801–2048, for microcomputer applications. The Nautical Almanac Office is developing software to utilize Kammeyer's ephemeris package. This Interactive Computer Ephemeris (ICE) is a set of Fortran subroutines, without a machine-dependent user interface, that returns rigorously calculated positions and phenomena of celestial objects for specified times and geographic locations. Star positions are calculated from mean places and proper motions from the FK5 Catalog.

The ICE code strictly adheres to Fortran-77 standards, so it can be compiled on a variety of computers. To run on a particular computer it is necessary to write a shell that serves as a user interface for ICE. Versions are being created for IBM PCs and clones running MS–DOS, Macintosh microcomputers, IBM mainframes running under VM, and VAX computers under VMS. The object code will be available to users who wish to write their own driver routines or incorporate the ICE routines into another program.

3.1 ICE ON MS–DOS

The MS–DOS version of the Interactive Computer Ephemeris requires a minimum of 384 Kbytes of RAM; a math coprocessor and hard disk are highly recommended. This version follows the general format and procedures of the annual Floppy Almanac, with improvements suggested by users' experience.

3.2 ICE ON THE MAC — MICE

The Macintosh Interactive Computer Ephemeris runs on Macintosh Plus, SE and II microcomputers. Use of a hard disk is highly recommended. Presenting ICE under the Macintosh user interface offers a new mode of interaction between a user and an almanac. We find that the Macintosh menu structure makes the program easy to learn and flexible to use. Background information, in the form of text files, can be instantaneously invoked. Using standard Cut and Paste facilities and common desk accessories, customized data sets can be constructed from disparate calculations and ancillary data can be prepared and introduced.

4. Satellite Almanac

The Satellite Almanac (Rohde, 1988) calculates ephemerides of the natural satellites of Mars, Jupiter, Saturn, Uranus, Neptune and Pluto for the time span 1988–1999. Running under MS–DOS, it produces to full precision the data contained in Section F of the Astronomical Almanac. Calculations are based on the best theories and orbital elements available at this time. For some of the outer satellites of Jupiter the program numerically integrates the orbits in real time. To facilitate practical applications the program also calculates sidereal time and the times at which the planets rise, set and transit. The program uses the general format and procedures of the Floppy Almanac. Use of a math coprocessor is highly recommended.

5. Future Electronic Almanacs

The form and substance of future computerized almanacs will depend on advances in science, computer hardware and computer software. By presenting an almanac covering nearly two and a half centuries we may appear to be establishing a permanent, gradually deteriorating standard. As orbital parameters and astronomical constants are improved, however, new versions must be introduced to supersede the old standards. The ICE code in particular is written to facilitate revision as new constants or theories become available.

In truth the rapid evolution of hardware decreases the likelihood that obsolete data will be unwittingly propagated. An ephemeris stored on today's floppy disk or PROM may not be readable by the next generation's machines. That this is not totally a blessing is illustrated by our recent and frustrating attempts to reconstruct the Improved Lunar Ephemeris from machine readable media of a decade ago. The successful completion of the project required careful term by term comparison with the printed book. Thus an almanac worth preserving is perhaps best preserved in printed form.

Development of new hardware and software raises anew the question of how a user should interact with a set of astronomical data. This requires consideration of both the needs of the user and the medium of presentation. For some purposes a skilled eye and traditional printed tables remain most efficient. At the other extreme, the user may be a machine controlled by a program. In an age when printed tables were the only form of presentation, considerable effort and ingenuity were devoted to selecting the functions to be tabulated, to finding optimal tabular intervals and effective page layouts, etc. Designing an almanac required a fusion of art and science. As data came to be used by computers, traditional formats were simply transferred to punched cards, magnetic tape and screens. This pattern was broken with the Almanac for Computers. The Floppy Almanac and now the Macintosh Interactive Computer Ephemeris are further attempts to utilize specific features of computer systems to put data at the service of the user.

It is scarcely a prediction to note that computers will become faster and more powerful. The possibilities of graphics and even sound are now ripe for exploitation. Keyboards, fingers and mice are cumbersome; and while the user-friendly metaphor of the desktop is useful for activities carried on at a desk, it is scarcely relevant to the subject of celestial mechanics. Thus we may hope for further breakthroughs in user interfaces. Just as the venerable science of celestial mechanics is flourishing, so too is the venerable art of producing almanacs.

6. References

Broucke, R. (1969) 'Perturbations in Rectangular Coordinates by Iteration', Celestial Mechanics 1, 110–126.

Bureau des Longitudes (1977) Connaissance des Temps, Nouvelle Série, 1978.

Carpenter, L. (1966) 'Planetary Perturbations in Chebyshev Series', NASA Technical Note D–3168.

Fox, L. and Parker, I.B. (1968) Chebyshev Polynomials in Numerical Analysis, Oxford University Press, London.

Kammeyer, P.C. (1989) 'Compressed Planetary and Lunar Ephemeris', these Proceedings.

Kaplan, G.H., Doggett, L.E. and Seidelmann, P.K. (1976) 'Almanac for Computers 1977', U.S. Naval Observatory Circular 155.

Kaplan, G.H., Carroll, T.S., Doggett, L.E. and Seidelmann, P.K. (1986) The Floppy Almanac User's Guide, U.S. Naval Observatory, Washington.

Rohde, J.R. (1988) The Satellite Almanac Users' Guide, U.S. Naval Observatory, Washington.

Yallop, B.D. (1981) 'Compact Data for Navigation and Astronomy for 1981 to 1985', Royal Greenwich Observatory Bulletins, Number 185.

LIST OF PARTICIPANTS

Mr. J. M. Alini
Paris, France

Dr. J. K. Andrade
EMBRATEL
Av. Presidente Vargis
1012 S. 609
Rio de Janerio Brazil

Dr. Arlot
Bureau Des Longitudes
77 Ave. Denfert-Rochereau
Paris 75014
France

R. Archinal
U S Naval Observatory
34th & Mass. Ave., N.W.
Washington, DC 20392-5100
U S A

Dr. E. Athanassoula
Observatoire de Marseille
2, Place le Verrier
13248 Marseille Cedex 04
France

J. A. Bangert
U S Naval Observatory
34th & Mass Ave., N.W.
Washington, DC 20392-5100
U S A

Mr. Kirk D. Borne
Space Tel Sci Inst
3700 San Martin Drive
Baltimore, MD 21218 USA

Mr. Alan P. Boss
DTM-Carnegie Inst. of Washington
5241 Broad Br. Rd. N.W.
Washington, DC 20015
U S A

Dr. Richard L. Branham
Centro Regional de Investigaciones Cientificas y Tecnologica
Casilla de Correro 131
Mendoza Argentina

Prof. R. A. Broucke
University of Texas
Aerospace Engineering
Austin, TX 78712
U S A

Dr. Bryan C. Brown
Computer Services Corp.
U S A

Dr. V. A. Brumberg
Institute of Applied Astronomy
Code 5531 Pulkovo
U S S R

Prof. Gene G. Byrd
University of Alabama
Dept. of Physics and Astronomy, Boc 1921
Tuscaloosa, al 35487
USA

Prof Mike Cahill
University of Wisconsin
400 University Drive
West Bend, WI 53095
U S A

Dr. Odile Calame
Cerga
Av. Copernic
06130 Grasse
FRANCE

T. S. Carroll
U S. Naval Observatory
34th & Mass Ave., N.W.
Washington, DC 20392-5100
U S A

Mr Jinyong Chen
Univ of South Florida
Dept. of Mathematics
Tampa, FL 33620
USA

Dr. Dimitris M. Christodoulou
Louisiana State Univ
Dept. of Physics
Baton Rouge, LA 70803
USA

Dr. Shannon Coffey
Naval Research Lab
4555 Overlook Avenue
Washington, DC 20375
USA

C. Cole
U. S. Naval Observatory
34th and Mass Ave. NW
Washington, DC 20392-5100 USA

Prof. B. Conway
University of Illinois
Dept. of Aero. & Astro. Engineering
Urbana, IL 61801
USA

Mr. Vincent T. Coppola
Cornell University
114 Thurston Hall
Ithaca, NY 14853-1503
U S A

Dr. Richard Cushman
Ryksuniversiteit Utrecht
Budapestlaan 6.
35 TA Utrecht
Netherland

Prof. Morris Davis
Univ of North Carolina
Physics & Astron Dept
CB# 3255 Phillips Hall
Chapel Hill, NC 27514
U S A

Dr. Andre Deprit
N I S T
Administration A302
Gaithersburg, MD 20899
USA

Etienne Deprit
Naval Research Lab
Code 8242
Washington, DC 20375
USA

J. DeYoung
U. S. Naval Observatory
34th and Mass Ave. NW
Washington, DC 20392-5100 U S A

L. E. Doggett
U. S. Naval Observatory
34th and Mass Ave NW
Washington, DC 20392-5100 U S A

Prof. R. L. Duncombe
University of Texas.
WRW 401 D, Dept. ASE/EM
Austin, TX 78712
USA

Dr. David W. Dunham
Computer Sciences Corp.
10110 Aerospace Road
Lanham, MD 20706
USA

Dr. Joan Bixby Dunham
Computer Sciences Corp
10110 Aerospace Road
Lanhan, MD 20706 USA

Mr. S. H. Dupree, Jr.
General Electric Co.
P.O. Box 8555
Bldg. C, Room 20C291
Philadelphia, PA 1910
USA

Dr. A. Elipe
University of Zaragoza
Dpto. Astronomia
Universidad de Zarogoza
Zaragoza 50009
SPAIN

Prof. S. Ferraz-Mello
Universidade de Sao Paulo
Caixa Postal 30627
Sao Paulo BRAZIL

A. D. Fiala
U. S. Naval Observatory
34th and Mass Ave NW
Washington, DC 20392-5100 USA

Dr. D. C. Froeschle
Observatory of Nice
BP139 06003
Nice FRANCE

Dr. E. Gaposchin
M.I.T. Lincoln Lab
Millstone Hill Radar
Grotan Road, RTE 40
Westford, MA 01886
USA

Dr. A. Ghaffari
7532 Royal Dominian Drive
Bethesda, MD 20817
USA

Dr. Asao Habe
Hokkaido University
Dept. of Physics
Sapporo, Hokkando 060
JAPAN

R. S. Harrington
U. S. Naval Observatory
34th and Mass Ave NW
Washington, DC 20392-5100 USA

Dr. Alan Harris
JPL - MS 183-501
4800 Oak Grove Drive
Pasadena, CA 91109
USA

W. H. Harris
U. S. Naval Observatory
34th and Mass Ave NW
Washington, DC 20392-5100 USA

Prof. Miaofu He
Shanghai Observatory
Academia Sinica
80 Nan Dan Road
Shanghai CHINA

Dr. Liam M. Healy
Naval Research Lab
Code 8242
Washington, DC 20375
USA

Prof. Jacques Henrard
Universitaires de Namur
9 Rempart de la Vierge
5000 Namur BELGIUM

Mr. Paul Hertz
Naval Research Laboratory
Code 4121.5
Washington, DC 20375
USA

J. L. Hilton
U. S. Naval Observatory
34th and Mass Ave NW
Washington, DC 20392-5100 USA

Sethanne Howard
Georgia State Univ
1230 Main Street
Canton, CA 30114
USA

Dr. R. A. Howland, Jr.
Univ of Notre Dame
Dept. of Aerospace and Mechanical Engineering
Notre Dame, IN 46556
USA

Prof. Piet Hut
Inst.for Advanced Study
Olden Lane
Princeton, NJ 08540
USA

Dr. Mohammad Ilyas
University of Science
School of Physics
Penang 11800 MALAYSIA
USA

P. M. Janiczek
U. S. Naval Observatory
34th & Mass Ave NW
Washington, DC 20392-5100 USA

Dr. F. C. Johnson
N I S T
Administration A302
Gaithersburg, MD 20899
USA

Dr. Bora Jovanovic
Belgrade Astronomical Observatory
Volgina 7
Belgrade YU-11050
YUGOSLAVIA

P. C. Kammeyer
U. S. Naval Observatory
34th and Mass Ave NW
Washington, DC 20392-5100 USA

G. H. Kaplan
U. S. Naval Observatory
34th and Mass Ave NW
Washington, DC 20392-5100

Prof. Tao Kiang
Dunsink Observatory
Castleknock
Dublin 15 IRELAND

Dr. Hiroshi Kinoshita
Tokyo Astronomical Observatory
2-21-1-Osawa, Mitaka
Tokyo 181 JAPAN

R. A. Kolvoord
Cornell University
Dept. of Theor. & Applied Mechanics
Thurston Hall
Ithaca, NY 14853 USA

Dr. Jean Kovalevsky
CERGA
Avenue Copernic
Grasse 06130

Dr. G. A. Krasinsky
Institute of Applied Astronomy
Pulkovo
196140 Leningrad
USSR

Dr. Myron Lecar
Center for Astrophysics
60 Garden St.
Cambridge, MA 02138
USA

Prof. Sang-Gak Lee
Seoul National University
SAN 56-1, Shimkim-Dong
Kwanaka-Ku
Seoul KOREA

Dr. Jay H. Lieske
J P L
MS 301-150
4800 Oak Grove Drive
Pasadena, CA 91109
USA

Prof. Lin Liu
Nanjing University
Dept. of Astronomy
Nanjing CHINA

Prof. Steve McMillan
Drexel University
32nd & Chestnut Sts.
Philadelphia, PA 19104
USA

Dr. Enrico Melchioni
University de Pisa
Pizza Duomo 57
Voohera 27058 ITALY

Dr. P. J. Message
University of Liverpool
Dept. of Applied Math &
Theoretical Physics
Liverpool L69 3BX UK

Dr. Kenneth Meyer
Univ of Cincinnati
Dept of Mathematics
Cincinnati, OH 45221
USA

Dr. Andrea Milani
Universita de Pisa
Instituto de Mathematica
via Buonarroti 2
I-56100 Pisa ITALY

Dr. B. R. Miller
N I S T
Administration, A302
Gaithersburg, MD 20899
USA

Dr. R. H. Miller
University of Chicago
Astronomy Dept.
5640 Ellis Ave.
Chicago, IL 60637
USA

Dr. E. F. Milone
The Univ of Calgary
Dept. of Physics
2500 Univ Drive NW
Calgary, Alberta T2N 1N4 CANADA

Dr. Edmundo Moreno
Instituto de Astronomia
Apartdo Postal 70-264
Mexico City, D.F. MEXICO

Stephane Morin
Observatoire de Marseille
2 Place Leverrier
Marseille Cedex 4 FRANCE

Mr. Marc A. Murison
Univ of WI at Madison
Dept of Astronomy
475 N. Charter St
Madison, WI 53706
U S A

Prof Juan Carlos Muzzio
Observatorio Astronomico
Passo del Bosque
1900 La Plata
Buenos Aires Argentina

Dr. Skip Newhall
JPL, 238-332
Pasadena, CA
91109-8099 U S A

Dr. William I. Newman
Cornell University
Department of Astronomy
Ithaca, NY 14853-6801
U S A

Dr. Anna Nobili
Universita de Pisa
Instituto de Mathematica
via Buonarroti 2
I-56100 Pisa Italy

Alexander Ollongren
Leiden University
Dept of Math & Comp. Sci.
Niels Bohrweg
Leiden Holand

S. P. Panossian
U S Naval Observatory
34th & Mass Ave
Washington, DC 20392-5100 U S A

D. Pascu
U S Naval Observatory
34th & Mass Ave., N.W.
Washington, DC 20392-5100 U S A

Dr. Thierry Pauwels
Kominkijke Serrenwacht van belgie
Ringlaan, 3
Brussel Belgium

Dr. Stanton J. Peale
Univ of California
Dept of Physics, UCSB
Santa Barbara, CA 93106
U S A

Prof. Kevin Prendergast
Columbia Univ. Box 77
New York, NY 10027
U S A

Mr. Thomas Quinn
CITA
60 St. George St.
Toronto Canada

Prof. D. L. Richardson
Univ of Cincinnati
Aerospace Engineering
ML #70
Cincinnati, OH 45221
U S A

R. Randall L. Rickelfs
University of Texas
Department of Astronomy
McDonald Observatory
Austin, TX 78712-1083
U S A

Mr. W. W. Roberts
Univ of Virginia
Dept of Allied Math
Thornton Hall
Charlottesville, VA 22901 U S A

J. R. Rohde
U S Naval Observatory
34th & Mass Ave., N.W.
Washington, DC 20392-5100 U S A

E. J. Santoro
U S Naval Observatory
34th and Mass Ave.,N.W.
Washington, DC 20392-5100 U S A

Prof. Dieter Schmidt
University of Cincinnati
Dept of Computer Science
Cincinnati, OH 45221-0008 U S A

R. E. Schmidt
U S Naval Observatory
34th and Mass Ave, N.W.
Washington, DC 20392-5100 U S A

Dr. Stevo Segan
Institute of Astronomy
Studentski TRG 16. PB550
Belgrade YU-1100
Yugoslavia

Dr. P. K. Seidelmann
U S Naval Observatory
34th & Mass Ave., N.W.
Washington, DC 20392-5100 U S A

Dr. M. L. Sein-Echaluce
University of Zaragoza
Dpto. Astronomia
Universidat de Zeragoza
Zaragoza 50009 Spain

Dr. Irwin I. Shapiro
Harvard-Smithsonian Ctr for Astrophysics
60 Garden Street
Cambridge, MA 02138
USA

Dr. J. W. Siry
NASA Goddard Space Flt. Ctr.
Greenbelt, MD 20071
U S A

Dr. V. I. Skripnichenko
Institute of Applied Astronomy
Pulkovo
196140 Leningrad
U S S R

Victor J. Slabinski
Intelsat
3457 South Utah Street
Arlington, VA 22206
U S A

Dr. Gary Smith
IBM - Houston
16307 Shady Elms
Houston, TX 77059
U S A

Dr. E. M. Standish
Jet Propulsion Laboratory
Pasadena, CA 91103
USA

W. T. Tangren
U S Naval Observatory
34th & Mass Ave, N.W.
Washington, DC 20392-5100 U S A

Prof. B. D. Tapley
Univ of Texas
Ctr for Space Research
Austin, TX 78731
U S A

Prof. Fu Tong
Purple Mountain Observatory
Nanjing China

Prof. Mauri Valtonen
University of Turku
Dept of Physical Science
IT Pitkakatu I
20520 Turku Finland

Mr. Ferenc Varadi
Univ. of CA at LA
Dept of Mathematics
Los Angeles, CA 90024
U S A

Prof. Harry Varvoglis
Univ of Thessaloniki
Dept of Physics
GR-54006 Thessaloniki
Greece

Dr. Christian Vellet
CERGA
Arcrue Caperuic
Grasse France

Dr. Rodolpho Vilhena de Moraes
ITA/CTA
San Jose doe Carnpos SP
Sao Paulo Brazil

Dr. Paul Weissman
Jet Propulsion Lab
Pasadena, CA 91109
U S A

Mr. Richard L. White
Space Telescope Science Institute
3700 San Martin Drive
Baltimore, MD 21218

Dr. C. A. Williams
Dept of Mathematics
Univ of South Florida
Tampa, FL 33620
USA

Dr. Harold A. Williams
Carnegie Inst. of
Washington - DTM
5241 Broad Br. Rd, N.W.
Washington, DC 20014-
1305 U S A

Prof. Jack Wisdom
Mass Inst. of
Technology
54-414 MIT
Cambridge, MA
U S A

Dr. Bernard Yallop
Royal Greenwich
Observatory
Herstmonceux Castle
Hailsham, East Sussex
United Kingdom

Prof. H. Yilmaz
105 Church Street
Winchester, MA 01890
U S A

INDEX

E R R A T A

Page 51 - This page is to be omitted: is duplicates page 50.

Page 156 - Lines 8 to 10 should read as follows :

"$$\overset{k/2}{\underset{n}{\mathscr{C}}}(x) = \frac{2n+k-2}{n} f(x) \overset{k/2}{\underset{n-1}{\mathscr{C}}}(x) - \frac{n+k-2}{n} \overset{k/2}{\underset{n-2}{\mathscr{C}}}(x) \qquad (12)$$

one can construct the expansions for Gegenbauer polynomials

$$\left(f(x), \overset{k/2}{\underset{0}{\mathscr{C}}}(x) = 1, \overset{k/2}{\underset{1}{\mathscr{C}}}(x) = kx \right)$$

Legendre polynomials (the "

GPSR Compliance

The European Union's (EU) General Product Safety Regulation (GPSR) is a set of rules that requires consumer products to be safe and our obligations to ensure this.

If you have any concerns about our products, you can contact us on ProductSafety@springernature.com

In case Publisher is established outside the EU, the EU authorized representative is:

Springer Nature Customer Service Center GmbH
Europaplatz 3
69115 Heidelberg, Germany

Batch number: 10607320

Printed by Printforce, the Netherlands